Transversity 2008

Proceedings of the Second Workshop on Transverse Polarization Phenomena in Hard Processes

Transversity 2008

Proceedings of the Second Workshop on Transverse Polarization Phenomena in Hard Processes

Ferrara, Italy   28 – 31 May 2008

edited by

Ciullo Giuseppe

INFN and Università degli Studi di Ferrara, Italy

Contalbrigo Marco

INFN sezione di Ferrara, Italy

Hasch Delia

INFN Laboratori Nazionali di Frascati, Italy

Lenisa Paolo

INFN and Università degli Studi di Ferrara, Italy

World Scientific

NEW JERSEY • LONDON • SINGAPORE • BEIJING • SHANGHAI • HONG KONG • TAIPEI • CHENNAI

Published by

World Scientific Publishing Co. Pte. Ltd.
5 Toh Tuck Link, Singapore 596224
USA office: 27 Warren Street, Suite 401-402, Hackensack, NJ 07601
UK office: 57 Shelton Street, Covent Garden, London WC2H 9HE

British Library Cataloguing-in-Publication Data
A catalogue record for this book is available from the British Library.

Fresco on the front cover:

Spin Game Fresco (Game Hall) in the Estense Castle of Ferrara
by Sebastiano Filippi (Bastianino), XVI secolo,
courtesy of Provincia di Ferrara.

The background:

Bugnato of the exterior walls — Palazzo dei Diamanti — Ferrara
(build from 1493, designed by Biagio Rossetti),
courtesy of Comune di Ferrara.

TRANSVERSITY 2008
Proceedings of the Second Workshop on Transverse Polarization Phenomena in Hard Processes

ISBN-13 978-981-4277-77-8
ISBN-10 981-4277-77-0

Printed in Singapore by World Scientific Printers

PREFACE

Transversity 2008, the second workshop on "Transverse polarization phenomena in hard processes",* follows the first one held in Como at a distance of three years. As in that case, the event comes at the end of a two years project financed by the Italian Ministry of Education (PRIN-2006). In the time interval between the two Workshops, decisive steps towards the revealing of transverse spin structure of the proton have been accomplished both on the theoretical and on the experimental sides. A particular mention deserves the milestone represented by the first extraction of Transversity and the Sivers function for the u- and d- quarks. In the same period hystoric experiments, that in the last decade contributed to the first pioneering measurements in the SIDIS sector, have concluded their data taking, and their place is being taken by upgrades of existing facilities or new facilities. These are the result of the new interesting phenomena which are appearing and call for additional experimental informations and novel experimental techniques.

It is in this framework of achievements and perspectives that the second Workshop on Transversity and related phenomena took place in Ferrara on May 28th–31st 2008. The site for the workshop has been chosen to be the Hall of the Camera di Commercio facing the XV century Castello Estense, in the heart of the town. The old "*imbarcadero*" of the Castle itself hosted most of the social events in a fancy medieval atmosphere. Even the picture for the conference poster, representing six "*putti*" playing the spinning-tops, is based on a XVI century fresco decorating the ceiling of one of the Castle halls and demonstrates that the interest for spinning objects has deep roots in history.

Over 80 physicists took part to the Workshop, 38 of them made presentations. The participation equally involved experimentalists and theoreticians

*Further information on the workshop and slides of contributions may be found at the following permanent website: http://www.fe.infn.it/transversity2008/

engaged in investigating the nature of transverse spin; the heterogeneous public favoured vivid discussions and fruitful exchange of up-to-date theoretical and experimental ideas on the subject.

The source of finance for the workshop, besides the Ministry funded PRIN, were the Universities of Ferrara, Insubria, Piemonte Orientale, Torino and Trieste; the INFN Frascati National Laboratories and the section of Ferrara; the Virtual Institute for spin-physics and QCD, the municipality and the province of Ferrara.

The success of the Workshop, both in terms of scientific merit and attendance, witnesses the vivacity and actuality of the field and the organizers do wish that it can become a regular appointment for the community. In view of this, we hope to meet all and more of you at the third workshop on Transverse polarization phenomena in hard processes.

4 December 2008

Ciullo Giuseppe
Università degli Studi di Ferrara and INFN
Contalbrigo Marco
INFN sezione di Ferrara
Hasch Delia
INFN Laboratori Nazionali di Frascati
Lenisa Paolo
Università degli Studi di Ferrara and INFN

ORGANIZING COMMITTEES

Local Organizing Committee

Paolo Lenisa (Chair)	Università di Ferrara and INFN
Vincenzo Barone	Università del Piemonte Orientale
Maria Pia Bussa	Università di Torino
Giuseppe Ciullo	Università di Ferrara and INFN
Marco Contalbrigo	INFN of Ferrara
Delia Hasch	INFN, Laboratori Nazionali di Frascati
Anna Martin	Università di Trieste
Philip G. Ratcliffe	Università dell'Insubria,

International Advisory Committee

Franco Bradamante (Chair)	Università degli Studi di Trieste
Mauro Anselmino	Università degli Studi di Torino
Stanley J. Brodsky	SLAC, USA
John C. Collins	Penn State University
Anatoli Efremov	JINR, Dubna
Elliot Leader	London University
Naomi Makins	Illinois University
Pieter Mulders	Vrije Universiteit, Amsterdam
Bernard Pire	CNRS
Frank Rathmann	IKP Forschungszentum Jülich
Klaus Rith	Universität Erlangen-Nürberg
Nahoito Saito	Kyoto University
Werner Vogelsang	BNL

ACKNOWLEDGMENT

The Organizers would like to acknowledge

- MIUR, VI-QCD and INFN for the finacial support,
- L. Barion, F. Giordano, L.L. Pappalardo, M. Stancari, M. Statera and the secretaries L. De Marco and P. Fabbri for their help in logistics and organization,
- Comune di Ferrara, Provincia di Ferrara, Camera di Commercio di Ferrara and Consorzio Ferrara Ricerche
- the companies Varian, Oerlikon-Leybold, Romete, Vacuum Science of L. Rimoldi, SpringerVerlag-Italia and World Scientific

04 December 2008

Ciullo Giuseppe
Università degli Studi di Ferrara and INFN

Contalbrigo Marco
INFN sezione di Ferrara

Hasch Delia
INFN Laboratori Nazionali di Frascati

Lenisa Paolo
Università degli Studi di Ferrara and INFN

CONTENTS

Introduction

PURPOSE AND STATUS OF THE ITALIAN TRANSVERSITY PROJECT

F. Bradamante
Dipartimento di Fisica dell'Università degli Studi di Trieste
and INFN, Sezione di Trieste,
Via A. Valerio 2, I-34127 Trieste, Italy

It is a great pleasure for me to open the second International Workshop on Transverse Polarization Phenomena in Hard Processes. The present Workshop is organized in the frame of an italian inter-university research project dedicated to transversity and transverse spin effects, which are presently some of the most debated issues in Hadron Physics. This project was proposed in 2006 to the Italian Ministry for Education, University and Research (MIUR, Ministero dell'Istruzione, dell'Università e della Ricerca) which each year supports two-year Research Projects of National Interest ("Progetti di Ricerca Interesse Nazionale", PRIN). The name of the project is "Study of transverse Spin Effects in the Nucleon". The proponers are some 35 theoretical and experimental physicists all involved in the physics of transversity. The proponents belonged to six Italian Institutes, located in Alessandria, Como, Ferrara, Frascati National Laboratory of INFN, Torino, and Trieste. The theoreticians are all well known physicists actively working in the field of transverse spin phenomena in hadronic physics. The experimentalists are active members of either the COMPASS or the HERMES Collaborations: RHIC experiments were not represented, since no Italian group is involved in RHIC. In spite of the tremendous competition among the PRIN projects, our proposal was approved, and the organisation of this workshop is part of the project. The Local Organizing Committee comprises one member from each participating Institute, and financial support is granted by all of them.

The project is the natural continuation of the PRIN2003 "Measurement of the Nucleon Transversity", which was proposed essentially by the same people in 2003, and which turned out to be very successful. Most important findings occurred during the years of that project, which were all presented and discussed in Como, from 7 to 10 September 2005 , when the first Transversity workshop was held.

To perform a global analysis was the goal of our PRIN2003, and it is most rewarding to acknowledge that now this goal has been achieved. The goals of PRIN 2006 are even more ambitious: we would like to create a general and homogenous framework for all the different processes that measure, in a direct or indirect manner, different combinations of the transverse-spin distribution and fragmentation functions, so as to converge towards a more and more definitive knowledge of transversity and towards more and more reliable predictions of new and as yet unmeasured processes. Another goal of this Research Program is to promote the measurement of h_1 at the new antiproton facility in the new large European project for hadronic physics FAIR, at the GSI laboratories in Darmstadt, via the measurement of the double-spin asymmetry of Drell-Yan pairs in the process $\bar{p}p \to l^+l^-X$.

It is a great pleasure and very rewarding for us to see that the community has responded to our call and that most of the actors in the field of transversity have come to Ferrara. In Como it was clear that the transversity phenomena had been definitely assessed, and that a new phase of investigation had to be opened up. In the past two years there have been very important advances in our field. But this field is really a new field, still in its infancy. As such, it is growing at a terrific rate, surprises are always possible, and we all must be ready for discussion. The precision of the measurements have improved, and the first signals which were observed before 2006 have been confirmed beyond any possible doubt. A first global analysis of all the data coming from SIDIS experiments and from e^+e^- annihilating into hadrons has already been published last year, allowing the first ever extraction of the transversity distributions of the u and d quarks.

During the workshop we will surely have many occasions to check how far we can go with our ambitious programme.

As a conclusion, let's look forward to a most interesting and useful Workshop, which has been organized thanks to MIUR support.

Contributions

TRANSVERSITY ASYMMETRIES

D. Boer

Department of Physics and Astronomy,
Vrije Universiteit Amsterdam,
De Boelelaan 1081, 1081 HV Amsterdam,
The Netherlands

Ways to access transversity through asymmetry measurements are reviewed. The recent first extraction and possible near future extractions are discussed.

Keywords: Transverse spin physics, cross section asymmetries

1. Transversity from 1978 to 2008

The year 1978 marks the birth of transversity *as a quark distribution* with the submission of the seminal paper[1] by Ralston and Soper on November 14, 1978. Transversity was a pre-existing term, but is meant here as the distribution of transversely polarized quarks inside a transversely polarized hadron. It depends on the lightcone momentum fraction x carried by the quark and is often denoted by $h_1(x)$. Theoretically it is defined as a hadronic matrix element of a nonlocal operator:

$$\int \frac{d\lambda}{2\pi} e^{i\lambda x} \langle P, S_T | \overline{\psi}(0) \mathcal{L}[0,\lambda] i\sigma^{i+}\gamma_5 \psi(\lambda n_-) | P, S_T \rangle = S_T^i \, h_1(x), \qquad (1)$$

where P and S_T denote the momentum and the transverse spin vector of the hadron, n_- is a lightlike direction, and $\mathcal{L}$ is a path-ordered exponential that renders the nonlocal operator color gauge invariant. Transversity is a chiral-odd or helicity-flip quantity, hence, in observables it always appears accompanied by another chiral-odd quantity, several of which will be discussed below. Ralston and Soper considered the double transverse spin asymmetry in the Drell-Yan process (reconsidered in detail at a later stage in Refs.[2–4]), i.e. the asymmetry in the azimuthal angular distribution of a produced lepton pair in the collision of two hadrons, in this case protons, with transverse spins parallel minus antiparallel:

$$A_{TT} = \frac{\sigma(p^\uparrow p^\uparrow \to \ell \bar{\ell} X) - \sigma(p^\uparrow p^\downarrow \to \ell \bar{\ell} X)}{\sigma(p^\uparrow p^\uparrow \to \ell \bar{\ell} X) + \sigma(p^\uparrow p^\downarrow \to \ell \bar{\ell} X)} \propto \sum_q e_q^2 \, h_1^q(x_1) \, h_1^{\bar{q}}(x_2). \quad (2)$$

However, polarized Drell-Yan is very challenging experimentally, as witnessed by the fact that even 30 years later it has not yet been performed. RHIC at BNL is at present the only place that can do double polarized proton-proton scattering, but A_{TT} is expected to be small at RHIC. It involves two unrelated transversity functions: the one for quarks and the one for antiquarks for which likely holds that $h_1^{\bar{q}} \ll h_1^q$. An upper bound on A_{TT} can be obtained by using Soffer's inequality,

$$|h_1(x)| \leq \frac{1}{2}\,[f_1(x) + g_1(x)]. \quad (3)$$

The upper bound on A_{TT} was shown[5] to be small at RHIC, of the percent level, requiring an accuracy that will not be reached soon.

Already before the advent of RHIC, people started to search for alternative ways of probing transversity. The first suggestion was made by Collins.[6] The idea was to exploit what is now referred to as the Collins effect, which is parameterized by the transverse momentum dependent fragmentation function $H_1^\perp(z, k_T^2)$. It is a spin-orbit coupling effect in the fragmentation of a transversely polarized quark, resulting in an asymmetric azimuthal angular distribution of produced hadrons around the quark polarization direction, a $\sin\phi$ distribution. $H_1^\perp$ is also a chiral-odd quantity.

Collins pointed out that there would be a $\sin(\phi_h + \phi_S)$ asymmetry in semi-inclusive DIS (SIDIS) proportional to $h_1 \otimes H_1^\perp$. Here ϕ_h and ϕ_S are the azimuthal angles of a final state hadron and the transverse spin of the initial state polarized hadron, respectively. The angles are measured w.r.t. the lepton scattering plane, which fixes the polarization state of the virtual photon such that the helicity flip state of the probed quark is selected.

The HERMES experiment at DESY was the first to measure a clearly nonzero, percent level, $\sin(\phi_h + \phi_S)$ Collins asymmetry in SIDIS.[7,8] This asymmetry has afterwards also been measured by the COMPASS experiment at CERN using a deuteron target[9,10] and, as first reported at this Transversity 2008 workshop, also on a proton target.[11] These measurements allow for an extraction of transversity once the Collins function $H_1^\perp$ is known. This exemplifies the crucial role played by electron-positron annihilation experiments. In Ref.[12] it was pointed out how $H_1^\perp$ can be extracted from a $\cos 2\phi$ asymmetry in $e^+ e^- \to \pi^+ \pi^- X$ (see also the recent, more extended Ref.[13]). This turned out to be the method that has actually been employed. The measurement of this $\cos 2\phi$ asymmetry has been performed at KEK using BELLE data.[14,15] This allowed for the first extraction of transversity by Anselmino *et al.*[16] in 2007, cf. Fig. 1.

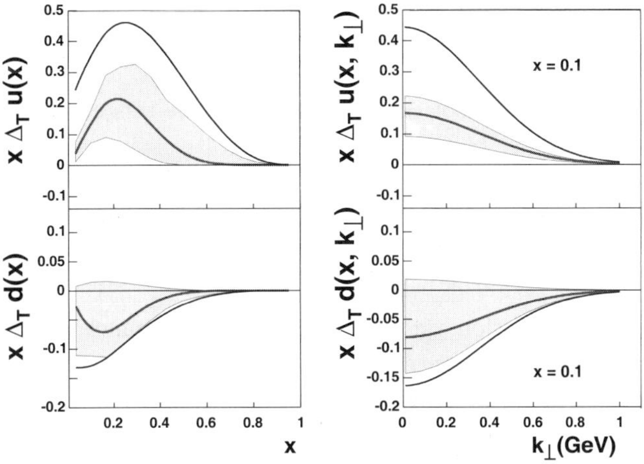

Fig. 1. First extraction of transversity by Anselmino *et al.*[16] Left: $xh_1(x)$ for u and d quarks, the red curves are the best fits, the blue curves the Soffer bounds. Right: k_T-dependence using a Gaussian Ansatz.

Next we turn to a discussion of the magnitude of the extracted transversity functions for u and d quarks. Often h_1 is compared with its Soffer bound in Eq. (3) or with g_1, which is interesting for theoretical reasons, but for phenomenology it is more relevant to compare it to f_1, since that is what determines the magnitude of asymmetries. The first extraction, the best fit, indicates that $h_1(x) \approx f_1(x)/3$, which means that transversity is not particularly small. Whether it is of the expected magnitude is a different matter though. One way of quantifying this is to compare it to expectations from lattice QCD and from models for the tensor charge,

$$\delta q = \int_0^1 dx \left[h_1^q(x) - h_1^{\bar{q}}(x)\right], \tag{4}$$

which is a fundamental charge, like the electric and the axial charge. Transversity is the only known way of obtaining the tensor charge experimentally. Using the central fit and assuming antiquark transversity to be small, the first extraction yields[17] (at $Q^2 = 2.4$ GeV^2)

$$\delta u \simeq +0.39, \; \delta d \simeq -0.16, \; \text{s.t.} \; \delta u - \delta d \simeq 0.55$$

A lattice determination with two dynamical quark flavors yields[18] (at $\mu^2 = 4$ GeV^2)

$$\delta u = +0.857 \pm 0.013, \;\; \delta d = -0.212 \pm 0.005, \;\; \text{s.t. } \delta u - \delta d = 1.068 \pm 0.016$$

The combination $\delta u - \delta d$ is given, because it has the advantage of cancelation of disconnected contributions which, although expected to be small, are not calculated.

Most models find tensor charges roughly in the range:

$$\delta u = +1.0 \pm 0.2, \quad \delta d = -0.2 \pm 0.2$$

All of this is consistent with the bounds derived by Soffer:[19]

$$|\delta u| \leq 3/2, \quad |\delta d| \leq 1/3$$

The recent extraction via the Collins effect asymmetries seems to indicate a u-quark tensor charge that is smaller than expected from lattice QCD and most models. However, at this workshop we learned that a new fit using newer and more accurate data yields a larger δu, which seems more in line with expectations.[20]

There is nevertheless another issue concerning the magnitude of the extracted transversity functions. The BELLE and SIDIS data are obtained at different scales: $Q^2 = 110$ GeV^2 and $\langle Q^2 \rangle = 2.4$ GeV^2, respectively. The extraction uses two Collins effect asymmetries, which are not like ordinary leading twist asymmetries. Both azimuthal asymmetries involve transverse momentum dependent functions (TMDs) and beyond tree level this becomes quite involved. The formalism that deals with TMDs beyond tree level is that of Collins-Soper (CS) factorization, initially considered for (almost) back-to-back hadron production in e^+e^- annihilation[21], and later for SIDIS and Drell-Yan.[22,23] In principle, CS factorization dictates how azimuthal asymmetries depend on Q^2, but in practice this has not been implemented in the h_1 extraction analysis.[16] Evolution is taken into account only partially in the following way. The Collins function is parameterized in terms of the unpolarized fragmentation function D_1,

$$H_1^{\perp}(z, k_T^2) \equiv D_1(z) F(z, k_T^2), \tag{5}$$

and the evolution is taken to be the one of the collinear function $D_1(z)$. This does not take into account that beyond tree level also the transverse momentum dependence requires modification with changing energy scale.

Collins effect asymmetries involve convolution integrals, for example the SIDIS asymmetry as a function of the observed transverse momentum $\boldsymbol{q}_T$

(with absolute value Q_T),

$$\frac{d\sigma(e\,p^\uparrow \to e'\,h\,X)}{d^2\boldsymbol{q}_T} \propto \frac{|\boldsymbol{S}_T|}{Q_T}\,\sin(\phi_h+\phi_S)\,\mathcal{F}\left[\frac{\boldsymbol{q}_T\cdot\boldsymbol{k}_T}{M}\,h_1 H_1^\perp\right], \tag{6}$$

involves a convolution that at tree level is of the form:

$$\mathcal{F}\left[w\,f\,D\right] \equiv \int d^2\boldsymbol{p}_T\,d^2\boldsymbol{k}_T\,\delta^{(2)}(\boldsymbol{p}_T+\boldsymbol{q}_T-\boldsymbol{k}_T)\,w(\boldsymbol{p}_T,\boldsymbol{q}_T,\boldsymbol{k}_T)$$
$$\times f(x,\boldsymbol{p}_T^2)D(z,z^2\boldsymbol{k}_T^2). \tag{7}$$

In general however it involves another factor U (called S in Refs.[22,23]):

$$\mathcal{F}\left[w\,f\,D\right] \equiv \int d^2\boldsymbol{p}_T\,d^2\boldsymbol{k}_T\,d^2\boldsymbol{l}_T\,\delta^{(2)}(\boldsymbol{p}_T+\boldsymbol{l}_T+\boldsymbol{q}_T-\boldsymbol{k}_T)\,w(\boldsymbol{p}_T,\boldsymbol{q}_T,\boldsymbol{k}_T)$$
$$\times f(x,\boldsymbol{p}_T^2)D(z,z^2\boldsymbol{k}_T^2)U(l_T^2). \tag{8}$$

In terms of diagrams the difference is expressed in Fig. 2. As a side remark,

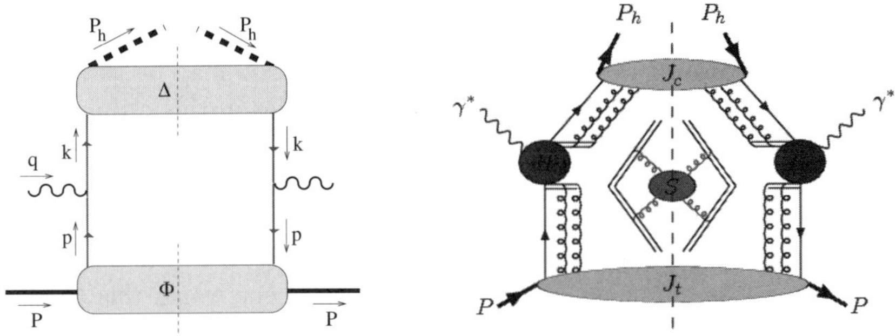

Fig. 2. The left figure shows pictorially the tree level expression ($H = 1$ and $U(l_T^2) \propto \delta(l_T^2)$), the right figure (by F. Yuan) shows the all-order expression (with $S = U$).

we note that it is possible to get rid of the convolutions by weighted integration over the observed transverse momentum of the asymmetry. This requires that the asymmetry is well-described for all values of the transverse momentum, which means that one has to connect the CS factorization expressions to the collinear factorization ones that are valid at large transverse momenta. For the Q_T-weighted Collins asymmetry in SIDIS this works, but for the Q_T^2-weighted $\cos 2\phi$ asymmetry in e^+e^- annihilation a direct extraction of the Collins function is not possible in this way.[13,24]

Beyond tree level the soft factor U dilutes the asymmetry, and increasingly so as Q^2 increases. Differently stated, for the same functions f and D and weight w, the quantity $\mathcal{F}\left[w\,f\,D\right]$ is smaller beyond tree level. This effect becomes stronger as Q increases and is referred to as Sudakov suppression. Conversely, if $\mathcal{F}[w\, f\, D]$ is obtained from experiment and if for instance f is extracted from it for given w and D, then f will be larger when using the expression beyond tree level. In Ref.[25] this was studied numerically and a rule of thumb for the Q^2 dependence of azimuthal asymmetries was put forward: asymmetries involving one k_T-odd TMD, such as the Collins effect asymmetry in SIDIS, approximately fall off as $1/\sqrt{Q}$; asymmetries involving two k_T-odd functions, such as the Collins effect asymmetry in e^+e^- annihilation, approximately fall off as $1/Q$. This behavior was obtained in the investigated range of $Q = 10 - 100$ GeV and is to a very large extent independent of model assumptions, even though the magnitude of the asymmetries does depend heavily on them.

This Sudakov suppression implies that tree level extractions of the Collins function from the $\cos 2\phi$ asymmetry at BELLE, leads to an underestimation of $H_1^\perp$ (since beyond tree level it will be larger). Hence, using that underestimated function to extract transversity from SIDIS data at a lower Q^2 (less Sudakov suppression), leads to an overestimation of h_1. Based on the results of Refs.[13,25], this overestimation may be as large as a factor of 2, although there are many uncertainties in this estimate and it does not take into account that in Ref.[16] some Q^2 dependence of $H_1^\perp$ is included through the scale dependence of D_1, as explained above.

To get clarity about the magnitude and about the reliability of the Collins effect extraction method, of course the best would be to do another independent measurement of transversity. Ideally one wants this to be a non-TMD, self-sufficient transversity measurement. These are the cleanest transversity asymmetries that consist of a single observable that only involves collinear distributions and do not require experimental input from other experiments done at different scales and/or using different processes.

Before addressing this topic in detail, it may be worth recalling that the scale dependence of $h_1(x)$ itself is quite well-known, i.e. to next-to-leading order.[26–28] The evolution of $h_1(x, Q^2)$ is very different from that of $g_1(x, Q^2)$, in part, because there is no gluon transversity distribution. h_1 grows with increasing Q^2 towards smaller x, to eventually become proportional to $\delta(x)$, but with a proportionality constant that decreases to zero as $Q^2 \to \infty$, hence $h_1(x, Q^2) \to 0$. Therefore, also the tensor charge decreases with Q^2, but it should be emphasized that it is only very mildly energy scale dependent. At the Planck scale the tensor charge is still only reduced by a factor 2 w.r.t. $Q^2 = 1$ GeV under next-to-leading order (NLO) evolution.

2. Transversity asymmetries

There is an obvious classification of transversity asymmetries into double and single transverse spin asymmetries, but from a theoretical point of view there is a more important distinction based on whether TMDs are involved or not. Cases where collinear factorization can be applied are much safer than cases for which CS factorization is expected to apply. The latter usually require some as yet unknown nonperturbative input and one has to resort to model assumptions, for instance about the transverse momentum shape of the TMDs, as was done for the first transversity extraction in Ref.[16]. For the analysis it also matters whether one has to combine information from several observables, either obtained under the same experimental conditions or different ones. Below these aspects will be discussed for the explicit routes to transversity.

2.1. *Double transverse spin asymmetries*

Almost no experiment aiming to extract h_1 will be self-sufficient. Most cleanly this requires experiments probing a single "$(h_1)^2$" observable. There are only two such processes:

- $\bar{p}^\uparrow \, p^\uparrow \to \ell \, \bar{\ell} \, X$
- $p^\uparrow \, p^\uparrow \to$ high-p_T jet $+ \, X$

Both processes were discussed by Artru and Mekhfi.[2] But the first process was only recently considered in detail, because of plans to use the future FAIR facility at GSI for its measurement. The second process was extensively discussed by Jaffe and Saito[29], who concluded that it is likely too challenging to be done at RHIC, because it leads to a permille level asymmetry (a result confirmed by Vogelsang[30]).

A somewhat less clean observable is $\bar{p}^\uparrow \, p^\uparrow \to \pi \, X$, which is $\propto (h_1)^2 D_1$, considered by Mukherjee, Stratmann and Vogelsang.[31] Also $\bar{p}^\uparrow \, p$ or $\bar{p} \, p^\uparrow$ Drell-Yan experiments are self-sufficient, but these involve TMDs and will be discussed in the next subsection on TMD single spin asymmetries.

First we look at double transverse spin asymmetries in $\bar{p}^\uparrow \, p^\uparrow$ collisions, in particular in Drell-Yan. It is ideally suited for h_1 extraction, because $h_1^{\bar{q}/\bar{p}} = h_1^{q/p}$, leading to:

$$A_{TT} = \frac{\sigma(\bar{p}^\uparrow \, p^\uparrow \to \ell \, \bar{\ell} \, X) - \sigma(\bar{p}^\uparrow \, p^\downarrow \to \ell \, \bar{\ell} \, X)}{\sigma(\bar{p}^\uparrow \, p^\uparrow \to \ell \, \bar{\ell} \, X) + \sigma(\bar{p}^\uparrow \, p^\downarrow \to \ell \, \bar{\ell} \, X)} \propto \sum_q e_q^2 \, h_1^q(x_1) \, h_1^q(x_2) \quad (9)$$

As said, this can perhaps be done at GSI-FAIR. Some of the considered options[32,33] are a collider mode at $\sqrt{s} = 14.5$ GeV (the currently preferred asymmetric collider option of 15 GeV antiprotons on 3.5 GeV protons) and a fixed target mode at $\sqrt{s} = 6.7$ GeV (usually quoted as $s = 45$ GeV^2).

A chiral quark soliton model calculation[34] of h_1 indicates that large asymmetries of 40-50% can be expected in the fixed target mode at $s = 45$ GeV^2. The asymmetry grows with increasing Q^2, but is generally smaller for the higher-$\sqrt{s}$ collider mode. Upper bounds on the asymmetry of approximately 17% at $Q = 2$ GeV to 38% at $Q = 12$ GeV for the collider mode of $\sqrt{s} = 14.5$ GeV have been obtained by Shimizu *et al.*[35] The first extraction of h_1 indicates that transversity is not much smaller than its upper bound, so asymmetries of order 10% at GSI kinematics should be expected. A Monte Carlo study regarding the feasibility of measuring A_{TT} at GSI-FAIR is promising.[36,37] But in the end the success of double polarized Drell-Yan at GSI-FAIR depends predominantly on whether significant polarization of the antiproton beam can be achieved.

From the study by Shimizu *et al.*[35] of the upper bound on A_{TT} it has also become clear that perturbative corrections hardly affect the asymmetry. The transition from leading to next-to-leading order pQCD is small and also resummation of large logs hardly has an effect. A similar robustness can be observed for a closely related asymmetry investigated in Ref.[38], the unintegrated asymmetry $A_{TT}(Q_T)$, which depends on the transverse momentum Q_T of the lepton pair w.r.t. the beam axis. Although resummation is essential for this observable, which is described within the CSS formalism[39] that derives from the CS formalism discussed before, resummation beyond the leading-logarithmic approximation (LL) has little effect on the asymmetry. As explained in Ref.[38] this is particular to $\bar{p}\,p$ scattering in the valence region.

In Ref.[38] the upper bound of the asymmetry $A_{TT}(Q_T)$ for GSI kinematics was shown to be of similar magnitude as the integrated asymmetry A_{TT} (which is obtained from $A_{TT}(Q_T)$ by integrating its numerator and denominator separately). Remarkably, $A_{TT}(Q_T)$ is very flat as a function of Q_T and remains flat under Q^2 evolution.

The asymmetry $A_{TT}(Q_T)$ for $p\,p$ scattering,[40] which is considerably smaller for RHIC ($\sqrt{s} = 200$ GeV) than J-PARC ($\sqrt{s} = 10$ GeV) kinematics, shows a very different behavior compared to $\bar{p}\,p$ scattering for potential GSI kinematics. The asymmetry is flat at LL level, but not at next-to-leading log. Resummation beyond LL clearly matters in $p\,p$ collisions.

In conclusion, the double transverse spin asymmetries in $\bar{p}^{\uparrow}\,p^{\uparrow}$ Drell-Yan offer clean, direct and unique probes of transversity and in the valence region they are very robust under perturbative corrections.

As mentioned, one could also consider $A_{TT}^{\pi^0}$ in $\bar{p}^\uparrow p^\uparrow \to \pi^0 X$, which is a slightly less clean observable as it requires input on the pion fragmentation function, which however is quite well-known. Upper bounds for assumed beam polarizations of 30% for $\bar{p}$ and 50% for p yield asymmetries of a few percent.[31] The difference between LO and NLO is a bit larger in this case.

2.2. *TMD single spin asymmetries*

What if one only has one polarized beam? This question is relevant for GSI if the antiproton beam cannot be polarized significantly. For one polarized beam there is a self-sufficient measurement of transversity which involves TMDs, namely the single spin asymmetry in $\bar{p}^\uparrow p \to \ell \bar{\ell} X$ or $\bar{p} p^\uparrow \to \ell \bar{\ell} X$. Both options are equally useful, there is no difference theoretically.

In the case of one transversely polarized hadron beam, there is a possible spin angle ϕ_S dependence of the differential cross section:

$$\frac{d\sigma}{d\Omega\, d\phi_S} \propto 1 + \lambda \cos^2\theta + \sin^2\theta \left[\frac{\nu}{2}\cos 2\phi - \rho\, |\boldsymbol{S}_T| \sin(\phi + \phi_S)\right] + \ldots$$

In a measurement of ρ (from $p^\uparrow - p^\downarrow$) also ν can be extracted from the same data (from $p^\uparrow + p^\downarrow$), i.e. under exactly the same experimental conditions. This is in contrast to the previously discussed Collins effect asymmetries. At tree level one has

$\nu \propto h_1^\perp\, h_1^\perp$ analogue of $\cos 2\phi$ asymmetry in e^+e^-

$\rho \propto h_1\, h_1^\perp$ analogue of Collins asymmetry in SIDIS

These two expressions involve the TMD distribution function[41] $h_1^\perp$, which in some respects is very similar to the Collins effect fragmentation function, but can be quite different from it. It is depicted in Fig. 3.

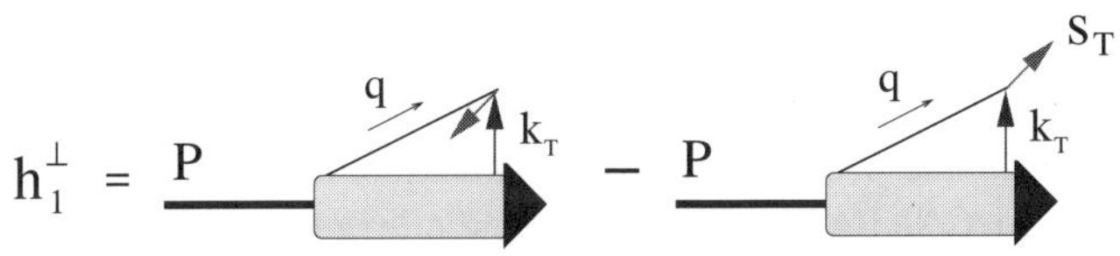

Fig. 3. Nonzero $h_1^\perp$ means that the transverse polarization S_T of quarks (with momentum $q \approx xP + k_T$) inside an unpolarized hadron (with momentum P) is nonzero. It is a k_T-odd and chiral-odd TMD.

The asymmetry ν has been measured in $\pi^- N \to \mu^+\mu^- X$ by the NA10 Collaboration[42,43] at CERN and the E615 Collaboration[44] at Fermilab, roughly 20 years ago. The data show an anomalously large asymmetry, which differs much from the perturbative QCD $\mathcal{O}(\alpha_s)$ Lam-Tung relation $\nu = (1-\lambda)/2$ and the $\mathcal{O}(\alpha_s^2)$ corrections to it.[45,46] Nonzero $h_1^\perp$ offers an explanation for this discrepancy.[47] Assuming u-quark dominance, Gaussian k_T-dependence for $h_1^\perp$ and x-dependence $\propto f_1(x)$, ρ can be related to ν:[48]

$$\rho = \frac{1}{2}\sqrt{\frac{\nu}{\nu_{\max}}}\,\frac{h_1^u}{f_1^u} \tag{10}$$

The result is displayed in Fig. 4.

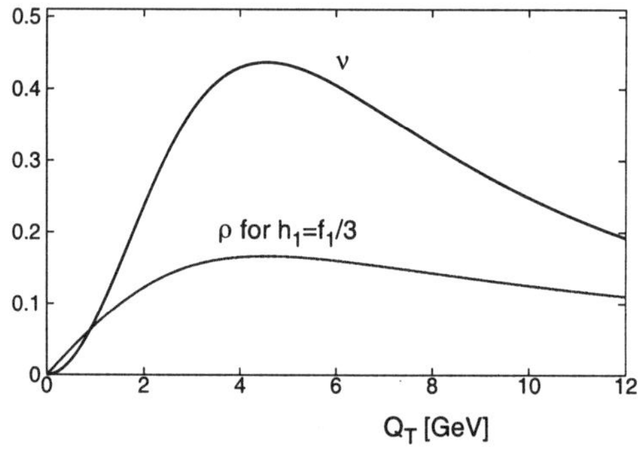

Fig. 4. Analyzing power ν of $\cos 2\phi$ asymmetry as fitted to NA10 data using a model Ansatz[47,48] for $h_1^\perp$ and the resulting prediction of the single spin asymmetry ρ using Eq. (10) for the case $h_1 = f_1/3$.

The asymmetry ν for pp (e.g. at RHIC, where also ρ can be measured) is expected to be smaller than for $\pi\, p$, due to absence of valence antiquarks. Preliminary pp data from Fermilab were shown at this workshop[49] and confirm this expectation. Earlier pd data[50] also show a small asymmetry, probably for the same reason.

The asymmetry ν for $\bar{p}p$ on the other hand is expected to be very similar to $\pi\, p$, due to the presence of valence antiquarks. Therefore, unpolarized $\bar{p}p$ Drell-Yan at GSI-FAIR will likely show a large anomalous $\cos 2\phi$ asymmetry, providing crucial information about its origin. As explained above, the measurement of ν and ρ at GSI-FAIR with only one polarized beam (either $\bar{p}^\uparrow$ or $p^\uparrow$) offers a probe of transversity. In this case predominantly $h_1^{\perp u/p}$ and $h_1^{u/p}$ are accessed, due to the charge-squared factor in Eq. (9).

The COMPASS experiment plans to do $\pi^\pm p^\uparrow$ Drell-Yan[51], which although not self-sufficient would provide valuable information on the flavor dependence of h_1 and $h_1^\perp$. Especially $\pi^+ p^\uparrow$ is of interest, as there is no data available on it yet and it provides information on the d-quark ratio $h_1^{\perp d/p}/h_1^{d/p}$, without suppression by a charge-squared factor. The ratio ν/ρ for $\pi^\pm p^\uparrow$ Drell-Yan in valence approximation namely provides the ratios $h_1^{\perp u/p}/h_1^{u/p}$ and $h_1^{\perp d/p}/h_1^{d/p}$ for π^- and π^+, respectively. Using the input on $h_1^\perp$ from for example unpolarized $p\bar{p}$ Drell-Yan (either from the Tevatron or from GSI-FAIR) would allow for an extraction of h_1 from $\pi^\pm p^\uparrow$ Drell-Yan at COMPASS.

The function $h_1^\perp$ may be extracted from ν at the Tevatron, but the high $\sqrt{s}$ leads to high Q^2 on average. This can result in considerable Sudakov suppression, which would be disadvantageous but interesting to verify. One may also probe $h_1^\perp$ via a $\cos 2\phi$ asymmetry in photon-jet production $p\,\bar{p} \to \gamma\,\text{jet}\,X$ at the Tevatron,[56]

$$\frac{d\sigma^{h_1\,h_2\to\gamma\,\text{jet}\,X}}{d\eta_\gamma\,d\eta_j\,d^2\boldsymbol{K}_{\gamma\perp}\,d^2\boldsymbol{q}_\perp} \propto \left(1+\nu_{\rm DY}\,R\cos 2(\phi_\perp-\phi_\gamma)\right) \tag{11}$$

where $\phi_\perp$ is the angle of the transverse momentum $\boldsymbol{q}_\perp$ of the photon-jet system and ϕ_γ is the angle of the transverse momentum $\boldsymbol{K}_{\gamma\perp}$ of the photon. The analyzing power consists of a proportionality factor R times $\nu_{\rm DY}$, the $\cos 2\phi$ asymmetry of Drell-Yan probed at the scale $|\boldsymbol{K}_{\gamma\perp}|$ which in general is different from Q, which might make a difference from the perspective of Sudakov suppression. The proportionality factor R is only a function of f_1. For typical Tevatron kinematics in the central region, recently investigated for the angular integrated case by the DØ Collaboration[57], $\nu_{\rm DY}\,R$ was estimated[56] to be $\sim 5-15\%$. That could be large enough to allow transversity related TMD studies at the Tevatron too.

Another "helper" process is the $\cos 2\phi$ asymmetry ν in unpolarized SIDIS $e\,p \to e'\,\pi\,X$, which would be proportional to $h_1^\perp H_1^\perp$. Given the Collins function it could in principle be used to extract $h_1^\perp$ too. The asymmetry in SIDIS turns out to be of quite different size compared to Drell-Yan. It has been investigated using model calculations in e.g. Refs.[52–54]. The asymmetries as a function of observed transverse momentum of the pion are typically on the percent level and are very similar in size for HERMES kinematics and JLab kinematics (the 12 GeV upgrade). Interestingly, the π^- asymmetries are positive and according to Ref.[52] roughly four times as large as the π^+ asymmetries which is of opposite sign. This factor of four is not related to the charge-squared factor ratio of u and d quarks.

The available data on the $\cos 2\phi$ asymmetry in unpolarized SIDIS are from EMC and COMPASS; the latter were presented at this workshop for the first time[55] (soon also HERMES data should become available). The data show that $\nu_{\rm SIDIS} \ll \nu_{\rm DY}$. The SIDIS data are obtained for not too large values of Q^2, where also higher twist contributions, such as the Cahn effect, can be relevant. The recent model calculation of Ref.[54] for instance shows this very clearly. This limits the usefulness of this observable for transversity related investigations. Nevertheless, it is interesting to study the importance of higher twist effects at HERMES and COMPASS energies through this observable.

It should be added that there is also high Q^2 data on the unpolarized azimuthal asymmetries in SIDIS from ZEUS ($\langle Q^2 \rangle = 750$ GeV^2). Within the sizeable errors the ZEUS data are consistent with pQCD expectations, but they have been presented with a lower cut-off on the transverse momentum of the final state hadron, cutting out contributions of interest here. Apart from that, high Q^2 is not favorable to probe the $h_1^\perp H_1^\perp$ contribution due to the Sudakov suppression discussed earlier.

2.3. *Non-TMD single spin asymmetries*

If one only has one transversely polarized proton beam, then there are two further possibilities to probe h_1 which do not involve TMDs, i.e. to use:

- transverse Λ polarization
- two hadron systems within a jet

Both options are not self-sufficient, at least not in a straightforward way; they involve unknown fragmentation functions, which most cleanly can be obtained from e^+e^- data. The big advantage is though that collinear factorization applies, therefore, one only deals with non-TMD functions.

Transverse Λ polarization enters with the transversity fragmentation function $H_1(z)$. It is still unknown, but can be measured in $e^+ \, e^- \to \Lambda^\uparrow \, \overline{\Lambda}^\uparrow X$: $\propto (H_1)^2$.[58] Subsequently, h_1 can be accessed via the spin transfer asymmetry $D_{NN} \propto h_1 H_1$ in either $e \, p^\uparrow \to e' \, \Lambda^\uparrow \, X$ or $p \, p^\uparrow \to \Lambda^\uparrow \, X$. The latter has been measured by the E704 Collaboration[59], yielding a D_{NN} of order 20-30% at a transverse momentum p_T of the Λ of around 1 GeV/c ($\sqrt{s} \approx 20$ GeV). However, because of the low p_T, this result can probably not be used to extract h_1 in a trustworthy manner. This should be different at RHIC. Upper bounds for D_{NN} calculated[60] for RHIC at $\sqrt{s} = 500$ GeV show promisingly large asymmetries at much larger p_T.

The other option is to use the Interference Fragmentation Function $H_1^{\sphericalangle}$, or more generally, chiral-odd two-hadron fragmentation functions. Consider for definiteness the final state $|(\pi^+ \, \pi^-)X\rangle$, i.e. a $\pi^+\pi^-$ pair inside a jet. The corresponding fragmentation correlation function $\Delta(z)$ of this final state can be parameterized as

$$\Delta(z) \propto \left[D_1 \not{P} + i H_1^{\sphericalangle} \frac{\not{R}_T \not{P}}{2M_\pi} \right], \tag{12}$$

where the two-hadron fragmentation functions D_1 and $H_1^{\sphericalangle}$ depend on the sum z of the momentum fractions $z^\pm$ of the $\pi^\pm$ and on the invariant mass of the two-pion system (not necessarily in a factorized way as assumed in Ref.[61]). The momenta appearing are $P = P_{\pi^+} + P_{\pi^-}$ and $R_T = (z^+ P_{\pi^-} - z^- P_{\pi^+})/z$. The k_T of the pion pair w.r.t. the fragmentating quark is integrated over. See Ref.[62] for details.

Nonzero $H_1^{\sphericalangle}$ can arise due to interference between different partial waves of the $(\pi^+ \pi^-)$ system and leads to single spin asymmetries $\sin(\phi^e_{S_T} + \phi^e_{R_T})$ in[61,63,64]

$$e\, p^\uparrow \to e' \,(\pi^+ \pi^-)\, X \qquad \propto h_1 \otimes H_1^{\sphericalangle}$$
$$p\, p^\uparrow \to (\pi^+ \pi^-)\, X \qquad \propto f_1 \otimes h_1 \otimes H_1^{\sphericalangle}$$

HERMES SIDIS data[65] below and above the ρ mass show a nonzero single spin asymmetry (with the same sign), which is another indication that transversity is nonzero. From the comparison[66] of the data to various model predictions for HERMES kinematics using different h_1 functions, we conclude that the two-hadron asymmetry data are compatible with $h_1 \approx f_1/3$, albeit with considerable room for other values too. COMPASS data could narrow this range down.

As said, both options discussed here have the advantage that one is dealing with collinear factorization. This means no Sudakov suppression and no process dependence. The latter topic will not be addressed here, but is intimately connected with the gauge invariant definition of TMDs, cf. e.g. Ref.[67]. Another advantage is that the evolution equations for $H_1(z)$ and $H_1^{\sphericalangle}(z)$ are known to next-to-leading order[68,69], they are in fact the same. Therefore, from a theoretical point of view exploiting the transversely polarized Λ or two-hadron fragmentation functions currently offers the safest and most straightforward way to extract transversity.

Like H_1, $H_1^{\sphericalangle}$ can most cleanly be extracted from electron-positron annihilation, in this case from $e^+ e^- \to (\pi^+ \pi^-)_{\text{jet}\,1} (\pi^+ \pi^-)_{\text{jet}\,2} X$ via a $\cos(\phi^e_{R_{1T}} + \phi^e_{R_{2T}})$ asymmetry[70] $\propto (H_1^{\sphericalangle})^2$. Since pions are easier to measure than polarized Λ's, this is probably the easiest route to transversity at this moment. BELLE can once again play a crucial role here (as BABAR could). Its data would allow for a non-TMD extraction of transversity in the not too far future. Therefore, all eyes are on BELLE again in this respect.

3. Routes to transversity

In the previous section several different routes to transversity were discussed. They can be classified into four types, which are summarized in Table 1 for the processes discussed before. On the one hand, there are the options that use collinear (non-TMD) functions, which are safer from a theoretical point of view. Some of these options are self-sufficient, but others require additional input, which most cleanly comes from e^+e^- collisions. On the other hand, there are the TMD options, which are theoretically challenging and it is somewhat ironic to note that what is theoretically the most challenging option, i.e. exploiting the Collins effect, is the one that has been done first experimentally.

Table 1. Summary of routes to transversity.

	non-TMD	TMD
self-sufficient	$\bar{p}^\uparrow p^\uparrow \to \ell\bar{\ell}X$	$p\bar{p}^\uparrow \to \ell\bar{\ell}X$
	$p^\uparrow p^\uparrow \to (\text{high-}p_T \text{ jet})\, X$	$\bar{p}p^\uparrow \to \ell\bar{\ell}X$
using external input	$e\,p^\uparrow \to e'\,\Lambda^\uparrow X$	$e\,p^\uparrow \to e'\,\pi\,X$
	$p\,p^\uparrow \to \Lambda^\uparrow X$	$\pi\,p^\uparrow \to \ell\bar{\ell}X$
	$e\,p^\uparrow \to e'\,(\pi^+\pi^-)\,X$	
	$p\,p^\uparrow \to (\pi^+\pi^-)\,X$	

Several remarks have to be added in relation to this table. As pointed out by Bacchetta and Radici[71] $H_1^{\sphericalangle}$ can also be extracted from $p\,p \to (\pi^+\pi^-)\,(\pi^+\pi^-)\,X$. Similarly, H_1 could be extracted from $p\,p \to \Lambda^\uparrow\,\bar{\Lambda}^\uparrow\,X$. This makes $p\,p^\uparrow$ experiments in principle self-sufficient too. But clearly this would be less clean and more involved than using e^+e^- extractions of $H_1^{\sphericalangle}$ and H_1, due to the appearance of additional distribution and fragmentation functions, and contributions from multiple partonic subprocesses. Here, different subprocesses can enter in numerator and denominator of the asymmetries, because of which the observables in $p\,p$ are likely to be considerably smaller than in e^+e^- annihilation. Nevertheless, it is important to keep in mind that one can make the experiments that require a separate extraction of an unknown fragmentation function, self-sufficient by considering more complicated $p\,p$ or $e\,p$ processes. This may not apply to the Collins function however. It is currently not clear whether $p\,p \to \pi\,\pi\,X$, where the two pions are in separate jets, can be used to safely extract the Collins function. Concerns regarding factorization have been raised in e.g. Refs.[72,73]

Note that the process $p\,p^\uparrow \to \pi\,X$ is absent from the table. This is because the single spin asymmetry A_N is described by a twist-three expression that consists of several contributions, not all proportional to transversity. Therefore, it is not clear how to safely extract it from this observable. Instead, $p\,p^\uparrow \to \gamma\,\pi\,X$ or $p\,p^\uparrow \to \pi\,\mathrm{jet}\,X$ (cf. also Ref.[74,75]) could be used, although also here factorization is yet to be established (which is the reason for not including them in the table).

Finally, it is worth adding that there is a special role for Drell-Yan at RHIC: $p^\uparrow\,p^\uparrow \to \ell\,\bar{\ell}\,X$. It offers a clean way to learn about transversity of antiquarks $h_1^{\bar{q}}$. Its contribution to the tensor charge is important to know. It would not be satisfactory to always have to assume that antiquark transversity is small, without knowing how small. Therefore, A_{TT} at RHIC is still worth measuring.

3.1. *Transversity GPD*

Now we turn to a completely independent way of accessing transversity that falls outside the framework of collinear functions and TMDs discussed thus far.

There are four chiral-odd Generalized Parton Distributions (GPDs),[76] which includes the transversity GPD[77] $H_T(x,\xi,t)$. In the forward limit: $H_T(x,0,0) = h_1(x)$, which offers an alternative way to access transversity and the tensor charge. Of course, the latter requires not only the extrapolation to the forward limit, but also integration over all x values. This will be quite challenging, but it may be worth pursuing this route too because it can be measured, for instance at JLab or a future electron-ion collider, without the need to polarize the proton.

Suggestions to probe H_T in exclusive electroproduction have been put forward, for instance, via the production of two vector mesons in particular polarization states[78,79], $\gamma^*\,p \to \rho_L^0\,\rho_T^+\,n$. Very recently it was suggested[80,81] that transversity could be measured via $\gamma^*\,p \to \pi^0\,p'$. In both cases the idea is that the spin states of the photon and the meson(s) enforce a helicity flip of the quarks inside the proton. In this way there is no need to polarize the proton. Helicity conservation requires the helicity flip on the proton side. This is similar to how the axial charge can be measured in unpolarized elastic *ep* or *νp* scattering.

It should be mentioned that in case of single vector meson production, e.g. $\gamma^*\,p \to \rho_T\,p'$, problems regarding factorization arise. Unfortunately this process cannot be used to extract transversity.[82,83]

Some information on chiral-odd GPDs has already been obtained from lattice QCD.[84] Besides yielding results for the tensor charge, they also show there to be nonzero transverse polarization of quarks inside unpolarized hadrons, hinting at nonzero $h_1^\perp$.

4. Conclusions

Although transversity is a very difficult quantity to measure, several transversity asymmetries have come within reach of present day and near-future experiments. Thanks to SIDIS data by HERMES and COMPASS, and e^+e^- annihilation data by BELLE the first extraction of transversity, exploiting the Collins effect, has been possible. This is an important step forward. The Collins effect asymmetries involve $\boldsymbol{k}_T$-dependent functions, TMDs, and are consequently more difficult to analyze theoretically. Therefore, an independent, preferably non-TMD extraction of transversity is desired. For this, $\bar{p}^\uparrow\, p^\uparrow$ Drell-Yan would be the ideal process, but two hadron fragmentation functions currently offer the most straightforward way. Many more observables could contribute to our knowledge of transversity, the tensor charge, and other chiral-odd quantities, such as $h_1^\perp$. Unpolarized Drell-Yan data and lattice QCD results strongly suggest that the transverse polarization of quarks inside unpolarized hadrons, which is encoded by $h_1^\perp$, is nonzero and large. If so, then especially $\bar{p}^\uparrow\, p$ or $\bar{p}\, p^\uparrow$ Drell-Yan offers another promising opportunity to probe transversity. Perhaps this will be possible at GSI-FAIR. It is a self-sufficient way of measuring transversity, in the sense that no information from other experiments or even other processes needs to be included in the analysis. Other, more demanding self-sufficient or nearly self-sufficient options that do not involve TMDs exist too. Amazingly most of these possibilities are in principle possible with existing accelerators. Further transversity measurements are therefore expected in the coming years, contributing valuably to our understanding of hadron spin.

Acknowledgments

I wish to thank the organizers for their kind invitation to give the opening talk at this exciting workshop, where so many new results were presented. I thank Markus Diehl for a useful discussion on vector meson production and Marco Contalbrigo for some feedback on the text.

References

1. J. P. Ralston, D. E. Soper, *Nucl. Phys. B* **152**, 109 (1979).

2. X. Artru, M. Mekhfi, *Z. Phys. C* **45**, 669 (1990).
3. R. L. Jaffe, X. Ji, *Nucl. Phys. B* **375**, 527 (1992).
4. J. L. Cortes, B. Pire, J. P. Ralston, *Z. Phys. C* **55**, 409 (1992).
5. O. Martin *et al.*, *Phys. Rev. D* **60**, 117502 (1999).
6. J. C. Collins, *Nucl. Phys. B* **396**, 161 (1993).
7. A. Airapetian *et al.* [HERMES Coll.], *Phys. Rev. Lett.* **94**, 012002 (2005).
8. M. Diefenthaler [HERMES Coll.], arXiv:0706.2242 [hep-ex].
9. V. Y. Alexakhin *et al.* [COMPASS Coll.], *Phys. Rev. Lett.* **94**, 202002 (2005).
10. E. S. Ageev *et al.* [COMPASS Coll.], *Nucl. Phys. B* **765**, 31 (2007).
11. S. Levorato [COMPASS Coll.], *these proceedings.*
12. D. Boer, R. Jakob, P. J. Mulders, *Nucl. Phys. B* **504**, 345 (1997).
13. D. Boer, *Nucl. Phys. B* (2008) *in press*, arXiv:0804.2408 [hep-ph].
14. R. Seidl *et al.* [BELLE Coll.], *Phys. Rev. Lett.* **96**, 232002 (2006).
15. R. Seidl *et al.* [BELLE Coll.], arXiv:0805.2975 [hep-ex].
16. M. Anselmino *et al.*, *Phys. Rev. D* **75**, 054032 (2007).
17. M. Wakamatsu, *Phys. Lett. B* **653**, 398 (2007).
18. M. Göckeler, Ph. Hägler *et al.* [QCDSF Coll.], *Phys. Lett. B* **627**, 113 (2005).
19. J. Soffer, *Phys. Rev. Lett.* **74**, 1292 (1995).
20. A. Prokudin, *these proceedings.*
21. J. C. Collins, D. E. Soper, *Nucl. Phys. B* **193**, 381 (1981).
22. X. Ji, J.-P. Ma, F. Yuan, *Phys. Rev. D* **71**, 034005 (2005).
23. X. Ji, J.-P. Ma, F. Yuan, *Phys. Lett. B* **597**, 299 (2004).
24. A. Bacchetta, D. Boer, M. Diehl, P. J. Mulders, *JHEP* **0808**, 023 (2008).
25. D. Boer, *Nucl. Phys. B* **603**, 195 (2001).
26. S. Kumano, M. Miyama, *Phys. Rev. D* **56**, 2504 (1997).
27. W. Vogelsang, *Phys. Rev. D* **57**, 1886 (1998).
28. A. Hayashigaki, Y. Kanazawa, Y. Koike, *Phys. Rev. D* **56**, 7350 (1997).
29. R. L. Jaffe, N. Saito, *Phys. Lett. B* **382**, 165 (1996).
30. W. Vogelsang, talk presented at the RBRC Workshop on "Future Transversity Measurements" held at BNL, September 18-20, 2000.
31. A. Mukherjee, M. Stratmann, W. Vogelsang, *Phys. Rev. D* **72**, 034011 (2005).
32. F. Rathmann, P. Lenisa, arXiv:hep-ex/0412078.
33. M. Nekipelov, *these proceedings.*
34. A. V. Efremov, K. Goeke, P. Schweitzer, *Eur. Phys. J. C* **35**, 207 (2004).
35. H. Shimizu *et al.*, *Phys. Rev. D* **71**, 114007 (2005).
36. A. Bianconi, M. Radici, *Phys. Rev. D* **71**, 074014 (2005).
37. A. Bianconi, M. Radici, *Phys. Rev. D* **72**, 074013 (2005).
38. H. Kawamura, J. Kodaira, K. Tanaka, *Phys. Lett. B* **662**, 139 (2008).
39. J. C. Collins, D. E. Soper and G. Sterman, *Nucl. Phys. B* **250**, 199 (1985).
40. H. Kawamura, J. Kodaira, K. Tanaka, *Nucl. Phys. B* **777**, 203 (2007).
41. D. Boer, P. J. Mulders, *Phys. Rev. D* **57**, 5780 (1998).
42. S. Falciano *et al.* [NA10 Coll.], *Z. Phys. C* **31**, 513 (1986).
43. M. Guanziroli *et al.* [NA10 Coll.], *Z. Phys. C* **37**, 545 (1988).
44. J. S. Conway *et al.*, *Phys. Rev. D* **39**, 92 (1989).
45. A. Brandenburg, O. Nachtmann, E. Mirkes, *Z. Phys. C* **60**, 697 (1993).
46. E. Mirkes, J. Ohnemus, *Phys. Rev. D* **51**, 4891 (1995).

47. D. Boer, *Phys. Rev. D* **60**, 014012 (1999).
48. D. Boer, *Nucl. Phys. B Proc. Suppl.* **79**, 638 (1999).
49. J.C. Peng [FNAL-E866/NuSea Coll.], *these proceedings.*
50. L. Y. Zhu *et al.* [FNAL-E866/NuSea Coll.], *Phys. Rev. Lett.* **99**, 082301 (2007).
51. M. Colantoni [COMPASS Coll.], *these proceedings.*
52. L. P. Gamberg, G. R. Goldstein, M. Schlegel, *Phys. Rev. D* **77**, 094016 (2008).
53. V. Barone, Z. Lu, B. Q. Ma, *Phys. Lett. B* **632**, 277 (2006).
54. V. Barone, A. Prokudin, B. Q. Ma, arXiv:0804.3024 [hep-ph].
55. W. Kafer [COMPASS Coll.], *these proceedings.*
56. D. Boer, P. J. Mulders, C. Pisano, *Phys. Lett. B* **660**, 360 (2008).
57. V. M. Abazov *et al.* [DØ Coll.], arXiv:0804.1107 [hep-ex].
58. A. P. Contogouris *et al.*, *Phys. Lett. B* **344**, 370 (1995).
59. A. Bravar *et al.* [E704 Coll.], *Phys. Rev. Lett.* **78**, 4003 (1997).
60. D. de Florian *et al.*, *Phys. Lett. B* **439**, 176 (1998).
61. R. L. Jaffe, X. Jin, J. Tang, *Phys. Rev. Lett.* **80**, 1166 (1998).
62. A. Bianconi, S. Boffi, R. Jakob, M. Radici, *Phys. Rev. D* **62**, 034008 (2000).
63. X. Ji, *Phys. Rev. D* **49**, 114 (1994).
64. J. C. Collins, S. F. Heppelmann, G. A. Ladinsky, *Nucl. Phys. B* **420**, 565 (1994).
65. A. Airapetian *et al.* [HERMES Coll.], *JHEP* **0806**, 017 (2008).
66. A. Bacchetta, M. Radici, *Phys. Rev. D* **74**, 114007 (2006).
67. C. J. Bomhof, P. J. Mulders, *Nucl. Phys. B* **795**, 409 (2008).
68. M. Stratmann, W. Vogelsang, *Phys. Rev. D* **65**, 057502 (2002).
69. F. A. Ceccopieri, M. Radici, A. Bacchetta, *Phys. Lett. B* **650**, 81 (2007).
70. X. Artru, J. C. Collins, *Z. Phys. C* **69**, 277 (1996).
71. A. Bacchetta, M. Radici, *Phys. Rev. D* **70**, 094032 (2004).
72. J. Collins, J. W. Qiu, *Phys. Rev. D* **75**, 114014 (2007).
73. J. Collins, arXiv:0708.4410 [hep-ph].
74. F. Yuan, *Phys. Rev. D* **77**, 074019 (2008).
75. F. Yuan, *Phys. Lett. B* **666**, 44 (2008).
76. M. Diehl, *Eur. Phys. J. C* **19**, 485 (2001).
77. P. Hoodbhoy, X. Ji, *Phys. Rev. D* **58**, 054006 (1998).
78. D. Y. Ivanov *et al.*, *Phys. Lett. B* **550**, 65 (2002).
79. R. Enberg, B. Pire, L. Szymanowski, *Eur. Phys. J. C* **47**, 87 (2006).
80. S. Ahmad, G. R. Goldstein, S. Liuti, arXiv:0805.3568 [hep-ph].
81. G. R. Goldstein, *these proceedings.*
82. M. Diehl, T. Gousset, B. Pire, *Phys. Rev. D* **59**, 034023 (1999).
83. J. C. Collins, M. Diehl, *Phys. Rev. D* **61**, 114015 (2000).
84. M. Göckeler *et al.* [QCDSF Coll.], *Phys. Rev. Lett.* **98**, 222001 (2007).

THE TRANSVERSE ANGULAR MOMENTUM SUM RULE

E. Leader
High Energy Physics, Imperial College London,
London, SW7 2AZ, UK

We explain the origin of the controversy about the existence of a transverse angular momentum sum rule, and show that it stems from utilizing an incorrect result in the literature, concerning the expression for the expectation values of the angular momentum operators. We demonstrate a new, short and direct way of obtaining correct expressions for these expectation values, from which a perfectly good transverse angular momentum sum rule can be deduced. We also introduce a new classification of sum rules into *primary* and *secondary* types. In the former all terms occurring in the sum rule can be measured experimentally; in the latter some terms cannot be measured experimentally.

Keywords: nucleon, angular momentum, sum rule.

1. Background

Shore and White's claim[1] that a_0 (which in the simple parton model is equal to the contribution to the nucleon's angular momentum arising from the quark spins) does *not* contribute to the nucleon's angular momentum, surprised us. Their analysis was based on a classical paper of Jaffe and Manohar,[2] who stressed the subtleties and warned that 'a careful limiting procedure has to be introduced'. Trying to understand this we became convinced that despite all the care, there are flaws. With the J-M result one cannot have a sum rule for a transversely polarized nucleon. With the correct version[3] one can!

2. Why the problem is non-trivial

What is the aim? We consider a nucleon with 4-momentum p^{μ} and covariant spin vector S^{μ} corresponding to some specification of its spin state e.g. helicity, transversity or spin along the Z-axis i.e. a nucleon in state $|p, S\rangle$. We require an expression for the expectation value of the angular momentum in this state i.e. for $\langle p, S|\boldsymbol{J}|p, S\rangle$ i.e. we require an expression in terms of p

and S. This can then be used to relate the expectation value of $\boldsymbol{J}$ for the nucleon to the angular momentum carried by its constituents.

2.1. *The traditional approach*

In every field theory there is an expression for the angular momentum density operator. The angular momentum operator $\boldsymbol{J}$ is then an integral over all space of this density. Typically the angular momentum density involves the energy-momentum tensor density $T^{\mu\nu}(x)$ in the form e.g.

$$\boldsymbol{J}_z = \boldsymbol{J}^3 = \int dV[xT^{02}(\boldsymbol{x}) - yT^{01}(\boldsymbol{x})]$$

Consider the piece $T^{02}(\boldsymbol{x})$. It is a local operator, so by translational invariance of the theory

$$T^{02}(\boldsymbol{x}) = e^{i\boldsymbol{P}.\boldsymbol{x}}T^{02}(0)e^{-i\boldsymbol{P}.\boldsymbol{x}}$$

where $\boldsymbol{P}$ are the linear momentum operators i.e. the generators of translations. Now the nucleon is in an eigenstate of momentum, so $\boldsymbol{P}$ acting on it just becomes $\boldsymbol{p}$. The numbers $e^{i\boldsymbol{p}.\boldsymbol{x}}e^{-i\boldsymbol{p}.\boldsymbol{x}}$ cancel out and we are left with:

$$\int dVx\, \langle p, S|\, T^{02}(0)\, |p, S\rangle$$

The matrix element is independent of x so we are faced with $\int dVx = \infty$? or $= 0$? Totally ambiguous! The problem is an old one: In ordinary QM plane wave states give infinities. The solution is an old one: Build a wave packet, a superposition of physical plane wave states. Now Jaffe and Manohar are generally very careful, but nonetheless there are errors in their derivation. They end up with the following expression for the matrix elements of the angular momentum operator:

$$\langle\langle \boldsymbol{p}, \boldsymbol{s}|\boldsymbol{J}_i|\boldsymbol{p}, \boldsymbol{s}\rangle\rangle_{JM} = \frac{1}{4mp_0}\left[(3p_0^2 - m^2)\boldsymbol{s}_i - \frac{3p_0 + m}{p_0 + m}(\boldsymbol{p}.\boldsymbol{s})\boldsymbol{p}_i\right]$$

where $p^\mu = (p^0, \boldsymbol{p})$ and $\boldsymbol{s}_i$ are the components of the rest frame spin vector. Recall that the parton picture is supposed to be valid when the nucleon is viewed in a frame where it is moving very fast. In other words to derive a sum rule involving partons we must take the limit $p^0 \to \infty$. If we consider *longitudinal* spin i.e $\boldsymbol{p}\, //\, \boldsymbol{s}$ one obtains:

$$\langle\langle \boldsymbol{p}, \boldsymbol{s}|\boldsymbol{J}_i|\boldsymbol{p}, \boldsymbol{s}\rangle\rangle_{JM} = \frac{1}{2}\boldsymbol{s}_i \qquad (1)$$

and there is no problem. But for transverse polarization one gets:

$$\langle\langle \boldsymbol{p}, \boldsymbol{s}|\boldsymbol{J}_i|\boldsymbol{p}, \boldsymbol{s}\rangle\rangle_{JM} = \frac{1}{4mp_0}\left[(3p_0^2 - m^2)\boldsymbol{s}_i\right] \tag{2}$$

which $\to \infty$ as $p_0 \to \infty$, so no sum rule is possible. We will see in a moment that the result for transverse spin is incorrect. The J-M reaction to our criticism was very gracious and positive!

"Better late than never. Aneesh and I finally found ourselves in the same place with the time to review the issues you raised by email and in your recent paper. We agree that there is an error in our eq. (6.9). It came from treating the quantity u(p', s)u(p, s) with insufficient care. Thanks for taking care and finding this mistake. It's good to get it cleared up. I have to add that I found your paper rather difficult to read. There is quite a bit of stuff that gets in the way of the relatively simple error..........."

2.2. *A new approach*

It is simple. It is short. It works for any spin. Previous methods only work for spin 1/2. We know how rotations affect states. If $|\boldsymbol{p}, m\rangle$ is a state with momentum $\boldsymbol{p}$ and spin projection m in the rest frame of the particle, and if $\hat{R}_z(\beta)$ is the operator for a rotation β about OZ, then

$$\hat{R}_z(\beta)|\boldsymbol{p}, m\rangle = |\boldsymbol{R}_z(\beta)\boldsymbol{p}, m'\rangle\, D^s_{m'm}[R_z(\beta)] \tag{3}$$

where the $D^s_{m'm}$ are the standard rotation matrices for spin s. But rotations are generated by the angular momentum operators! i.e.

$$\hat{R}_i(\beta) = e^{-i\beta \boldsymbol{J}_i}$$

so that

$$\boldsymbol{J}_i = i\frac{d}{d\beta}\hat{R}_i(\beta)\big|_{\beta=0}$$

From Eq. (3) we know what the matrix element of $\hat{R}_i(\beta)$ looks like. So we simply differentiate and put $\beta = 0$. Thus we have

$$\langle \boldsymbol{p}', m'|\boldsymbol{J}_i|\boldsymbol{p}, m\rangle = i\frac{\partial}{\partial\beta}\langle \boldsymbol{p}', m'|R_i(\beta)|\boldsymbol{p}, m\rangle|_{\beta=0}$$

One technical point: you have to know that the derivative of the rotation matrix for spin s at $\beta = 0$ is just the spin matrix for that spin. e.g. for spin 1/2 just $\sigma_i/2$.

3. Comparison of results

For the expectation values we find, for *any* spin configuration (longitudinal, transverse etc) the remarkably simple result (suppressing a delta-function term):

$$\langle\langle \boldsymbol{p}, \boldsymbol{s} | \boldsymbol{J}_i | \boldsymbol{p}, \boldsymbol{s} \rangle\rangle = \frac{1}{2} \boldsymbol{s}_i$$

This agrees precisely with the JM result for longitudinal spin Eq. (1). But for transverse polarization our result differs from the JM Eq. (2), which implied no possibility of a transverse sum rule. With our correct result there is no fundamental distinction between the transverse and longitudinal cases.

4. Sum rules

Consider a nucleon moving along OZ with momentum $\boldsymbol{p}$ and spin projection m along OZ. We expand the nucleon state as a superposition of n-parton Fock states.

$$|\boldsymbol{p}, m\rangle = \sum_n \sum_{\{\sigma\}} \int d^3\boldsymbol{k}_1 \ldots d^3\boldsymbol{k}_n \, \psi_{\boldsymbol{p},m}(\boldsymbol{k}_1, \sigma_1, \ldots \boldsymbol{k}_n, \sigma_n)$$
$$\times \, \delta^{(3)}(\boldsymbol{p} - \boldsymbol{k}_1 \ldots - \boldsymbol{k}_n) \, |\boldsymbol{k}_1, \sigma_1, \ldots \boldsymbol{k}_n, \sigma_n\rangle.$$

where σ_j labels the spin state of the parton, either a projection along OZ for quarks, or helicity for gluons.

There are two independent cases:

(a) *Longitudinal polarization* i.e. the nucleon rest frame spin vector $\boldsymbol{s}$ is along OZ. The sum rule for $\boldsymbol{J}_z$ yields the well known result

$$1/2 = 1/2\, \Delta\Sigma + \Delta G + \langle L_z^q \rangle + \langle L_z^G \rangle \tag{4}$$

(b) *Transverse polarization* i.e. $\boldsymbol{s} = \boldsymbol{s}_T$ where $\boldsymbol{s}_T \perp \boldsymbol{p}$. The sum rule for $\boldsymbol{J}_x$ or $\boldsymbol{J}_y$ yields a a *new* sum rule

$$1/2 = 1/2 \sum_{q,\bar{q}} \int dx \, \Delta_T q(x) + \sum_{q,\bar{q},G} \langle L_{\boldsymbol{s}_T} \rangle \tag{5}$$

Here $L_{\boldsymbol{s}_T}$ is the component of $\boldsymbol{L}$ along $\boldsymbol{s}_T$.

The structure functions $\Delta_T q^a(x) \equiv h_1^q(x)$ are known as the quark transversity or transverse spin distributions in the nucleon. As mentioned no such parton model sum rule is possible with the J-M formula for the expectation value of $\boldsymbol{J}_i$ because for $i = x, y$ the matrix elements diverge as $p \to \infty$.

It is absolutely crucial to note that the sum rule Eq. (5) involves a *sum* of quark and antiquark densities. Not realizing this has led to some misunderstandings.

The *tensor charge* of the nucleon involves the *difference* of the first moments of quark and antiquark contributions. Thus the transverse spin sum rule, although it involves the transverse spin or transversity quark and antiquark densities, does *not* involve the nucleon's tensor charge. The tensor charge operator is *not* related to the angular momentum.

The structure functions $\Delta_T q(x) \equiv h_1^q(x)$ are most directly measured in doubly polarized Drell-Yan reactions

$$p(\boldsymbol{s}_T) + p(\boldsymbol{s}_T) \to l^+ + l^- + X$$

where the asymmetry is proportional to

$$\sum_f e_f^2 [\Delta_T q_f(x_1) \Delta_T \bar{q}_f(x_2) + (1 \leftrightarrow 2)].$$

They can also be determined from the asymmetry in semi-inclusive hadron-hadron interactions like

$$p + p\,(\boldsymbol{s}_T) \to H + X$$

where H is a detected hadron, typically a pion, and in semi-imclusive lepton-hadron reactions (SIDIS) with a transversely polarized target, like

$$\ell + p\,(\boldsymbol{s}_T) \to \ell + H + X.$$

The problem here is that in these semi-inclusive reactions $\Delta_T q(x)$ always occurs multiplied by the Collins fragmentation function, about which we are only at present gathering information.

5. A new classification of sum rules

Part of the reason that there are claims and counter-claims about the existence of certain sum rules is that different people have a different interpretation as to what a sum rule really implies. To clarify this we propose a new classification into *primary* and *secondary* sum rules.

- A *primary* sum rule is one in which every term occurring in the sum rule can be measured experimentally. If the derivation of the sum rule is rigorous and if it fails experimentally, one can conclude that the theory behind it is incorrect. Examples are the Bjorken sum rule,[4] the ELT sum rule[5] and the Ji sum rule.[6]
- A *secondary* sum rule is one in which not every term occurring can be measured experimentally. Examples are Eq. (4) and Eq. (5), where we do not know how to measure the orbital angular momentum terms experimentally. Consequently a secondary sum rule can't test the validity of a theory, but this does not mean the sum rule is vacuous. It can tell us about the terms which we cannot measure, and that can be of value in model building or in understanding the structure of say the nucleon. Do not forget that the renaissance of spin dependent deep inelastic scattering, both theory and experiment, is a direct consequence of using a secondary sum rule i.e. Eq. (4) to proclaim the existence of a "spin crisis in the parton model".[7]

Of course the above is an idealization. I do not know of a single case where literally everything is measurable. So in the Bjorken and ELT sum rules one has to extrapolate $g_1(x)$ and $x[g_1(x) + 2g_2(x)]$ respectively to $x = 0$, and in the Ji case one must extrapolate $E(x, \xi, \Delta^2)$ to $\Delta^2 = 0$. Nonetheless I think the classification is useful.

6. Conclusions

In order to derive angular momentum sum rules we need an expression for the matrix elements of the angular momentum operators $\boldsymbol{J}$ in terms of the momentum $\boldsymbol{p}$ and spin $\boldsymbol{s}$ of the particle. Such matrix elements are divergent and ambiguous in the traditional approach. The infinities and ambiguities can be handled using wave packets, but the calculations are long and unwieldy. and the results, in some classic papers, are incorrect for a transversely polarized nucleon. Consequently it was claimed that no angular momentum rule was possible for a transversely polarized nucleon.

We have found a simple, direct method for evaluating these matrix elements, which is free of infinities and ambiguities. It uses the facts that we know how states transform under rotations, and that the rotation operators are exponentials of the generators of rotations i.e. of the angular momentum operators. It leads quickly and relatively painlessly to correct results.

The great success of the correct approach is that it allows the derivation of a sum rule also for transversely polarized nucleons.

Finally, we have proposed a classification of sum rules into *primary* and *secondary* sum rules, according to whether all, or not all, the terms in a sum rule can be measured experimentally. Whereas the former could, in principle, disprove a theory, the latter can only give us information about quantities which we cannot measure directly. *Both* the longitudinal and the transverse angular momentum sum rules are secondary.

Acknowledgements

This work was carried out in collaboration with Ben Bakker, Vrije Universiteit, Amsterdam and Larry Trueman, Brookhaven National Laboratory, USA.

References

1. G. M. Shore and B. E. White, *Nucl. Phys. B* **581**, 409 (2000).
2. R. L. Jaffe and A. Manohar, *Nucl. Phys. B* **337**, 509 (1990).
3. B. L. Bakker, E. Leader and T. L. Trueman, *Phys. Rev. D* **70**, 114001 (2004).
4. J. D. Bjorken, *Phys. Rev. D*1, 1376 (1971).
5. A. V. Efremov, E. Leader and O. V. Teryaev, *Phys. Rev. D*55, 4307 (1997).
6. X. Ji, *Phys. Rev. Lett.*78, 610 (1997).
7. E. Leader and M. Anselmino, *Z. Phys. C* **41**, 239 (1988).

MEASUREMENT OF COLLINS AND SIVERS ASYMMETRIES AT HERMES

L.L. Pappalardo
(On behalf of the HERMES Collaboration)

INFN and Università degli studi di Ferrara
Dipartimento di Fisica
Via Saragat, 1 44100 Ferrara

Preliminary results on azimuthal single-spin asymmetries in semi-inclusive electro-production of pions and charged kaons at the HERMES experiment are presented. Significant amplitudes for both the Collins and the Sivers mechanisms are extracted from the full data set collected with a transversely polarized target. A new method for the evaluation of the acceptance effects, developed in view of a future extraction of the $P_{h\perp}$-weighted moments, is also discussed.

Keywords: Transversity, Sivers function, Collins mechanism, Sivers mechanism.

1. The Collins and Sivers mechanisms

After integrating over the quark transverse momentum p_T, three parton distribution functions are needed at leading twist for a complete description of the momentum and spin distributions of the quarks within the nucleon. Two of these have been experimentally explored in some detail: the *momentum distribution* $q\left(x,Q^2\right)$, and the *helicity distribution* $\Delta q\left(x,Q^2\right)$. In contrast, the third one, known as *transversity*,[1,2] remained unmeasured until very recently.[3] In a basis of transverse spin eigenstates, it reflects the difference in probabilities to find, in a transversely polarized nucleon, quarks with their spin aligned and anti-aligned to the spin of the nucleon. This quantity, however, has no probabilistic interpretation in the helicity basis, in which the other two distribution functions are naturally defined. In this basis, indeed, it is related to a forward scattering amplitude involving helicity flip of both quark and target nucleon ($N^{\Rightarrow}q^{\leftarrow}\rightarrow N^{\Leftarrow}q^{\rightarrow}$), which is suppressed by chirality conservation. Being a chiral-odd object, transversity is not measurable in inclusive Deep Inelastic Scattering (DIS).

At HERMES azimuthal Single Target-Spin Asymmetries (SSA) in Semi-Inclusive Deeply Inelastic Scattering (SIDIS) on a transversely polarized proton target are investigated. In such events the scattered lepton is detected in coincidence with at least one of the hadrons produced in the fragmentation of the struck quark. This allows, in particular, to access the so-called *Collins moments*, in which transversity is convoluted with the chiral-odd *Collins fragmentation function*, describing the correlation between the transverse spin of the struck quark and the transverse momentum $P_{h\perp}$ of the produced hadron.[4] The transverse polarization of the struck quark can indeed influence the transverse (with respect to the virtual photon direction) component of the hadron momentum, leading to an azimuthal asymmetry in the momentum distribution of the produced hadrons (*Collins mechanism*). Similar asymmetries might also arise from a completely different mechanism involving a correlation between the transverse polarization of the target nucleon and the transverse momentum p_T of quarks (*Sivers mechanism*).[5] This correlation is accounted for by the naïve T-odd *Sivers function* $f_{1T}^{\perp}$, which, being related to a forward scattering amplitude involving helicity flip of only the target nucleon ($N^{\Rightarrow}q^{\leftarrow}\rightarrow N^{\Leftarrow}q^{\leftarrow}$), must involve orbital angular momentum of the quarks.[6,7] The so-called *Sivers moments*, which are proportional to a convolution of the Sivers function with the spin-independent fragmentation function, are also accessible at HERMES in SIDIS.

2. The HERMES experiment

The data analyzed was recorded during the 2002–2005 running period of the HERMES experiment using a transversely nuclear-polarized hydrogen gas target internal to the $E = 27.6$ GeV HERA positron/electron storage ring at DESY. Being interested in the extraction of azimuthal moments which are independent on the beam polarization, the two beam helicity states were combined resulting in a vanishing net beam polarization. The open-ended target cell was fed by an atomic-beam source[8] based on Stern-Gerlach separation and RF transitions of hyperfine states. The nuclear polarization of the atoms was flipped at $1-3$ min time intervals. The average value of the proton polarization was $\langle P_z \rangle = 0.73 \pm 0.05$.

Scattered leptons and any coincident hadrons were detected by the HERMES spectrometer,[9] whose acceptance spans the vertical and horizontal scattering-angle ranges $40 < |\theta_y| < 140$ mrad and $|\theta_x| < 170$ mrad. Leptons are identified with an efficiency exceeding 98% and a hadron contamination of less than 1%. A dual radiator RICH allows to identify the charged hadrons ($\pi^{\pm}, K^{\pm}, p$) in the momentum range 2 GeV $< P_h <$ 15 GeV.

3. Extraction of Collins and Sivers moments

Events were selected subject to the kinematic requirements $W^2 > 10$ GeV^2, $0.1 < y < 0.95$ and $Q^2 > 1$ GeV^2, where W is the invariant mass of the photon-nucleon system and y is the fractional beam energy transfer to the target. Coincident hadrons were only accepted in the semi-inclusive range $0.2 < z < 0.7$, where z is the hadron energy fraction. The cross section asymmetry with respect to the target polarization was evaluated as:

$$A^h_{UT}(\phi,\phi_S) = \frac{1}{\langle P_z\rangle}\frac{N_h^{\Uparrow}(\phi,\phi_S)+N_h^{\Downarrow}(\phi,\phi_S)}{N_h^{\Uparrow}(\phi,\phi_S)-N_h^{\Downarrow}(\phi,\phi_S)}\,, \tag{1}$$

were $N_h^{\Uparrow(\Downarrow)}$ represents the yield in the target spin state "$\Uparrow$ ($\Downarrow$)" for a hadron type h, and ϕ and ϕ_S are two azimuthal angles, defined with respect to the lepton scattering plane, as shown in Fig. 1.

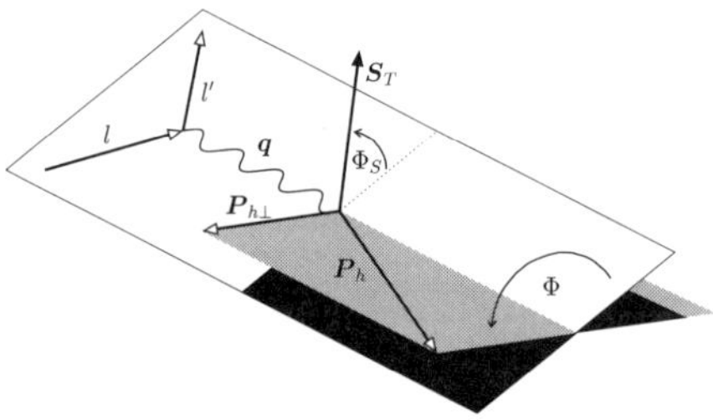

Fig. 1. Kinematics of semi-inclusive DIS on a transversely polarized target.

The asymmetry (1) can be expanded in terms of azimuthal moments. Each of these moments is characterized by a peculiar modulation in the azimuthal angles ϕ and ϕ_S. In particular, the Collins and Sivers moments are modulated by $\sin(\phi+\phi_S)$ and $\sin(\phi-\phi_S)$, respectively. Their different azimuthal modulation allows to extract them separately.

In 2005 the HERMES Collaboration published a first evidence of non-zero Collins and Sivers moments for charged pions.[10] The results, based on a limited data sample, were extracted in a least-squares fit of the asymmetry (1). The Collins and Sivers moments have now been extracted using the full HERMES transverse data set, resulting in a substantial increase of the statistical precision. Furthermore, these moments have also been extracted for neutral pions and charged kaons. Due to the relatively poor statistics for kaons, which would have produced non reliable fit results in some azimuthal (ϕ,ϕ_S) bins, an unbinned Maximum Likelihood fit was adopted, based on the following probability density function (p.d.f.):

$$F(P_z,\phi,\phi_S) = 1 + P_z \cdot \Big[2\langle\sin(\phi+\phi_S)\rangle^h_{UT}\ \sin(\phi+\phi_S) +$$
$$2\langle\sin(\phi-\phi_S)\rangle^h_{UT}\ \sin(\phi-\phi_S)\ + 2\langle\sin(3\phi-\phi_S)\rangle^h_{UT}\ \sin(3\phi-\phi_S) +$$
$$2\langle\sin(2\phi-\phi_S)\rangle^h_{UT}\ \sin(2\phi-\phi_S)\ + 2\langle\sin(\phi_S)\rangle^h_{UT}\ \sin(\phi_S)\Big]\ . \qquad (2)$$

Here P_z denotes the value of the target polarization and the indices U and T stand for Unpolarized beam and Transversely polarized target, respectively. For completeness, three additional terms are included in the fit besides the Collins and Sivers ones: the leading twist $\langle\sin(3\phi-\phi_S)\rangle^h_{UT}$ moment and two twist-3 $\langle\sin(2\phi-\phi_S)\rangle^h_{UT}$ and $\langle\sin(\phi_S)\rangle^h_{UT}$ moments.

The preliminary results for the Collins moments for charged pions are reported in Fig. 2 as a function of x, z and $P_{h\perp}$. The shaded bands represent the systematic uncertainty, which include contributions from acceptance effects, instrumental smearing, QED radiation and hadron misidentification. A common 8.1% scale uncertainty arises from the uncertainty on the target polarization.

Significantly positive (negative) Collins amplitudes are observed for π^+ (π^-). These results confirm the previously published ones[10] and demonstrate that both transversity and Collins function are non-zero.

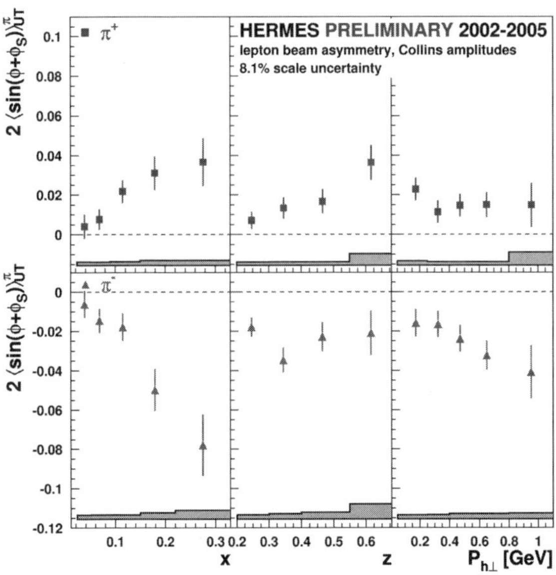

Fig. 2. Collins moments for π^+ (upper panels) and π^- (lower panels).

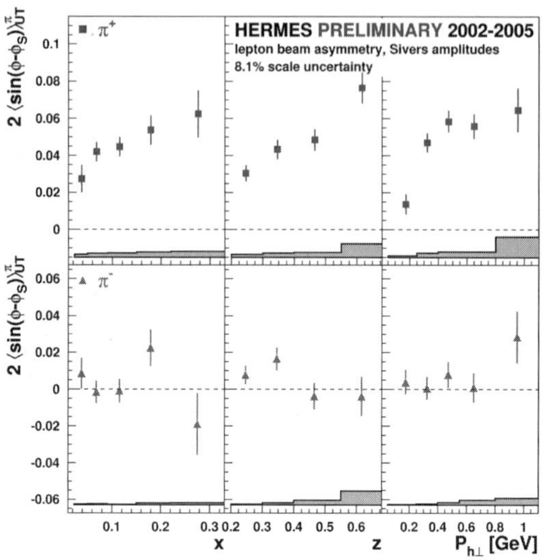

Fig. 3. Sivers moments for π^+ (upper panels) and π^- (lower panels).

The fragmentation of an u-quark is said to be *favored* if the produced hadron contains an u-quark as a valence quark (e.g. π^+) and *unfavored* in the opposite case (e.g. π^-). Assuming that the scattering off u-quarks is the dominating subprocess (u-quark dominance), the unexpectedly high negative Collins amplitude for π^- suggests a unfavored Collins function with a magnitude similar to that of the favored one but with opposite sign.

The Sivers moments, reported in Fig. 3, show a significantly positive amplitude for π^+. This result demonstrates that the Sivers function is non-zero and implies the existence of non-zero orbital angular momentum of the quarks, which is one of the still unmeasured contributions to the nucleon spin.[7] An amplitude consistent with zero is measured for π^-.

The Collins and Sivers moments were also extracted for π^0, as shown in Fig. 4. The results show a Collins amplitude consistent with zero and a Sivers amplitude slightly positive.

The measured Collins and Sivers moments are in agreement with the following relation based on the isospin symmetry for the π-meson triplet[a]:

$$\langle\sin(\phi\pm\phi_S)\rangle_{UT}^{\pi^+}+C\cdot\langle\sin(\phi\pm\phi_S)\rangle_{UT}^{\pi^-}-(1+C)\cdot\langle\sin(\phi\pm\phi_S)\rangle_{UT}^{\pi^0}=0\ , \quad (3)$$

[a]The coefficient C represents the unpolarized cross-section ratio for semi-inclusive negative and positive pion production ($C=\sigma_{UU}^{\pi^-}/\sigma_{UU}^{\pi^+}$).

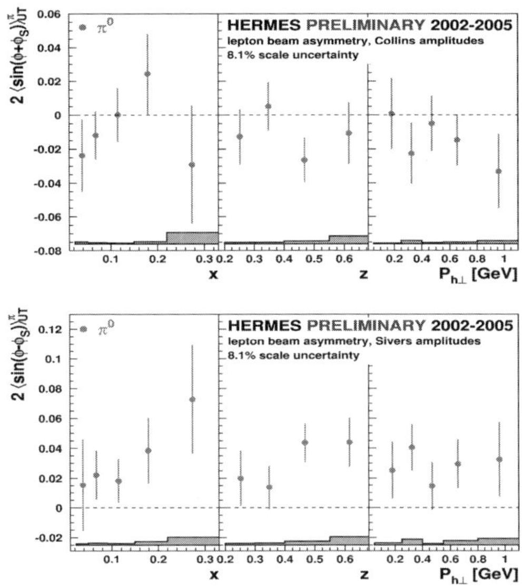

Fig. 4. Collins (upper panel) and Sivers (lower panel) moments for π^0.

which predicts, for the π^0, amplitudes intermediate between those of π^+ and π^-.

To probe the contribution from the sea quarks, the Collins and Sivers moments were also extracted for the charged kaons. The corresponding results are compared in Figs. 3 and 5 with those obtained for the charged pions.

The Collins moments for π^+ and K^+ are compatible within the statistical uncertainty, while those for K^- and π^- are of opposite sign. However, there is in principle no reason to expect a similar amplitude for K^- and π^- since, differently from the π^-, the K^- has no valence quarks in common with the target proton. Unexpectedly, the Sivers amplitude for K^+ is found to be roughly twice as big as that for π^+. Since the valence content of these two mesons differs only in the anti-quark involved, this observation suggests a significant Sivers function for the proton sea quarks.

4. A method for the estimation of the acceptance effects

The extraction of the transversity and the Sivers function from the measured Collins and Sivers moments reported above is not straightforward since these distribution functions are embedded within convolution integrals over the intrinsic transverse momenta. In order to solve these integrals one

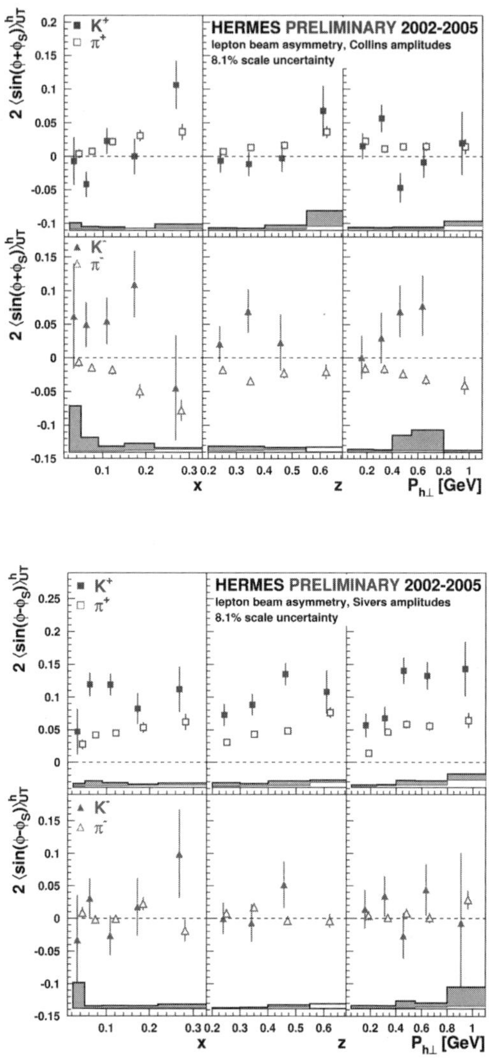

Fig. 5. Collins (upper plot) and Sivers (lower plot) moments for K^+ (upper panels) and K^- (lower panels). The pion results (open dots) are also shown for comparison.

needs to make an assumption (model) for the distribution of the transverse momenta. The most popular one is the so-called Gaussian ansatz.[11]

A model-independent extraction of the transversity and the Sivers function is possible from the so-called $P_{h\perp}$-weighted Collins and Sivers moments. Differently from the 'unweighted' moments presented above, these moments have an extra weight, given by the ratio $P_{h\perp}/z$, which allows to solve the convolution integral without the need of any assumption on the intrinsic transverse momenta distribution. However, $P_{h\perp}$-weighted moments can be affected by severe acceptance effects due to the acceptance in $P_{h\perp}$ of the spectrometer.

An innovative method for the estimation of the acceptance effects has been developed. This method consists of two main steps. As a first step the full kinematic dependence of the Collins and Sivers moments is extracted from the data in a Maximum Likelihood fit based on a fully differential p.d.f.:

$$F = 1 + P_z \cdot \Big[A_{\text{Col}}(x, Q^2, z, P_{h\perp}; a_C)\ \sin(\phi + \phi_S) + A_{\text{Siv}}(x, Q^2, z, P_{h\perp}; a_S)\ \sin(\phi - \phi_S)\Big] \,. \tag{4}$$

Here the Collins and Sivers amplitudes A_{Col} and A_{Siv} are parameterized in terms of a Taylor expansion on the four relevant kinematic variables x, Q^2, z, $P_{h\perp}$ through the sets of coefficients a_C and a_S, extracted as fit parameters. In the second step, these parameterizations are folded with the unpolarized cross section σ_{UU}, evaluated either in 4π and within the HERMES acceptance. The difference between the amplitudes folded in 4π and those folded within the acceptance then provides an estimate, based on the data signal itself, of the acceptance effects affecting the measured Collins and Sivers moments. This method has been successfully tested using a Monte Carlo simulation based on the GMC_TRANS generator, which allows to generate events based on phenomenological models for the Collins and Sivers mechanisms. In the future this method will be applied to the real data for the extraction of the $P_{h\perp}$-weighted moments.

References

1. J.P. Ralston and D.E. Soper, Nucl. Phys. **B152**, 109 (1979).
2. V. Barone, A. Drago and G. Ratcliffe, Physics Reports **359**, 1-168 (2002).
3. M. Anselmino *et al.*, arXiv:hep-ph/0701006 (2007), Revised version to appear in Phys. Rev. D.
4. J.C. Collins, Nucl. Phys. **B396**, 161 (1993).
5. D.W. Sivers, Phys. Rev D **41**, 83 (1990).
6. S.J. Brodsky, D.S. Hwang and I. Schmidt, Phys. Lett. **B530**, (2002).
7. M. Burkardt, Phys. Rev. **D66**, 114005 (2002).
8. A. Airapetian *et al.* (HERMES), Nucl. Instr. and Meth. **A540**, 68 (2005).
9. K. Ackerstaff *et al.* (HERMES), Nucl. Inst. & Meth. **A 417**, 230 (1998).
10. A. Airapetian *et al.* (HERMES), Phys. Rev. Lett. **94**, 012002 (2005).
11. A. Bacchetta, PhD Thesis (2002), arXiv:hep-ph/0212025.

REVIEW OF SSA RESULTS ON DEUTERON AT COMPASS

A. Richter
(On behalf of the COMPASS Collaboration)
Physikalisches Institut, Universität Erlangen-Nürnberg,
Erwin-Rommel-Str. 1, 91058 Erlangen, Germany

The measuring of transverse single spin asymmetries (SSA) is part of the physics program at COMPASS, a fixed target experiment at CERN SPS. In 2002-04 the COMPASS experiment has collected data with a transversely polarised ^{6}LiD target using a 160 GeV/c polarised μ^+ beam. By measuring transverse SSA one has access to the transversity distribution function $\Delta_T q(x)$. This is one of the three quark distribution functions, which are needed to fully describe the spin structure of the nucleon at leading twist. At COMPASS three different quark polarimeters have been used to access transversity: the Collins effect, which produces an azimuthal asymmetry in the single hadron distribution, the azimuthal target spin asymmetries of charged hadron pairs and the transverse polarisation of Λ hyperons.

In addition the SSA arising from the correlation between the transverse nucleon spin and the quark intrinsic transverse momentum (Sivers effect), was measured, together with six more transverse target spin asymmetries.

All the asymmetries measured on the deuteron target are small and compatible with zero.

Keywords: Transversity; Transverse Spin Asymmetry; Collins Asymmetry; Sivers Asymmetry; Interference Fragmentation Function; COMPASS.

1. Introduction

To fully describe the quark spin structure of the nucleon at leading twist three distribution functions are needed: The momentum distribution $q(x)$, which describes the probability of finding a quark with a fraction x of the nucleon momentum, the helicity distribution $\Delta q(x)$, which describes the probability in a longitudinally polarised nucleon (w.r.t. the direction of motion) of finding a quark with spin parallel to the nucleon spin and the transversity distribution $\Delta_T q(x)$, which describes the probability in a transversely polarised nucleon (w.r.t. the direction of motion) of finding a quark with spin parallel to the nucleon spin. While the first two are known, transversity is now under intensive examination.

Since transversity is chiral-odd, it decouples from inclusive deep inelastic scattering (DIS). It is possible to measure it in semi-inclusive deep inelastic scattering (SIDIS) in combination with chiral-odd fragmentation functions, like the Collins fragmentation function $\Delta_T^0 D_q^h$ for single hadron production or the interference fragmentation function $H_1^{\sphericalangle}$ for the production of hadron pairs, which give rise to an azimuthal single spin asymmetry (SSA) in the final state hadrons.

A third channel, in which transversity was measured at COMPASS, is the transverse polarisation of Λ hyperons, which is not part of this contribution.

A different mechanism, where SSA are created from the correlation between the nucleon spin and the quark intrinsic transverse momentum k_T is the Sivers effect.[1] It is described by the so-called Sivers distribution function $\Delta_T^0 q(x, k_T)$.

In semi-inclusive DIS of polarised leptons on a transversely polarised target eight azimuthal modulations appear in the cross section. Within QCD parton model four azimuthal asymmetries can be interpreted at leading order, two of them being the already discussed Collins and Sivers asymmetries. The other two leading twist asymmetries, related to different transverse momentum dependent quark distribution functions, and also additional four asymmetries, which can be interpreted as twist-three contributions, have also been measured for the first time at COMPASS.

The COMPASS experiment[2] is a fixed target experiment, which uses a polarised μ^+ beam with a momentum of 160 GeV/c at the CERN SPS M2 beamline. The main goal of its spin physics program is the study of the spin structure of the nucleon. In 2002-04 a polarised ^{6}LiD target was used, whereas the target nucleons had a polarisation transverse to the beam direction for nearly 20% of the beam time in this years.

2. One hadron asymmetries

2.1. *Collins asymmetry*

The Collins mechanism[3] leads to a modulation in the azimuthal distribution of the inclusively produced hadrons given by:

$$N = N_0 \cdot (1 + f \cdot P_T \cdot D_{NN} \cdot A_C \cdot \sin(\Phi_{Coll})) \qquad (1)$$

where P_T is the target polarisation ($\sim 50\%$), f the target dilution factor (~ 0.38), i.e. the fraction of the target nuclei, which can be polarised, $D_{NN} = (1-y) / \left(1 - y + y^2/2\right)$ the spin transfer coefficient. Φ_{Coll} is the

Collins angle, defined as $\Phi_h - \Phi_{S'} = \Phi_h + \Phi_S - \pi$, where Φ_h is the azimuthal angle of the hadron momentum, $\Phi_{S'}$ is the azimuthal angle of the quark spin after scattering and Φ_S is the azimuthal angle of the nucleon spin in the $\gamma - N$ system.

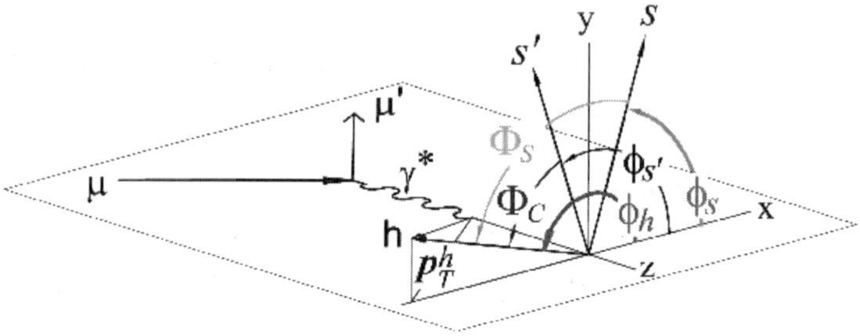

Fig. 1. Definition of Collins and Sivers angle.

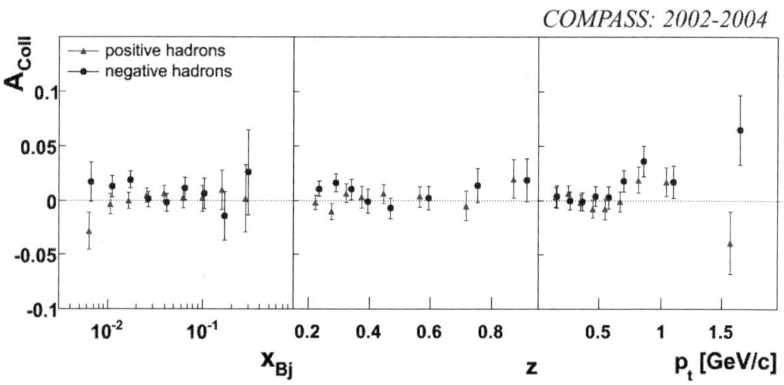

Fig. 2. Collins asymmetries for unidentified hadrons as a function of x_{bj}, z and p_t.

A_C is the Collins asymmetry and is given by:

$$A_C = \frac{\sum_q e_q^2 \Delta_T q(x) \Delta_T^0 D_q^h \left(z, p_T^h\right)}{\sum_q e_q^2 q(x) D_q^h \left(z, p_T^h\right)} \tag{2}$$

The Collins asymmetry is proportional to the convolution of the transversity distribution $\Delta_T q(x)$ and the Collins fragmentation function $\Delta_T^0 D_q^h$ divided by the momentum distribution q and the unpolarised fragmentation function D_q^h. $z = E_h/(E_l - E_{l'})$ is the fraction of the virtual photon energy carried by the hadron and p_T^h is the transverse momentum of the hadron with respect to the direction of the virtual photon.

The event selection requires standard DIS cuts: photon virtuality $Q^2 > 1\ (\mathrm{GeV/c})^2$, $0.1 < y < 0.9$ and mass of the final hadronic state $W > 5\ \mathrm{GeV/c^2}$. Additional cuts on the detected hadrons are: $z > 0.2$ and $p_T^h > 0.1$.

The Collins asymmetry of unidentified hadrons[4,5] is shown in fig. 2 as a function of x_{bj}, z and p_T^h; the asymmetries are small and compatible with zero. Only the statistical error is shown in the error bars, while several studies have shown that the systematic errors are much smaller than the statistical ones. The most energetic (leading) hadrons show a similar behavior to that of all detected hadrons.

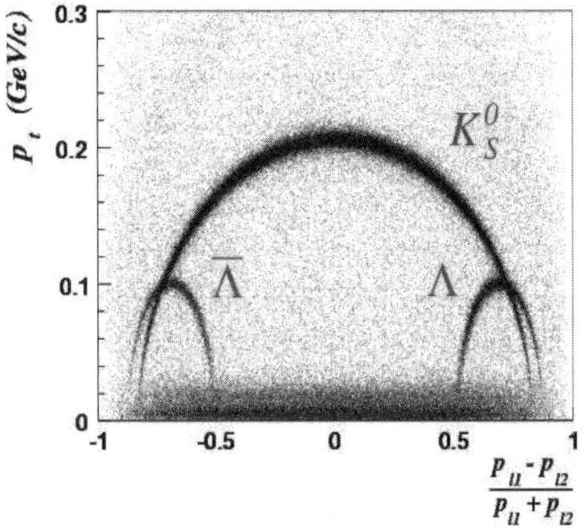

Fig. 3. Armenteros plot of the hadron pair.

The same analysis done for the charged hadrons has been repeated to extract the Collins and Sivers asymmetry for $\pi^{\pm}$ and $K^{\pm}$ for the data collected in 2003 and 2004, using the information of the Ring Imaging Cherenkov Detector detector (RICH) to identify the hadrons, and for K_S^0 for the whole data from 2002-04. The COMPASS RICH1 detector uses C_4F_{10} as radiator gas, with a refractive index of ~ 1.0015 for 7 eV Cherenkov photons, that gives a threshold of 2 (9) GeV for pions (kaons) and allows to separate pions from kaons up to 45-50 GeV. More than 80% of the hadrons

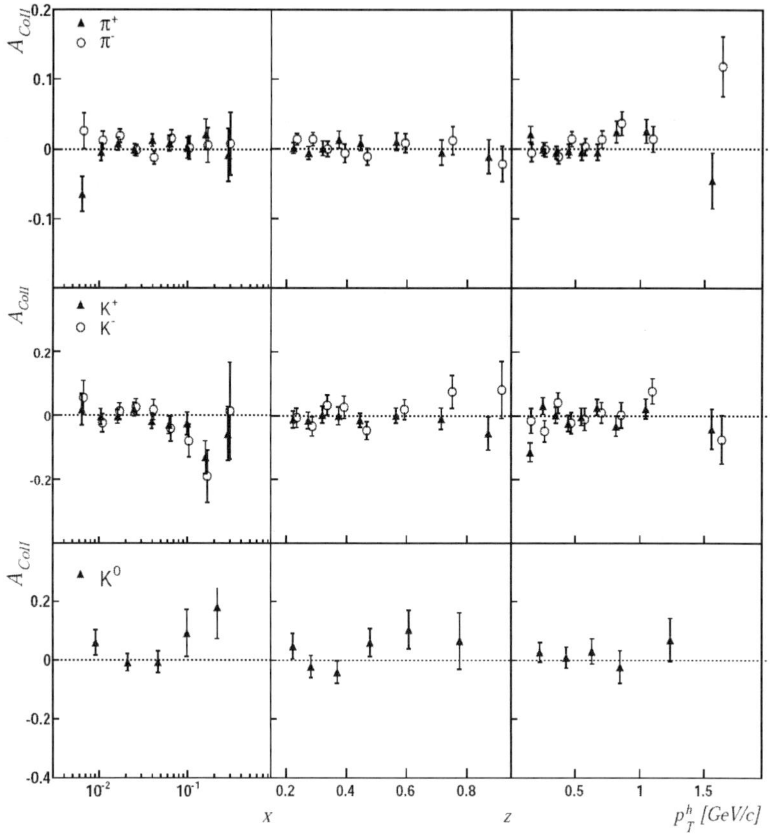

Fig. 4. Collins asymmetries for identified hadrons (pions, kaons) as a function of x_{bj}, z and p_t.

in this energy interval identified by the RICH are pions and the rest are kaons.

K_S^0 have been identified by using secondary vertices with two outgoing and no incoming particle and calculating the invariant mass of the decay products assumed to be pions. Further cuts are on the transverse momentum of one of the hadrons relative to the hadron momentum sum ($p_t > 0.025$ GeV/c, see Armenteros plot: fig. 3), on the primary vertex pointing of the K_S^0 and on a minimum distance between the primary vertex and the forward secondary vertex at least 10 cm. With this cuts a signal to background ratio of 15 for the K_S^0 sample is reached.

The Collins asymmetry of the identified hadrons[6] are shown in fig. 4.

Like for the unidentified hadrons, the asymmetries are small and compatible with zero.

Although there were some predictions for small asymmetries by assuming an opposite contribution from u and d quarks, causing a large cancellation for the isoscalar deuteron target at COMPASS regimes, there was no clear expectation that the asymmetries are so small. On the (u-dominated) proton the HERMES collaboration has measured a non-zero Collins asymmetry,[7] leading to the conclusion that both, the transversity distribution $\Delta_T u(x)$ and the Collins fragmentation function $\Delta_T^0 D_u^h$, are non-zero.

As the Collins fragmentation function is measured at the Belle experiment,[8] it is possible to do an extraction of the transversity distributions.[9] Hereby the COMPASS deuteron data allow, in conjunction to the HERMES data on proton and the Belle data in e^+e^- hadron production, to extract $\Delta_T d(x)$.

2.2. *Sivers asymmetry*

The Sivers asymmetry gives a modulation in the azimuthal distribution of $\Phi_{Siv} = \Phi_h - \Phi_S$, which is independent of the Collins angle and can therefore be disentangled in SIDIS.

COMPASS has published results for the unidentified[4,5] as well as for the identified hadrons ($\pi^\pm$, $K^\pm$, K_S^0).[6] The asymmetries are small and compatible with zero (see fig. 5). Given the non-zero asymmetry measured by HERMES for Sivers, also there might be a cancellation between u and d, i.e. $\Delta_T^0 d \sim -\Delta_T^0 u$. For a summary concerning a simultaneous description of the COMPASS and HERMES data see:[10] Furthermore the vanishing of the Sivers asymmetry on deuteron can be interpreted as the absence of gluon orbital angular momentum in the nucleon.[11]

2.3. *Other single spin asymmetries*

In the SIDIS cross-section there are alltogether eight structure functions, which depend on the transverse target polarisation and cause eight spin dependent asymmetries with different, independent azimuthal modulations. All of them were being measured at COMPASS. Two of them are the Collins and the Sivers asymmetry. Another two are also twist-2 asymmetries and can therefore be interpreted in the QCD parton model. They will allow to extract two further quark distribution functions. The remaining six can be interpreted as twist-3 contributions. All those asymmetries are found to be very small and compatible with zero.

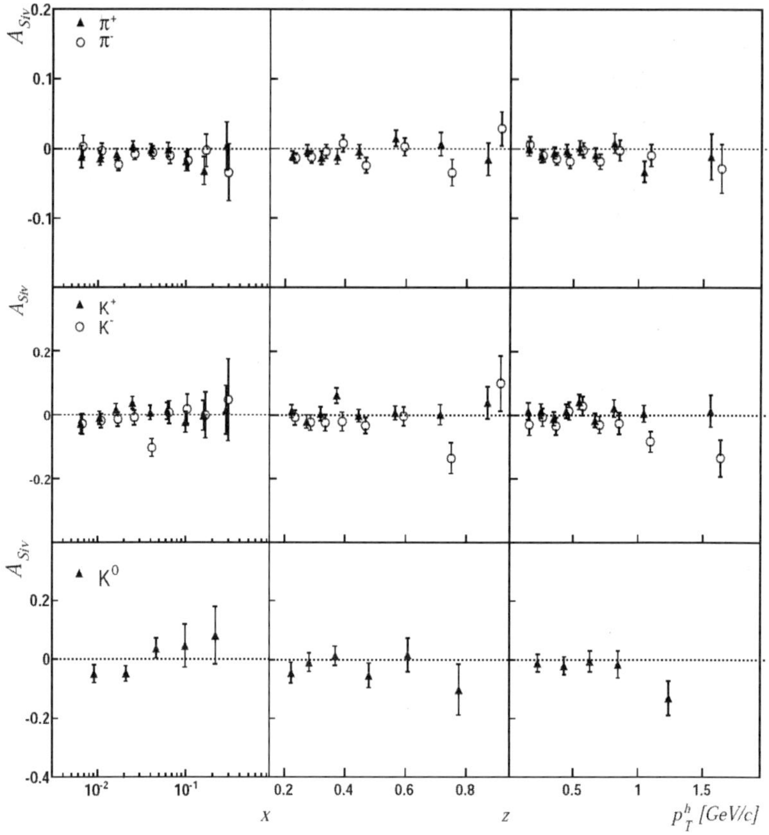

Fig. 5. Sivers asymmetries for identified hadrons (pions, kaons) as a function of x_{bj}, z and p_t.

3. Two hadron asymmetries

$\Delta_T q(x)$ can also be accessed via the production of pairs of two unpolarised hadrons.[12,13] The two hadron production counting rates are given by:

$$N = N_0 \cdot (1 + f \cdot P_T \cdot D_{NN} \cdot A_{RS} \cdot \sin(\Phi_{RS})) \qquad (3)$$

The azimuthal angle Φ_{RS} is defined as $\Phi_{RS} = \Phi_R - \Phi_S - \pi$, with Φ_R the azimuthal angle of the plane containing the two hadrons and Φ_S the azimuthal angle of the target spin vector with respect to the lepton scattering plane. Φ_R is defined at COMPASS as the azimuthal angle of the transverse component R_T of the vector $\frac{z_2\mathbf{P}_1 - z_1\mathbf{P}_2}{z_1+z_2}$. Hereby the indices 1 and 2 refer to the two final state hadrons.

The asymmetry A_{RS} is given by:

$$A_{RS} = \frac{\sum_q e_q^2 \Delta_T q(x) H_1^{\sphericalangle}\left(z, M_h^2\right)}{\sum_q e_q^2 q(x) D_q^h\left(z, M_h^2\right)} \qquad (4)$$

M_h is here the invariant mass of the hadron pair. $D_q^h\left(z, M_h^2\right)$ is the also unknown unpolarised fragmentation function into two hadrons. The asymmetry is proportional to the convolution of the transversity distribution $\Delta_T q(x)$ and the interference fragmentation function $H_1^{\sphericalangle}\left(z, M_h^2\right)$ that can be measured at the Belle experiment.

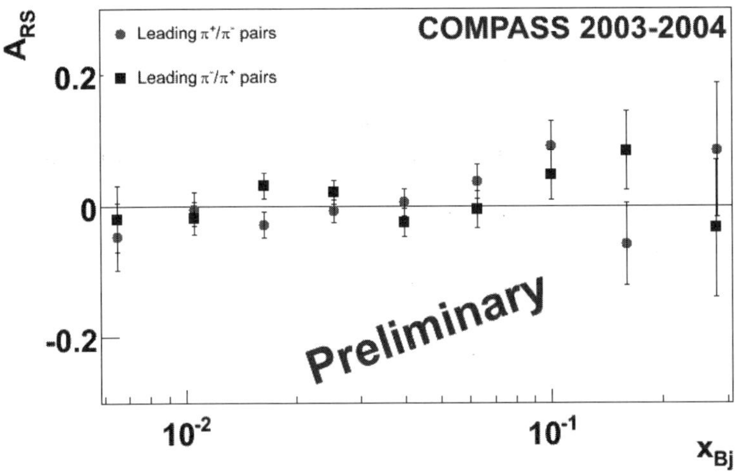

Fig. 6. Asymmetries of leading $\pi^+\pi^-$ hadron pairs as a function of x_{bj}.

There were two different analysis of the data: first all hadron pairs with a positive and a negative hadron were considered, giving four combinations for identified hadrons: $\pi^+\pi^-$, π^+K^-, $K^+\pi^-$ and K^+K^-.

Secondly, a z ordered data sample was selected, using only the leading and the sub-leading hadron (in z) of the event. Here also combinations of hadrons with the same charge are possible.

The event selection is basically the same as for the single hadron asymmetries, with the exception of the z-cut, which was $z_{1,2} > 0.1$ for each hadron and $z_1 + z_2 < 0.9$ for the sum to remove exclusive production.

The asymmetries were studied as function of x_{bj}, z and M_h. No clear signal variant from zero was seen (for one example see fig. 6). For its proton target the HERMES collaboration has measured non-zero results.[14]

4. Summary

In 2002-04 the COMPASS collaboration has measured a full set of single spin asymmetries on a transversely polarised deuterium target. The asymmetries on the deuteron were all found to be small and compatible with zero, which implicates a cancellation of the u and d quark contribution to the asymmetries in a deuterium target.

In 2007 COMPASS has also measured transverse asymmetries on a polarised proton target.[15]

References

1. D. W. Sivers, *Phys. Rev. D* **41**, 83 (1990).
2. P. Abbon *et al.*, [COMPASS collaboration] *Nucl. Instrum. Meth. A* **577**, 455 (2007).
3. J. C. Collins, *Nucl. Phys. B* **396**, 161 (1993).
4. V. Y. Aleaxakhin *et al.*, [COMPASS collaboration] *Phys. Rev. Lett.* **94**, 202002 (2005).
5. E. S. Ageev *et al.*, [COMPASS collaboration] *Nucl. Phys. B* **765**, 31 (2007).
6. M. Alexeev *et al.*, [COMPASS collaboration]; arXiv:0802.2160 [hep-ex] (2008).
7. M. Diefenthaler *et al.*, [HERMES collaboration] *proceedings of DIS 2007*, 579 (2007); arXiv:0706.2242 [hep-ex].
8. K. Abe *et al.*, [Belle collaboration] *Phys. Rev. Lett.* **96**, 232002 (2006).
9. M. Anselmino *et al.*; arXiv:0701006v3 [hep-ph] (2007).
10. M. Anselmino *et al.*, *proceedings of Transversity 2005*; arXiv:0511017v1 [hep-ph] (2005).
11. S. J. Brodsky, S. Gardner, *Phys. Lett. B* **643**, 22 (2006).
12. A. Bacchetta, M. Radici, *proceedings of DIS 2004*; arXiv:0407345 [hep-ph] (2004).
13. X. Artru, *proceedings of 10th Rhodanien Seminar: The Spin in Physics*; arXiv:0207309 [hep-ph] (2002).
14. A. Airapetian *et al.*, [HERMES collaboration]; arXiv:0803.2367v2 [hep-ex] (2008).
15. S. Levorato, [COMPASS collaboration], these proceedings.

SINGLE SPIN ASYMMETRIES ON A TRANSVERSELY POLARISED PROTON TARGET AT COMPASS

S. Levorato
(On behalf of COMPASS Collaboration)
Physics Department, INFN Trieste and Trieste University,
Trieste, 34100, Italy

COMPASS is a running fixed-target experiment at the CERN SPS with a rich physics program focused on nucleon spin structure and on hadron spectroscopy. One of the main goals of the spin program is the measurement of the transverse spin effects in semi-inclusive DIS off transversely polarised nucleons. In the years 2002, 2003 and 2004 data have been taken using a 160 GeV/c naturally polarised μ^+ beam and a deuterium target (^6LiD) transversely polarised respect to the beam direction. In 2007 the run year has been devoted to collect data with a proton (NH_3) target. The preliminary results for the Collins and Sivers asymmetries, extracted from the 2007 data with transverse target polarisation, are presented here. Results are also compared with existing model predictions.

Keywords: Semi-Inclusive Deep Inelastic Scattering; Transversely polarised proton target; Collins asymmetry; Sivers asymmetry; COMPASS.

1. Single Spin asymmetries

The cross-section for polarised deep inelastic scattering[1] of leptons off spin 1/2 hadrons can be expressed, at leading twist, as a function of three independent quark distribution functions: $q(x)$, $\Delta q(x)$ and $\Delta_T q(x)$. The latter, which describes the probability of finding a quark with spin parallel to the nucleon spin in a transversely polarised nucleon, is chiral-odd and can be measured in transversely polarised nucleon (anti)-nucleon hard scattering or in semi-inclusive deep inelastic scattering (SIDIS) on a transversely polarised target.

In SIDIS $\Delta_T q(x)$ can be measured in combination with the chiral odd Collins fragmentation function $\Delta_0^T D_q^h$, via azimuthal single spin asymmetries (SSA) in single hadron production. According to Collins,[2] the fragmentation of a transversely polarised quark in unpolarised hadrons presents an azimuthal modulation with respect to the plane defined by the quark momentum and the quark spin. In SIDIS the hadron yield can be written then as:

$$N = N_0 \cdot (1 + f \cdot P_t \cdot D_{nn} \cdot A_C \cdot \sin(\Phi_C)) \tag{1}$$

where f is the target dilution factor, P_t the target polarisation and $D_{nn} = (1-y)/(1-y+y^2/2)$ the transverse spin transfer coefficient from the initial to the struck quark. The angle Φ_C is known as "Collins angle" and is conveniently defined in the system where the z-axis is the virtual photon direction and the x-z plane is the muon scattering plane. In this frame $\Phi_C = \Phi_h + \Phi_S - \pi$, where Φ_h is the hadron azimuthal angle, and ϕ_S the azimuthal angle of the transverse spin of the initial nucleon. Finally A_C is the Collins asymmetry resulting from the convolution between the Collins fragmentation function and the transverse spin distribution:

$$A_C = \frac{\sum_q e_q^2 \, \Delta_T q(x) \, \Delta_T^0 D_q^h(z, p_T^h)}{\sum_q e_q^2 \, q(x) \, D_q^h(z, p_T^h)} \tag{2}$$

where e_q is the quark charge, $D_q^h(z, p_T^h)$ is the unpolarised fragmentation function, $z = E_h/(E_l - E_{l'})$ is the fraction of available energy carried by the hadron, and p_T^h is the hadron transverse momentum with respect to the virtual photon direction. As is clear from eq.1, the Collins asymmetry A_C is revealed as a $\sin\Phi_C$ modulation in the number of produced hadrons.

A second source of azimuthal asymmetry is related to the Sivers effect,[3] arising from a possible coupling of the intrinsic transverse momentum $\vec{k}_T$ of unpolarised quarks to the spin of a transversely polarised nucleon. In this case the number of produced hadrons can be written as:

$$N = N_0 \cdot (1 + f \cdot P_t \cdot A_S \cdot \sin(\Phi_S)) \tag{3}$$

where the Sivers angle Φ_S is defined as $\Phi_h - \Phi_s$, and the asymmetry A_S probes the so called Sivers distribution function $\Delta_0^T q$:

$$A_S = \frac{\sum_q e_q^2 \, \Delta_0^T q(x, \vec{k}_T) \, D_q^h(z)}{\sum_q e_q^2 \, q(x) \, D_q^h(z)}. \tag{4}$$

In this case the asymmetry A_S is revealed as a $\sin\Phi_S$ modulation in the number of produced hadrons.

It has to be noted here that since the Collins and the Sivers angles are independent,[4] it is possible to measure from the same data both the Collins and the Sivers asymmetries.

2. Data sample and analysis

The Collins and the Sivers asymmetries have been measured by COMPASS from 2002 to 2004 on a transversely polarised deuterium target; both the asymmetries turned out to be small, compatible with zero.[5–7] In 2007 COMPASS took data with a proton target (NH_3) and a 50% sharing between longitudinal and transverse target configuration, accumulating $\sim 40 \cdot 10^{12}$ and $\sim 42 \cdot 10^{12}$ μ on tape respectively.

A new target with a larger diameter (4 cm) and three cells (upstream, central, downstream, 30, 60, and 30 cm long respectively) have been used: the tree cells setup is meant to reduce the difference in acceptance between the two polarisations. The upstream and the downstream cells are polarised along the same direction, which is opposite to the central cell. The Ammonia material is characterized by a dilution factor f $\sim$ 0.15 and a very high polarisation: $\sim$ 90%. As in 2006 the target is inside the new large acceptance Superconducting magnet. The transverse run has been divided in 12 "periods", each of them corresponding to about 5 full days of data taking. Consecutive periods correspond to opposite polarisation of each of the tree target cells. Asymmetries are extracted using at the same time the informations coming from cells in two consecutive periods with opposite configuration.

Almost all the data collected in the transverse polarisation configuration of the target have been processed a first time. For the results presented here about 20% of the whole collected data have been used. These data have been selected requiring a good stability of the spectrometer and of the reconstruction between consecutive periods. In the analysis the events are considered only if one primary vertex is found in the target region. To select DIS events the photon virtuality Q^2 is taken greater than 1 (GeV/c)^2, y between 0.1 and 0.9, and the invariant mass of the final hadronic state $W > 5$ GeV/c^2. The hadron sample on which the asymmetries are computed consists of all the charged hadrons coming from the reaction vertex with $p_T^h > 0.1$ GeV/c and $z > 0.2$. In table 1 the final statistics entering the asymmetries extraction is given for the 6 used periods, separately for positive and negative hadrons. Figure 1 illustrates the Q^2-x Bjorken phase-space covered by the COMPASS experiment after the $Q^2 > 1$ selection. As it can be seen the high energy of the muon beam allow to reach the 10^{-3} region of x, in the DIS regime, moreover most of the statistic is at low x values: $0.008 \div 0.02$.

A number of systematic studies have been performed in order to determine the systematic errors. Extensive tests both to measure false asym-

Table 1. Number of hadrons used for this analysis.

Period	Positive hadrons:	Negative hadrons
W39/W40	2742704	2149343
W41/W42a	2199513	1752684
W42b/W42	761014	603061
Total:	5703231	4505088

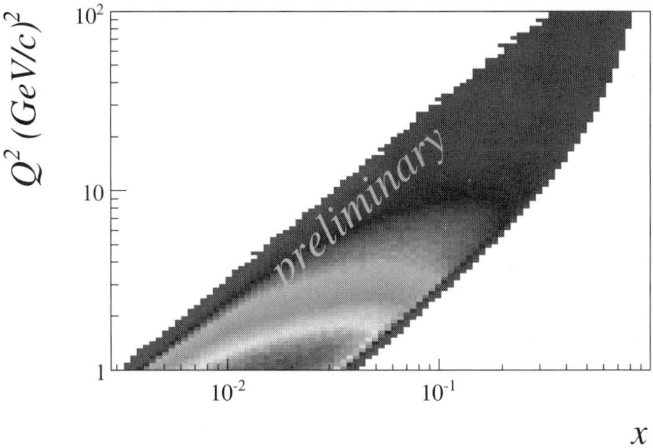

Fig. 1. Scatter plot of Q^2 as function of x Bjorken for the events after all the kinematic cuts. Most of the statistics is concentrated at low x Bjorken values.

metries and to investigate the stability of the physics results, in each data taking period, were done combining cells with the same polarisation and using different splitting combination of the target cells (top-bottom, left-right). The results obtained with different asymmetries extraction methods have been compared too.

From all these tests the systematic errors have been estimated to be $0.3 \cdot \sigma_{stat}$ for the Collins asymmetries and $0.5 \cdot \sigma_{stat}$ for the Sivers asymmetries.

2.1. *Results*

The Collins and Sivers asymmetries were evaluated as a function of x, p_T^h, and z dividing the corresponding kinematical range in bins (with variable width, in order to have a comparable statistics in each of them), and integrating over the other two variables. In total, the asymmetries were evaluated in 9 x-bins, 9 p_T^h-bins, and 8 z-bins. The method used for extraction is based on a two dimensional (8 times 8) binning in Φ_h and Φ_S. Combining the information of the opposite target polarisation cells a non-linear system of equations for the cross section amplitude modulations can be written, and the extraction of the amplitudes is then obtained via maximum likelihood method. The results have been checked with several other statistical methods described in Ref. 6. In figure 2 the preliminary results

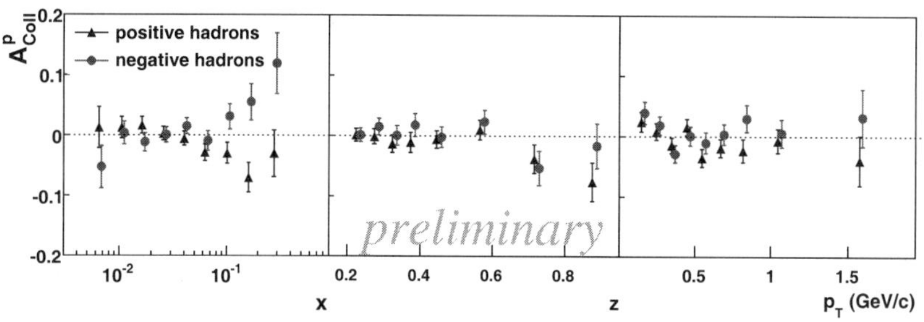

Fig. 2. Collins asymmetries for positive unidentified hadrons (triangles) and for negative unidentified hadrons (circles) as function of x, p_T^h, and z.

for the Collins asymmetries as function of x, p_T^h, and z are shown both for positive and for negative unidentified hadrons. The asymmetry is small, basically statistically compatible with zero up to $x = 0.05$ while a signal is visible in the last points: the asymmetry then increases in module up to 10% and with oppoiste sign for positive and negative hadrons. For p_T^h and z the asymmetry amplitude is compatible with zero due to the fact that most of the statistical sample is in the low x region. In fact requiring $x > 0.05$ the asymmetry signal becomes more evident both in p_T^h and z bins. This can be seen in fig. 3, showing Collins asymmetry for the data in the valence region, i.e. with $x > 0.05$. It is clear from the plots that there is not an appreciable z or p_T^h dependence.

Figure 4 shows the theoretical expectations of the Collins asymmetries in the COMPASS kinematical range[8] both for positive and negative hadrons. The predictions are obtained from a global analysis of the HERMES[9] data, COMPASS[10] deuterium data and the BELLE[11] data. The good agreement for all the different kinematic variables is manifested in the figure. In

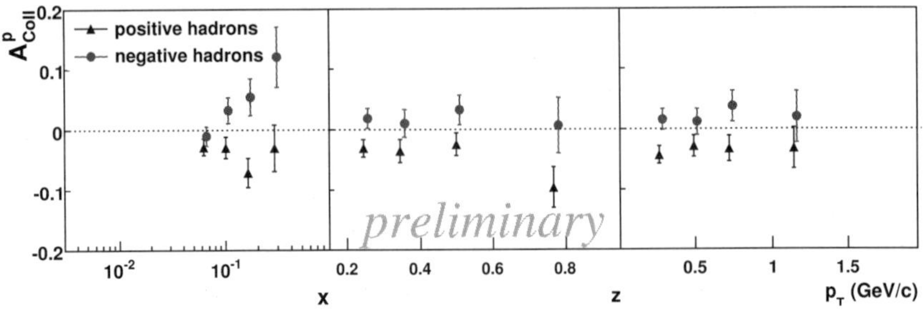

Fig. 3. Collins asymmetries for positive unidentified hadrons (triangles) and for negative unidentified hadrons (circles) with the cut $x > 0.05$ as function of x, p_T^h, and z.

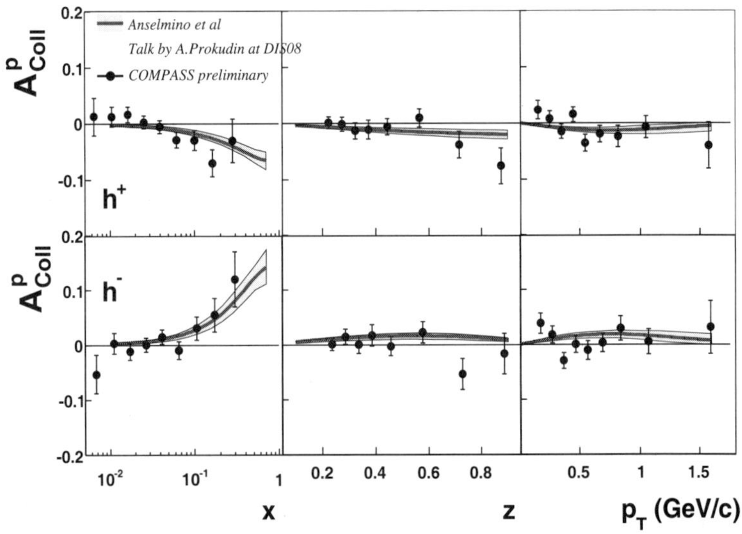

Fig. 4. Compass preliminary Collins asymmetries on proton with the latest prediction of Anselmino *et al.* (A. Prokudin at DIS 08).

figure 5 the preliminary results of the Sivers asymmetries are shown as function of x, p_T^h and z. At variance with A_C, the Sivers asymmetry is small and statistically compatible with zero for both positive and negative hadrons over all the measured x range. The result for positive hadrons is at variance from what has been measured by HERMES,[9] and the two data

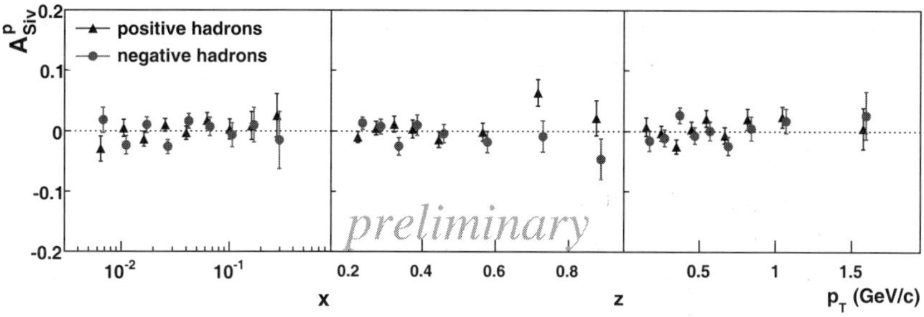

Fig. 5. Sivers asymmetries for positive unidentified hadrons (black) and for negative unidentified hadrons (circles) as function of x, p_T^h, and z.

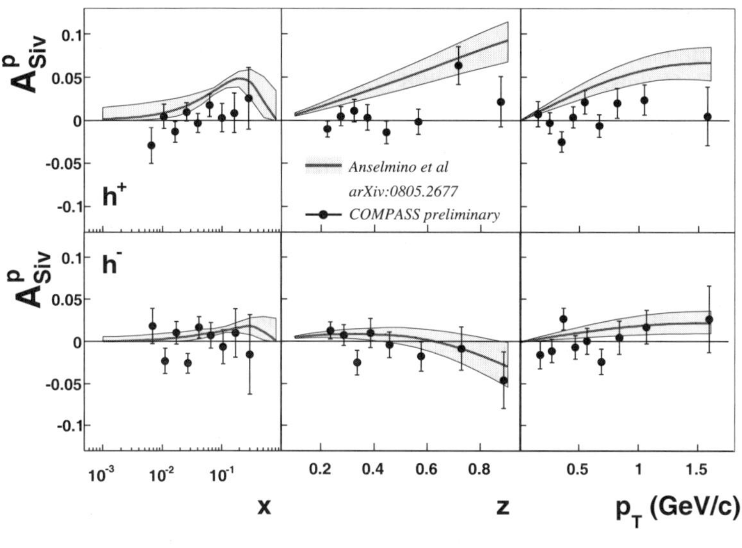

Fig. 6. COMPASS Sivers asymmetry on proton for positive and negative hadrons, with the latest prediction of Anselmino *et al.*

samples are marginally compatible. A possible explanation of this result may come from the very different kinematic range of the two experiments.

Figure 6 compares the COMPASS Sivers asymmetry on proton data for positive and negative hadrons with the latest prediction of Anselmino *et al.*[12] For positive hadrons the agreement is not satisfactory. As in the Collins case the predictions are driven by the HERMES data.

Other theoretical models as for example the one proposed by S. Arnold *et al.*[13] show the same disagreement for the Sivers results.

3. Summary

Preliminary results of Collins and Sivers asymmetries for 2007 COMPASS proton data have been presented. Collins asymmetries for positive and negative hadrons are different from zero and of opposite sign for the two charges and agree with the previous HERMES results. In the Sivers case the measured asymmetries are compatible with zero, within the present statistics, both for positive and negative hadrons, at variance with the HERMES result.

References

1. V. Barone, A. Drago and P. G. Ratcliffe, Phys. Rept. 359 (2002) 1.
2. J. C. Collins, Nucl. Phys. B **396** (1993) 161.
3. D. W. Sivers, Phys. Rev. D **41** (1990) 83.
4. D. Boer, P. J. Mulders, Phys. Rev. D **57** (1998) 5780.
5. The COMPASS Collaboration, hep-ex/0501073.
6. E. S. Ageev *et al.*, hep-ex/0610068.
7. M. Alekseev *et al.*, "Collins and Sivers Transverse Spin Asymmetries for Pions and Kaons on deuterons" hep-ex/0802.2160.
8. A. Prokudin DIS 2008, 7–11 April 2008.
9. M. Diefenthaler [HERMES Collaboration], proceedings of DIS 2007, page 579; arXiv:0706.2242 [hep-ex].
10. The Compass Experiment at CERN Nucl. Instrum. Meth. A **577** (2007) 455–518.
11. A. Abashian *et al.* (Belle): Nucl. Instrum. Meth. A **479** (2002) 117.
12. M. Anselmino *et al.*, "Sivers Effect for Pion and Kaon Production in Semi-Inclusive Deep Inelastic Scattering," arXiv:0805.2677 [hep-ph].
13. S. Arnold, A. V. Efremov, K. Goeke, M. Schlegel and P. Schweitzer, "Sivers effect at HERMES, COMPASS and CLAS12", arXiv:0805.2137 [hep-ph].

NEW PRELIMINARY RESULTS ON THE TRANSVERSITY DISTRIBUTION AND THE COLLINS FRAGMENTATION FUNCTIONS

M. Anselmino[1], M. Boglione[1], U. D'Alesio[2,3], A. Kotzinian[4], F. Murgia[3], A. Prokudin[1,5*], S. Melis[1], C. Türk[1]

1- Dipartimento di Fisica Teorica, Università di Torino and INFN, Sezione di Torino, Via P. Giuria 1, I-10125 Torino, Italy

2- Dipartimento di Fisica, Università di Cagliari, Cittadella Universitaria di Monserrato, I-09042 Monserrato (CA), Italy

3- INFN, Sezione di Cagliari, C.P. 170, I-09042 Monserrato (CA), Italy

4- CEA-Saclay, IRFU/Service de Physique Nuclaire, 91191 Gif-sur-Yvette, France; Yerevan Physics Institute, 375036 Yerevan, Armenia; JINR, 141980 Dubna, Russia

5- Di.S.T.A., Università del Piemonte Orientale "A. Avogadro", Alessandria, Italy

First results from a new global analysis of the experimental data on azimuthal asymmetries in SIDIS, from HERMES and COMPASS collaborations, and in $e^+e^- \to h_1 h_2 X$ processes, from the Belle collaboration, are presented. The new data allow for a more precise determination of the Collins fragmentation function and of the transversity distribution function for u and d quarks, in comparison with the results of our previous analysis. Estimates for the single spin asymmetry $A_{UT}^{\sin(\phi_h+\phi_S)}$ at JLab and COMPASS, operating on a transversely polarized proton target, are given.

The transversity distribution function, usually denoted as h_1 or $\Delta_T q$, together with the unpolarized distribution functions q and the helicity distributions Δq, contains basic information about the spin structure of nucleons.

SIDIS processes represent a testing ground for the transversity distribution, as this chiral-odd function couples to another chiral-odd quantity, the Collins fragmentation function,[1] giving rise to an azimuthal asymmetry, the so-called Collins asymmetry. Another azimuthal asymmetry produced in the $e^+e^- \to h_1 h_2 X$ process[2] results from a convolution of two Collins fragmentation functions for quarks and antiquarks.

---

*presenting author

The first global analysis of data on azimuthal asymmetries in SIDIS and in e^+e^- annihilations, resulting in the first extraction of the transversity distribution and the Collins fragmentation function, was presented in Ref. [3].

Recently, much higher statistics data on the azimuthal asymmetries for SIDIS have become available: the HERMES Collaboration have presented pion and kaon azimuthal asymmetries;[4] the COMPASS Collaboration have presented their measurements for separated charged pion and kaon asymmetries;[5] the Belle collaboration have issued new high precision data of the Collins asymmetry in e^+e^- annihilation.[6] It is therefore timely to reconsider the results of Ref. [3] using the new data in order to improve our knowledge of the Collins fragmentation function and of the transversity distribution.

We consider the $\sin(\phi_S + \phi_h)$ moment,

$$A_{UT}^{\sin(\phi_S+\phi_h)} = 2\,\frac{\int d\phi_S\, d\phi_h\, [d\sigma^\uparrow - d\sigma^\downarrow]\, \sin(\phi_S + \phi_h)}{\int d\phi_S\, d\phi_h\, [d\sigma^\uparrow + d\sigma^\downarrow]}\,, \tag{1}$$

measured by the HERMES[4] and COMPASS[5] Collaborations.

Belle collaboration[6] measures the Collins asymmetry in $e^+e^- \to h_1 h_2 X$ processes by two different methods, either the so-called "$\cos(2\varphi_0)$ method", using the Gottfried-Jackson c.m. frame where one of the produced hadrons identifies the $\hat{z}$ direction, or the "$\cos(\varphi_1+\varphi_2)$ method" in the Collins-Soper frame where the quark thrust axis is used as the $\hat{z}$ direction (for details and definitions see for example Refs. [3,6,7]).

We assume the usual gaussian and factorized parameterization of unpolarized parton distribution (PDF) and fragmentation (FF) functions:[3]

$$f_{q/p}(x,k_\perp) = f_{q/p}(x)\,\frac{e^{-k_\perp^2/\langle k_\perp^2\rangle}}{\pi\langle k_\perp^2\rangle}\,, \quad D_{h/q}(z,p_\perp) = D_{h/q}(z)\,\frac{e^{-p_\perp^2/\langle p_\perp^2\rangle}}{\pi\langle p_\perp^2\rangle}\,,$$

with $\langle k_\perp^2\rangle$ and $\langle p_\perp^2\rangle$ fixed to the values found in Ref. [8] by analyzing unpolarized SIDIS: $\langle k_\perp^2\rangle = 0.25\ (\mathrm{GeV}/c)^2$, $\langle p_\perp^2\rangle = 0.20\ (\mathrm{GeV}/c)^2$. Integrated parton distribution and fragmentation functions $f_{q/p}(x)$ and $D_{h/q}(z)$ are available in the literature; in particular we use the GRV98 PDF set[9] and the DSS fragmentation function set.[10]

As in our previous analysis[3] we adopt the following parameterizations for the transversity distribution, $\Delta_T q(x,k_\perp)$, and the Collins FF,

$\Delta^N D_{h/q^\uparrow}(z, p_\perp)$:

$$\Delta_T q(x, k_\perp) = \frac{1}{2}\,\mathcal{N}_q^T(x)\,[f_{q/p}(x) + \Delta q(x)]\;\frac{e^{-k_\perp^2/\langle k_\perp^2\rangle_T}}{\pi\langle k_\perp^2\rangle_T}\,, \tag{2}$$

$$\Delta^N D_{h/q^\uparrow}(z, p_\perp) = 2\,\mathcal{N}_q^C(z)\,D_{h/q}(z)\,h(p_\perp)\,\frac{e^{-p_\perp^2/\langle p_\perp^2\rangle}}{\pi\langle p_\perp^2\rangle}\,, \tag{3}$$

with

$$\mathcal{N}_q^T(x) = N_q^T\,x^\alpha(1-x)^\beta\,\frac{(\alpha+\beta)^{(\alpha+\beta)}}{\alpha^\alpha\beta^\beta}\,, \tag{4}$$

$$\mathcal{N}_q^C(z) = N_q^C\,z^\gamma(1-z)^\delta\,\frac{(\gamma+\delta)^{(\gamma+\delta)}}{\gamma^\gamma\delta^\delta}\,, \tag{5}$$

$$h(p_\perp) = \sqrt{2e}\,\frac{p_\perp}{M_h}\,e^{-p_\perp^2/M_h^2}\,, \tag{6}$$

and $-1 \le N_q^T \le 1$, $-1 \le N_q^C \le 1$. We assume $\langle k_\perp^2\rangle_T = \langle k_\perp^2\rangle$. The helicity distributions $\Delta q(x)$ are taken from Ref. [11].

Using these parameterizations we obtain[3] the following expression for $A_{UT}^{\sin(\phi_S+\phi_h)}$:

$$A_{UT}^{\sin(\phi_S+\phi_h)} = \frac{\dfrac{P_T}{M_h}\,\dfrac{1-y}{sxy^2}\,C\displaystyle\sum_q e_q^2\,\mathcal{N}_q^T(x)\,\left[f_{q/p}(x)+\Delta q(x)\right]\,\mathcal{N}_q^C(z)\,D_{h/q}(z)}{\dfrac{e^{-P_T^2/\langle P_T^2\rangle}}{\langle P_T^2\rangle}\,\dfrac{[1+(1-y)^2]}{sxy^2}\displaystyle\sum_q e_q^2\,f_{q/p}(x)\,D_{h/q}(z)}\,, \tag{7}$$

where

$$C \equiv \sqrt{2e}\,\frac{\langle p_\perp^2\rangle_C^2}{\langle p_\perp^2\rangle}\,\frac{e^{-P_T^2/\langle P_T^2\rangle_C}}{\langle P_T^2\rangle_C^2}\,, \qquad \langle p_\perp^2\rangle_C = \frac{M_h^2\langle p_\perp^2\rangle}{M_h^2+\langle p_\perp^2\rangle}\,,$$
$$\langle P_T^2\rangle = \langle p_\perp^2\rangle + z^2\langle k_\perp^2\rangle\,, \qquad \langle P_T^2\rangle_C = \langle p_\perp^2\rangle_C + z^2\langle k_\perp^2\rangle\,. \tag{8}$$

For the case of the $\cos(\varphi_1+\varphi_2)$ method the cross section for the process $e^+e^- \to h_1 h_2 X$ reads:

$$\frac{d\sigma^{e^+e^-\to h_1h_2X}}{dz_1\,dz_2\,d\cos\theta\,d(\varphi_1+\varphi_2)} = \frac{3\alpha^2}{4s}\sum_q e_q^2\Big\{(1+\cos^2\theta)\,D_{h_1/q}(z_1)\,D_{h_2/\bar q}(z_2)$$
$$+\frac{1}{4}\,\sin^2\theta\,\cos(\varphi_1+\varphi_2)\,\Delta^N D_{h_1/q^\uparrow}(z_1)\,\Delta^N D_{h_2/\bar q^\uparrow}(z_2)\Big\} \tag{9}$$

where we have defined

$$\int d^2\boldsymbol{p}_\perp\,\Delta^N D_{h/q^\uparrow}(z, p_\perp) \equiv \Delta^N D_{h/q^\uparrow}(z)\,. \tag{10}$$

Normalizing to the azimuthal averaged unpolarized cross section one has:

$$A_{12}(z_1, z_2, \theta, \varphi_1 + \varphi_2) \equiv \frac{1}{\langle d\sigma \rangle} \frac{d\sigma^{e^+e^- \to h_1 h_2 X}}{dz_1\, dz_2\, d\cos\theta\, d(\varphi_1 + \varphi_2)}$$
$$= 1 + \frac{1}{4} \frac{\sin^2\theta}{1 + \cos^2\theta} \cos(\varphi_1 + \varphi_2)$$
$$\times \frac{\sum_q e_q^2\, \Delta^N D_{h_1/q^\uparrow}(z_1)\, \Delta^N D_{h_2/\bar{q}^\uparrow}(z_2)}{\sum_q e_q^2 D_{h_1/q}(z_1)\, D_{h_2/\bar{q}}(z_2)}. \tag{11}$$

For the $\cos(2\varphi_0)$ method the analogue of Eq. (11) reads

$$A_0(z_1, z_2, \theta_2, \varphi_0) = 1 + \frac{1}{\pi} \frac{z_1 z_2}{z_1^2 + z_2^2} \frac{\sin^2\theta_2}{1 + \cos^2\theta_2} \cos(2\varphi_0)$$
$$\times \frac{\sum_q e_q^2\, \Delta^N D_{h_1/q^\uparrow}(z_1)\, \Delta^N D_{h_2/\bar{q}^\uparrow}(z_2)}{\sum_q e_q^2 D_{h_1/q}(z_1)\, D_{h_2/\bar{q}}(z_2)}. \tag{12}$$

To eliminate false asymmetries, the Belle Collaboration[6] considers the ratio of unlike-sign to like-sign pion pair production, A_U and A_L.

The evolution of transversity is known up to NLO; in the present paper we consistently use its LO evolution.[12] On the other hand the scale dependence of the Collins FF $\Delta^N D_{h/q^\uparrow}$ is still quite unknown; in our analysis we tentatively assume the same Q^2 evolution as for the unpolarized FF $D_{h/q}$.

We can now pursue our strategy of gathering simultaneous information on the transversity distribution function $\Delta_T q(x, k_\perp)$ and the Collins fragmentation function $\Delta^N D_{h/q^\uparrow}(z, p_\perp)$. To such a purpose we perform a global best fit analysis of experimental data involving these functions, namely the data from the SIDIS measurements by the HERMES[4] and COMPASS[5] Collaborations, and the data from $e^+e^- \to h_1 h_2 X$ unpolarized processes by the Belle Collaboration.[6]

The two sets of Belle data are not indepedent, so we include only one set of data in the fit, either A_0 or A_{12} data. In this paper we report the results we obtained by using A_{12} data the $\cos(\varphi_1 + \varphi_2)$ method. The consequences of fitting A_0 instead of A_{12} are presently under investigation.

Table 1 contains the results of our best fit to the new data sets.[4–6] In Fig. 1, our present results for the transversity distribution and the Collins fragmentation function are compared to those of our previous extraction;[3] Fig. 2 shows the fits to the HERMES[4] and COMPASS[5] data, while Fig. 3 shows the fit to the Belle A_{12} asymmetry data and the predictions for the Belle A_0 asymmetry.[6]

Table 1. Best values of the free parameters for the u and d transversity distribution functions and for the favoured and unfavoured Collins fragmentation functions. We obtain χ^2/d.o.f. $= 1.3$. Notice that the errors generated by MINUIT are strongly correlated, and should not be taken at face value. The significant fluctuations in our results are shown by the shaded areas in Figs. 2 and 3.

$N_u^T = 0.79 \pm 0.11$	$N_d^T = -1.00 \pm 0.15$	$\alpha = 0.62 \pm 0.18$
$N_{fav}^C = 0.43 \pm 0.05$	$N_{unf}^C = -1.00 \pm 0.17$	$\gamma = 0.96 \pm 0.06$
$M_h^2 = 0.91 \pm 0.46\ (\text{GeV}/c)^2$	$\beta = 0.31 \pm 0.27$	$\delta = 0.01 \pm 0.03$

Finally, in Fig. 4, we show our estimates for COMPASS and JLab experiments operating on a proton target. Notice that JLab results will give important information on the range of large x values, which is left basically

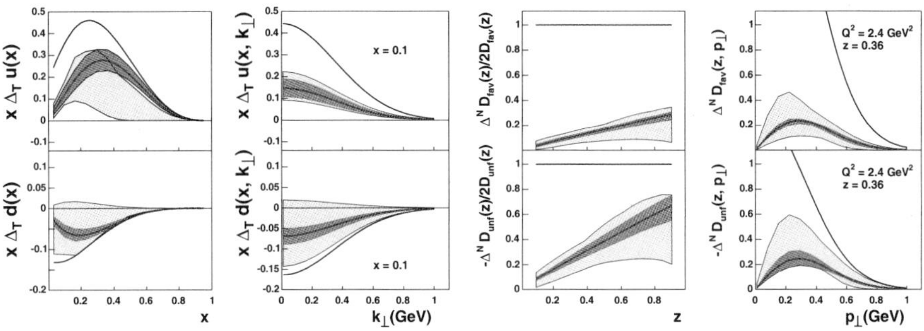

Fig. 1. Left panel: the transversity distribution functions for u and d flavours as determined by our global fit; we also show the Soffer bound (highest or lowest lines) and the (wider) bands of our previous extraction.[3] Right panel: favoured and unfavoured Collins fragmentation functions as determined by our global fit; we also show the positivity bound and the (wider) bands as obtained in Ref. [3].

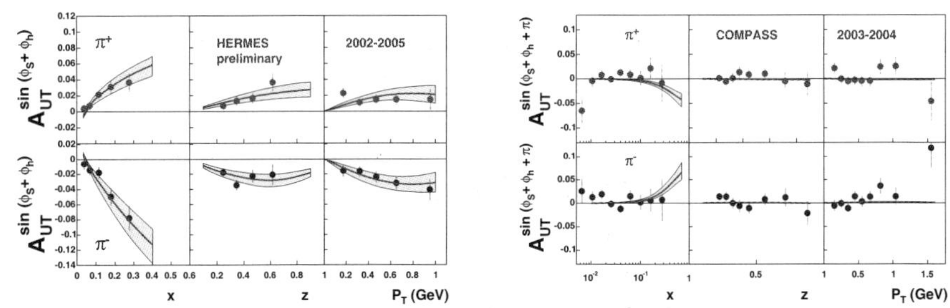

Fig. 2. Fits of HERMES[4] and COMPASS[5] data. The shaded area corresponds to the uncertainty in the parameter values, see Ref. [3].

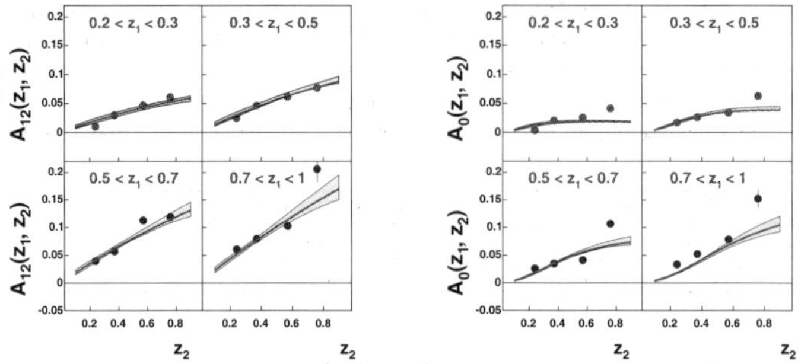

Fig. 3. Left panel: fit of the Belle[6] data on the A_{12} asymmetry (the $\cos(\varphi_1 + \varphi_2)$ method). Right panel: predictions for the A_0 Belle asymmetry (the $\cos(2\varphi_0)$ method).

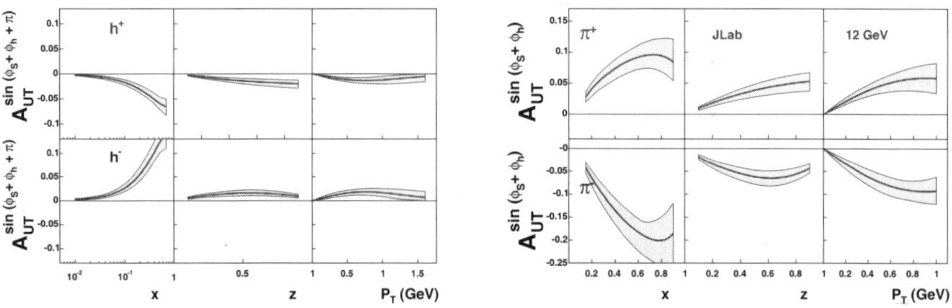

Fig. 4. Estimates of the single spin asymmetry $A_{UT}^{\sin(\phi_h+\phi_S)}$ for COMPASS (left panel) and JLab (right panel) operating with proton target.

unconstrained by the present SIDIS data from HERMES and COMPASS. Our large x estimates are based on an extrapolation of the transversity distribution function into an unexplored region of x, and consequently must be taken with some care. To parametrize the large x behaviour of transversity we use the same parameter β for $\Delta_T u$ and $\Delta_T d$, as the experimental data do not cover the large x region. Indeed this lack of information has strong influence. The same is true for the Collins fragmentation functions, whose large z behaviour is driven by the same parameter δ for favoured and unfavoured Collins FF. The small to medium x region is covered by the experimental data of COMPASS and HERMES and the data support the choice of a universal behaviour x^α for $\Delta_T u$ and $\Delta_T d$. The future JLab measurements will shed light onto the validity of this approximations.

The resulting transversity distribution for u quark is slightly bigger than that of our previous analysis,[3] while d quark transversity now turns out to be definitely negative and non compatible with zero (see Fig. 1).

The Collins fragmentation functions are determined with higher precision, as can be seen in Fig. 1. The unfavored Collins FF turns out to be comparable in size with the favoured Collins FF, thus corroborating previous results.[3,13]

The first x-moments of the transversity distribution – related to the tensor charge[a], and defined as $\Delta_T q \equiv \int_0^1 dx \Delta_T q(x)$ – are found to be $\Delta_T u = 0.59^{+0.14}_{-0.13}$, $\Delta_T d = -0.20^{+0.05}_{-0.07}$ at $Q^2 = 0.8$ GeV^2. Such values are quite close to various model predictions[14–17] for tensor charges which span $0.5 \lesssim \Delta_T u \lesssim 1.5$ and $-0.5 \lesssim \Delta_T d \lesssim 0.5$.

From this preliminary analysis, we conclude that the inclusion of the new data sets[4–6] improves our determination of the transversity distribution and Collins FFs, as can be seen by the reduced size of the present uncertainty bands (Fig. 1), and enables us to give more precise predictions for forthcoming experiments (see Fig. 4). The new distributions are compatible with the extraction of Ref. [3] and close to some model predictions for the transversity distribution (see for example Ref. [14]).

References

1. J. C. Collins, Nucl. Phys. **B396**, 161 (1993).
2. D. Boer, R. Jakob, and P. J. Mulders, Nucl. Phys. **B504**, 345 (1997).
3. M. Anselmino *et al.*, Phys. Rev. **D75**, 054032 (2007).
4. HERMES, M. Diefenthaler, (2007), arXiv:0706.2242.
5. COMPASS, M. Alekseev *et al.*, (2008), arXiv:0802.2160.
6. Belle, R. Seidl *et al.*, Phys. Rev. **D78**, 032011 (2008).
7. D. Boer, (2008), arXiv:0804.2408.
8. M. Anselmino *et al.*, Phys. Rev. **D71**, 074006 (2005).
9. M. Gluck, E. Reya, and A. Vogt, Eur. Phys. J. **C5**, 461 (1998).
10. D. de Florian, R. Sassot, and M. Stratmann, Phys. Rev. **D75**, 114010 (2007).
11. M. Gluck, E. Reya, M. Stratmann, and W. Vogelsang, Phys. Rev. **D63**, 094005 (2001).
12. O. Martin, A. Schafer, M. Stratmann, and W. Vogelsang, Phys. Rev. **D57**, 3084 (1998).
13. A. V. Efremov, K. Goeke, and P. Schweitzer, Phys. Rev. **D73**, 094025 (2006).
14. I. C. Cloet, W. Bentz, and A. W. Thomas, Phys. Lett. **B659**, 214 (2008).
15. M. Wakamatsu, Phys. Lett. **B653**, 398 (2007), 0705.2917.
16. QCDSF, M. Gockeler *et al.*, Phys. Lett. **B627**, 113 (2005), hep-lat/0507001.
17. H.-x. He and X.-D. Ji, Phys. Rev. **D54**, 6897 (1996), hep-ph/9607408.

[a]We have assumed $\Delta_T q_{sea} = 0$.

SIVERS EFFECT IN SIDIS PION AND KAON PRODUCTION

M. Anselmino[1], M. Boglione[1], U. D'Alesio[2,3], A. Kotzinian[4], S. Melis[1], F. Murgia[3], A. Prokudin[1,5], C. Türk[1]

1 - Dipartimento di Fisica Teorica, Università di Torino and INFN, Sezione di Torino, Via P. Giuria 1, I-10125 Torino, Italy

2 - Dipartimento di Fisica, Università di Cagliari, Cittadella Universitaria di Monserrato, I-09042 Monserrato (CA), Italy

3 - INFN, Sezione di Cagliari, C.P. 170, I-09042 Monserrato (CA), Italy

4 - CEA-Saclay, IRFU/Service de Physique Nucléaire, 91191 Gif-sur-Yvette, France; Yerevan Physics Institute, 375036 Yerevan, Armenia; JINR, 141980 Dubna, Russia

5 - Di.S.T.A., Università del Piemonte Orientale "A. Avogadro", Alessandria, Italy

We study the Sivers effect in transverse single spin asymmetries (SSA) for pion and kaon production in Semi-Inclusive Deep Inelastic Scattering (SIDIS) processes. We perform a fit of $A_{UT}^{\sin(\phi_h-\phi_S)}$ data from HERMES and COMPASS Collaborations, which allow a new determination of the Sivers distribution functions for quark and anti-quark with u, d and also s flavours.

High statistics data on the $A_{UT}^{\sin(\phi_h-\phi_S)}$ azimuthal asymmetries for SIDIS have recently become available: the COMPASS collaboration has presented separated charged pion and kaon asymmetries, together with some data for K_S^0 production,[1,2] while the HERMES Collaboration has measured neutral pion and charged kaon azimuthal asymmetries, and reanalyzed higher precision data on charged pion asymmetries.[3] Soon after their release, HERMES data on charged kaon asymmetries have raised a new puzzle. In fact, the single spin asymmetry corresponding to K^+ production is much larger (about a factor two) than the analogous asymmetry for π^+, while one would naively expect, on the basis of u quark dominance, K^+ and π^+ asymmetries to be roughly the same. Actual attempts to reproduce the K^+ asymmetry confirm this naive expectation: while most of the available sets of fragmentation functions describe rather well the pion multiplicities observed at

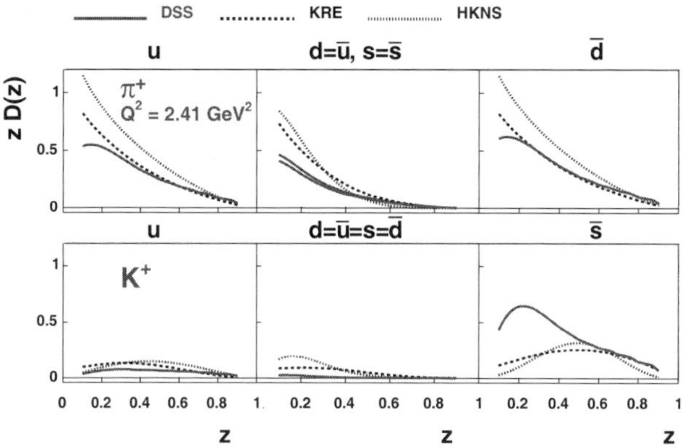

Fig. 1. The LO unpolarized fragmentation functions for u, d and s flavours, at $Q^2 = 2.4$ $(\mathrm{GeV}/c)^2$, as given by Kretzer[5] (dashed lines), by Hirai, Kumano, Nagai and Sudoh[6] (dotted lines) and by De Florian, Sassot and Stratmann[4] (solid lines).

HERMES, many of them fail to reproduce the kaon multiplicities in SIDIS production. The main reason is the role of the strange quarks: for example, one expects that K^+ mesons can be abundantly produced by $\bar{s}$ quarks, via creation from the vacuum of a light $u\bar{u}$ pair, rather than by u quarks, via creation from the vacuum of a heavy $s\bar{s}$ pair. Such a feature is particularly emphasized in the set recently obtained by de Florian, Sassot and Stratmann (DSS),[4] which has $D_{\bar{s}}^{K^+}(z) \gg D_u^{K^+}(z)$ over the whole z range. This is shown in Fig. 1, where the LO DSS fragmentation functions are compared with those extracted from Kretzer (KRE)[5] and from Hirai, Kumano, Nagai and Sudoh (HKNS).[6] In the DSS set the presence of a large $D_{\bar{s}}^{K^+}$ FF can help to understand the "puzzle" of the K^+ asymmetry. Indeed, if a non-negligible $\bar{s}$ Sivers function exists, then its action combined with a large $D_{\bar{s}}^{K^+}$ fragmentation function can give rise to a significant difference between K^+ and π^+ Sivers asymmetries.

The SIDIS transverse single spin asymmetry $A_{UT}^{\sin(\phi_h-\phi_S)}$ measured by HERMES and COMPASS is defined as

$$A_{UT}^{\sin(\phi_h-\phi_S)} = 2\,\frac{\int d\phi_S\, d\phi_h\, [d\sigma^\uparrow - d\sigma^\downarrow]\, \sin(\phi_h-\phi_S)}{\int d\phi_S\, d\phi_h\, [d\sigma^\uparrow + d\sigma^\downarrow]}\,, \qquad (1)$$

and shows the azimuthal modulation triggered by the correlation between the nucleon spin $\boldsymbol{S}$ and the quark intrinsic transverse momentum $\boldsymbol{k}_\perp$, embodied in the Sivers distribution function $\Delta^N f_{q/p^\uparrow}(x, k_\perp)$:

$$\hat{f}_{q/p^\uparrow}(x, \boldsymbol{k}_\perp) = f_{q/p}(x, k_\perp) + \frac{1}{2}\,\Delta^N f_{q/p^\uparrow}(x, k_\perp)\; \boldsymbol{S}\cdot(\hat{\boldsymbol{P}} \times \hat{\boldsymbol{k}}_\perp)\,, \qquad (2)$$

where $f_{q/p}(x, k_\perp)$ is the unpolarized x and $k_\perp$ dependent parton distribution, and the mixed product $\boldsymbol{S} \cdot (\hat{\boldsymbol{P}} \times \hat{\boldsymbol{k}}_\perp)$ explicitly gives the azimuthal dependence mentioned above. The "weighting" factor $\sin(\phi_h - \phi_S)$ in Eq. (1) is appropriately chosen to single out, among the various azimuthal dependent terms appearing in $[d\sigma^\uparrow - d\sigma^\downarrow]$,[7,8] only the contribution of the Sivers mechanism.

At order $(k_\perp/Q)$, the denominator of Eq. (1) is simply twice the unpolarized cross section, while the numerator is a subtle convolution of the Sivers distribution function, the unpolarized fragmentation function and the elementary cross section with some specific azimuthal phases, crucial when taking into account the full non-collinear nature of the process.[9]

In our analysis we shall consider u, d and s flavours for quarks and antiquarks. The Sivers function is parameterized in terms of the unpolarized distribution function, in the following factorized form:[10]

$$\Delta^N f_{q/p^\uparrow}(x, k_\perp) = 2\,\mathcal{N}_q(x)\, h(k_\perp)\, f_{q/p}(x, k_\perp)\,, \tag{3}$$

with

$$\mathcal{N}_q(x) = N_q\, x^{\alpha_q}(1-x)^{\beta_q}\, \frac{(\alpha_q+\beta_q)^{(\alpha_q+\beta_q)}}{\alpha_q^{\alpha_q}\,\beta_q^{\beta_q}}\,, \qquad h(k_\perp) = \sqrt{2e}\,\frac{k_\perp}{M_1}\, e^{-k_\perp^2/M_1^2}\,, \tag{4}$$

where N_q, α_q, β_q and M_1 (GeV/c) are free parameters to be determined by fitting the experimental data. Since $h(k_\perp) \le 1$ for any $k_\perp$ and $|\mathcal{N}_q(x)| \le 1$ for any x (notice that we allow the constant parameter N_q to vary only inside the range $[-1, 1]$), the positivity bound for the Sivers function is automatically fulfilled. We adopt the usual (and convenient) gaussian factorization for the unpolarized distribution and fragmentation functions:

$$f_{q/p}(x, k_\perp) = f_q(x)\,\frac{1}{\pi\langle k_\perp^2\rangle}\, e^{-k_\perp^2/\langle k_\perp^2\rangle}\,, \quad D_q^h(z, p_\perp) = D_q^h(z)\,\frac{1}{\pi\langle p_\perp^2\rangle}\, e^{-p_\perp^2/\langle p_\perp^2\rangle}, \tag{5}$$

with the values of $\langle k_\perp^2\rangle$ and $\langle p_\perp^2\rangle$ fixed to the values found by analysing the Cahn effect in unpolarized SIDIS:[11] $\langle k_\perp^2\rangle = 0.25$ (GeV/c)^2 and $\langle p_\perp^2\rangle =$ 0.20 (GeV/c)^2 .

The recent SIDIS experimental data on Sivers asymmetries for pion *and kaon* production give us the opportunity to study sea-quark Sivers functions for $\bar{u}$, $\bar{d}$, s and $\bar{s}$ quarks. In order to evaluate the significance of the sea-quark Sivers contributions we first perform a fit of the SIDIS data using flavour independent ratios of the sea-quark Sivers functions with the corresponding unpolarized PDF's: that is, for $\bar{u}$, $\bar{d}$, s and $\bar{s}$ flavours we attempt an "unbroken sea" ansatz:

$$\Delta^N f_{q/p^\uparrow}(x,k_\perp) = 2\,\mathcal{N}_{sea}(x)\,h(k_\perp)\,f_{q/p}(x,k_\perp)\,,$$
$$\mathcal{N}_{sea}(x) = N_{sea}\,x^{\alpha_{sea}}(1-x)^{\beta_{sea}}\,\frac{(\alpha_{sea}+\beta_{sea})^{(\alpha_{sea}+\beta_{sea})}}{\alpha_{sea}^{\alpha_{sea}}\,\beta_{sea}^{\beta_{sea}}}\,, \qquad (6)$$

where $N_{sea}, \alpha_{sea}, \beta_{sea}$ are the same for all sea quarks, $q = \bar{u}$, $\bar{d}$, s and $\bar{s}$.

As the SIDIS data from HERMES and COMPASS have a limited coverage in x, typically $x < 0.3 - 0.4$, the experimental asymmetries we are fitting contain very little information on the large x tail of the Sivers functions. Therefore, as a first attempt, we assume the same value of β for all Sivers functions, setting $\beta_{sea} = \beta_u = \beta_d \equiv \beta$. Thus for the "unbroken sea" ansatz we have 8 free parameters.

For the purposes of our fit, we use the GRV98LO[12] unpolarized parton distribution functions $f_q(x,Q^2)$ and the DSS[4] fragmentation functions $D_q^h(z,Q^2)$ – all evolved at the appropriate Q^2 values – with the additional $k_\perp$ gaussian dependences of Eq. (5). While the choice of the DSS fragmentation functions allows for the best description of the large asymmetries observed for K^+, the use of different sets of distribution functions, including the most recent analysis of s quark distributions from HERMES,[13] would not affect our results significantly. For the Sivers function, we use the functional form of Eq. (3). Notice that the (unknown) Q^2 evolution of these functions is assumed to be the same as for the unpolarized PDF's, $f_q(x,Q^2)$.

By fitting simultaneously pion and kaon production data from HERMES[3] and COMPASS[1] we obtain an acceptable overall description of the experimental data, with $\chi^2_{dof} = 1.16$. The new experimental data give clear indications of the need of a non negligible sea-quark Sivers function, with $N_{sea} = -0.13 \pm 0.03$ sensitively different from zero. Although the total χ^2_{dof} is definitely good, a more careful examination of the results shows that while we achieve a perfect description of the π^+ production data at HERMES,[3] with $\chi^2 \simeq 1$ per data point, the description of kaon production data is rather poor, with $\chi^2 \simeq 3$ per data point for K^+ production at HERMES.[3] This indicates that the "unbroken sea" ansatz fails to reproduce the differences between pion and kaon production, and clearly suggests the need of a parameterization which should allow for a more structured flavour dependence of the sea-quark Sivers functions.

Including four new functions in our analysis would result in a substantial growth of the number of parameters and would consequently limit the predictive power of the model. To keep the number of parameters under control,

Table 1. Best values of the free parameters for the "broken sea" ansatz.

$N_u = 0.35^{+0.08}_{-0.08}$	$N_d = -0.90^{+0.43}_{-0.10}$	$N_s = -0.24^{+0.62}_{-0.50}$
$N_{\bar{u}} = 0.04^{+0.22}_{-0.24}$	$N_{\bar{d}} = -0.40^{+0.33}_{-0.44}$	$N_{\bar{s}} = 1^{+0}_{-0.0001}$
$\alpha_u = 0.73^{+0.72}_{-0.58}$	$\alpha_d = 1.08^{+0.82}_{-0.65}$	$\alpha_{sea} = 0.79^{+0.56}_{-0.47}$
$\beta = 3.46^{+4.87}_{-2.90}$	$M_1^2 = 0.34^{+0.30}_{-0.16}$ (GeV/c)^2	

we define a simple "broken sea" ansatz by introducing four free parameters, $N_{\bar{u}}$, $N_{\bar{d}}$, N_s, and $N_{\bar{s}}$ which give different sizes to the sea-quark Sivers functions, while keeping the same functional forms ($\alpha_{\bar{u}} = \alpha_{\bar{d}} = \alpha_s = \alpha_{\bar{s}} \equiv \alpha_{sea}$ and $\beta_{sea} = \beta_u = \beta_d \equiv \beta$). For the "broken sea" ansatz fit we then have 11 parameters.

The results we obtain for these parameters by fitting simultaneously the four experimental data sets on $A_{UT}^{\sin(\phi_h-\phi_S)}$, corresponding to pion and kaon production at HERMES[3] and COMPASS,[1] are presented in Table 1, together with their corresponding errors.[9] The fit performed under the "broken sea" ansatz shows a remarkable improvement, especially concerning the description of kaon data. We now obtain $\chi^2 = 1.20$ per data point for K^+ production at HERMES,[3] while for pions we have $\chi^2 = 0.94$ per data point, and a total $\chi^2_{dof} = 1.00$.

The quality of our results is shown in Fig. 2 where our best fits to the SSA's are compared with the experimental data[1,3] (the SSA's are plotted as a function of one variable at a time, either z or x or P_T, while an integration over the other variables has been performed consistently with the cuts of the corresponding experiment). The shaded areas in Fig. 2 corresponds to 95.45% Confidence Level.

Our computation of the K^0_S asymmetry at COMPASS can be compared with the available data,[2] as shown in Fig. 2. Notice that these curves are not best fits, but simple estimates, based on the extracted Sivers functions.

The Sivers functions generated by our best fit procedure are presented in Fig. 3, where we plot, on the left panel, the first $\boldsymbol{k}_\perp$ moment and, on the right panel, the $k_\perp$ dependence of $\Delta^N f_{q/p^\uparrow}$ at a fixed value of $x = 0.1$. The highest and lowest dashed lines show the positivity limits $|\Delta^N f| = 2f$.

The HERMES data on kaon asymmetries, surprisingly large for K^+, cannot be explained without a sea-quark Sivers distribution. In particular, we definitely find $\Delta^N f_{\bar{s}/p^\uparrow} > 0$, and confirm the previous findings for valence flavours $\Delta^N f_{u/p^\uparrow} > 0$ and $\Delta^N f_{d/p^\uparrow} < 0$.[10,14–16] There are simple reasons for the above results. The Sivers distribution function for $\bar{s}$ quarks turns out to be definitely positive, due to the large positive value of $A_{UT}^{\sin(\phi_h-\phi_S)}$ for

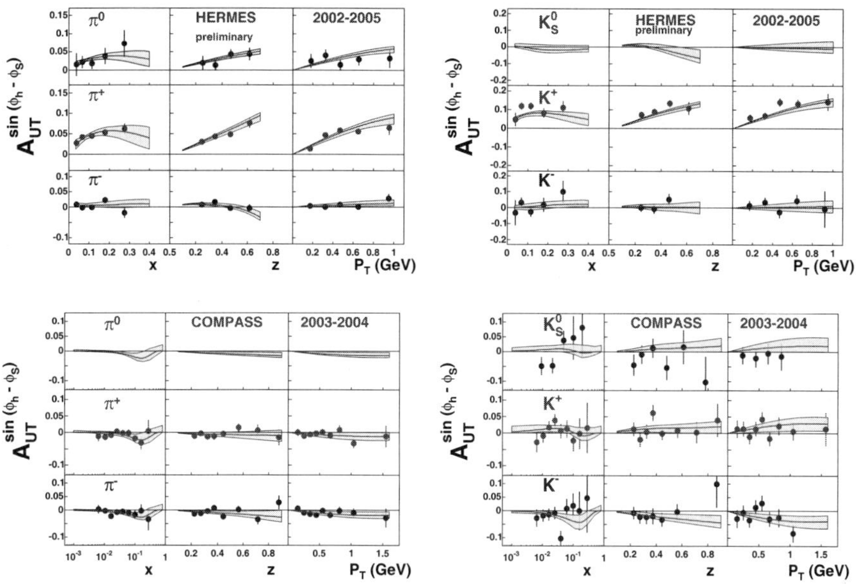

Fig. 2. The results obtained from our fit (solid lines) are compared with the HERMES[3] and COMPASS[1] (deuteron target) measurements of $A_{UT}^{\sin(\phi_h - \phi_S)}$ for pion and kaon production. The shaded area corresponds to the theoretical uncertainty of the parameters.

K^+; notice that the value of $N_{\bar{s}}$ saturates the positivity bound $|N_q| \leq 1$. Similarly, the positive sign of $\Delta^N f_{u/p^\uparrow}$ is, essentially, driven by the positive π^+ SSA and the opposite sign of $\Delta^N f_{d/p^\uparrow}$ by the small SSA measured by COMPASS on a deuteron target. The Sivers functions for $\bar{u}$, $\bar{d}$ and s quarks, instead, turn out to have much larger uncertainties.

We notice that the Burkardt sum rule[17] is almost saturated by u and d quarks alone at $Q^2 = 2.4$ (GeV/c)^2, leaving little room for a gluon Sivers function, in agreement with other similar results.[18,19]

Using the Sivers functions determined through our fit, we can give predictions for other transverse single spin asymmetries $A_{UT}^{\sin(\phi_h - \phi_S)}$ which will be measured in the near future. Fig. 4 shows the results we obtain for the COMPASS experiment operating with a hydrogen target, adopting the same experimental cuts which were used for the deuterium target .[11]

Forthcoming measurements at the energies of 6 and 12 GeV are going to be performed at JLab, on transversely polarized proton, neutron and deuteron targets. The obtained data will have a very strong impact as they will cover a kinematical region corresponding to large values of x, a region which is so far unexplored from other SIDIS measurements. In particular, a combined analysis of HERMES, COMPASS and JLab SIDIS data will allow

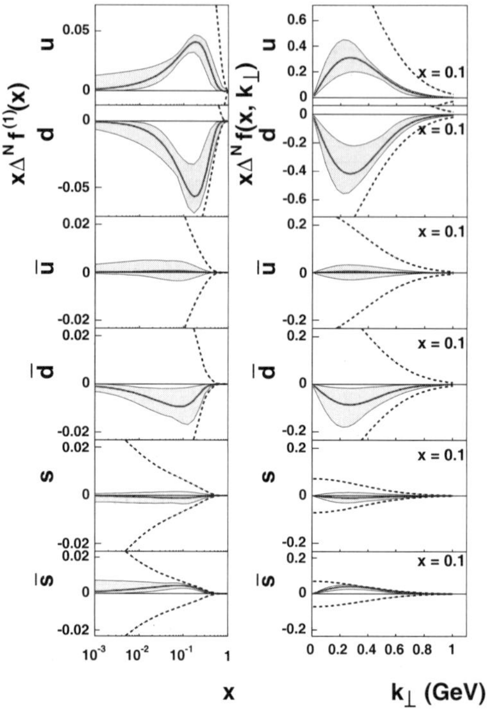

Fig. 3. The Sivers distribution functions for u, d and s flavours as determined by our simultaneous fit of HERMES and COMPASS data (see text for details). On the left panel, the first moment $x\,\Delta^N f^{(1)}(x)$ is shown as a function of x for each flavour, on the right panel, the Sivers distribution $x\,\Delta^N f(x, k_\perp)$ is shown as a function of $k_\perp$ at a fixed value of $x = 0.1$ for each flavour.

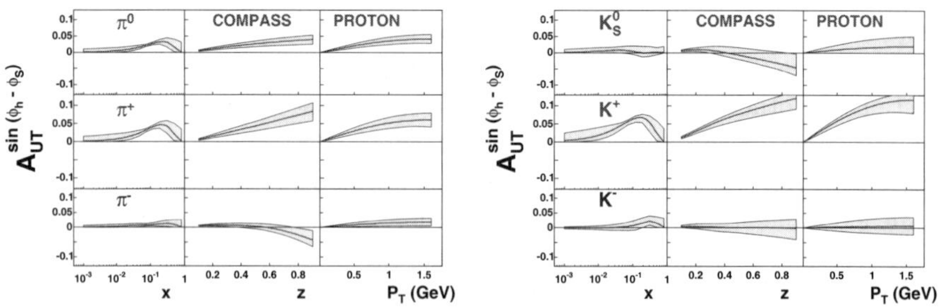

Fig. 4. Predictions for the single spin asymmetry $A_{UT}^{\sin(\phi_h-\phi_S)}$ for pion and kaon production off a hydrogen target, which will be measured by the COMPASS Collaboration.

a much better determination of the β parameters, which control the large x behaviour of the Sivers distribution functions. In addition, the combined analysis of proton and neutron target events will help flavour disentangling and a more precise determination of u and d quark contributions. We have also presented predictions for the JLab SSA's, for pion and kaon production off proton, neutron and deuteron targets, at 6 and 12 GeV.[9] Notice that these predictions, while well constrained by the available SIDIS data at small x values, might be less stringent at large x values: for example, relaxing the assumption of a unique β value for all flavours, would only marginally affect our present fit of HERMES and COMPASS data, but would much widen the uncertainty band above $x \simeq 0.4$.

References

1. A. Martin, *Czech. J. Phys.* **56**, F33 (2006).
2. M. Alekseev *et al.*, arXiv:0802.2160 [hep-ex] (2008).
3. M. Diefenthaler, arXiv:0706.2242 [hep-ex] (2007).
4. D. de Florian, R. Sassot and M. Stratmann, *Phys. Rev.* **D75**, 114010 (2007).
5. S. Kretzer, *Phys. Rev.* **D62**, 054001 (2000).
6. M. Hirai, S. Kumano, T. H. Nagai and K. Sudoh, *Phys. Rev.* **D75**, 094009 (2007).
7. A. Bacchetta *et al.*, *JHEP* **02**, 093 (2007).
8. M. Anselmino, M. Boglione, U. D'Alesio, F. Murgia and A. Prokudin, *in preparation.*
9. M. Anselmino *et al.*, arXiv:0805.2677 [hep-ph] (2008).
10. M. Anselmino *et al.*, *Phys. Rev.* **D72**, 094007 (2005).
11. M. Anselmino *et al.*, *Phys. Rev.* **D71**, 074006 (2005).
12. M. Gluck, E. Reya and A. Vogt, *Eur. Phys. J.* **C5**, 461 (1998).
13. A. Airapetian *et al.*, arXiv:0803.2993 [hep-ex] (2008).
14. W. Vogelsang and F. Yuan, *Phys. Rev.* **D72**, 054028 (2005).
15. J. C. Collins *et al.*, *Phys. Rev.* **D73**, 014021 (2006).
16. M. Anselmino *et al.*, hep-ph/0511017 (2005).
17. M. Burkardt, *Phys. Rev.* **D69**, 091501 (2004).
18. M. Anselmino, U. D'Alesio, S. Melis and F. Murgia, *Phys. Rev.* **D74**, 094011 (2006).
19. S. J. Brodsky and S. Gardner, *Phys. Lett.* **B643**, 22 (2006).

SPIN-ORBIT CORRELATIONS

M. Burkardt

Physics Department, New Mexico State University,
Las Cruces, NM 88003, U.S.A.

We summarize the intuitive connection between deformations of parton distributions in impact parameter space and single-spin asymmetries. Lattice results for the x^2-moment of the twist-3 polarized parton distribution $g_2(x)$ are used to estimate the average transverse force acting on the active quark in SIDIS in the instant after being struck by the virtual photon.

1. Impact Parameter Dependent PDFs and SSAs

The Fourier transform of the GPD $H_q(x,0,t)$ yields the distribution $q(x,\mathbf{b}_\perp)$ of unpolarized quarks, for an unpolarized target, in impact parameter space [1]

$$q(x,\mathbf{b}_\perp) = \int \frac{d^2\mathbf{\Delta}_\perp}{(2\pi)^2} H_q(x,0,-\mathbf{\Delta}_\perp^2)\, e^{-i\mathbf{b}_\perp\cdot\mathbf{\Delta}_\perp}, \tag{1}$$

with $\mathbf{\Delta}_\perp = \mathbf{p}'_\perp - \mathbf{p}_\perp$. For a transversely polarized target (e.g. polarized in the $+\hat{x}$-direction) the impact parameter dependent PDF $q_{+\hat{x}}(x,\mathbf{b}_\perp)$ is no longer axially symmetric and the transverse deformation is described by the gradient of the Fourier transform of the GPD $E_q(x,0,t)$ [2]

$$q_{+\hat{x}}(x,\mathbf{b}_\perp) = q(x,\mathbf{b}_\perp) - \frac{1}{2M}\frac{\partial}{\partial b_y}\int \frac{d^2\mathbf{\Delta}_\perp}{(2\pi)^2} E_q(x,0,-\mathbf{\Delta}_\perp^2)\, e^{-i\mathbf{b}_\perp\cdot\mathbf{\Delta}_\perp} \tag{2}$$

$E_q(x,0,t)$ and hence the details of this deformation are not very well known, but its x-integral, the Pauli form factor F_2, is. This allows to relate the average transverse deformation resulting from Eq. (2) to the contribution from the corresponding quark flavor to the anomalous magnetic moment. This observation is important in understanding the sign of the Sivers function.

In a target that is polarized transversely (*e.g.* vertically), the quarks in the target nucleon can exhibit a (left/right) asymmetry of the distribution

$f_{q/p^\uparrow}(x_{\rm B}, \boldsymbol{k}_T)$ in their transverse momentum $\boldsymbol{k}_T$ [3,4]

$$f_{q/p^\uparrow}(x_{\rm B}, \boldsymbol{k}_T) = f_1^q(x_{\rm B}, k_T^2) - f_{1T}^{\perp q}(x_{\rm B}, k_T^2)\frac{(\hat{\boldsymbol{P}} \times \boldsymbol{k}_T) \cdot \boldsymbol{S}}{M}, \tag{3}$$

where $\boldsymbol{S}$ is the spin of the target nucleon and $\hat{\boldsymbol{P}}$ is a unit vector opposite to the direction of the virtual photon momentum. The fact that such a term may be present in (4) is known as the Sivers effect and the function $f_{1T}^{\perp q}(x_{\rm B}, k_T^2)$ is known as the Sivers function. The latter vanishes in a naive parton picture since $(\hat{\boldsymbol{P}} \times \boldsymbol{k}_T) \cdot \boldsymbol{S}$ is odd under naive time reversal (a property known as naive-T-odd), where one merely reverses the direction of all momenta and spins without interchanging the initial and final states. The momentum fraction x, which is equal to $x_{\rm B}$ in DIS experiments, represents the longitudinal momentum of the quark *before* it absorbs the virtual photon, as it is determined solely from the kinematic properties of the virtual photon and the target nucleon. In contradistinction, the transverse momentum $\boldsymbol{k}_T$ is defined in terms of the kinematics of the final state and hence it represents the asymptotic transverse momentum of the active quark *after* it has left the target and before it fragments into hadrons. Thus the Sivers function for semi-inclusive DIS includes the final state interaction between struck quark and target remnant, and time reversal invariance no longer requires that it vanishes.

Fig. 1. In SIDIS (a) the ejected (red) quark is attracted by the (anti-red) spectators. In contradistinction, in DY (b), before annihilating with the (red) active quark, the approaching (anti-red) antiquark is repelled by the (anti-red) spectators.

Indeed, as time reversal not only reverses the signs of all spins and momenta, but also transforms final state interactions (FSI) into initial state interactions (ISI), it has been shown that the Sivers function relevant for SIDIS and that relevant for Drell-Yan (DY) processes must have opposite signs [5],

$$f_{1T}^{\perp}(x_{\rm B}, k_T^2)_{SIDIS} = -f_{1T}^{\perp}(x_{\rm B}, k_T^2)_{DY}, \tag{4}$$

where the asymmetry in DY arises from the ISI between the incoming antiquark and the target. The experimental verification of this relation would provide a test of the current understanding of the Sivers effect within QCD. It is instructive to elucidate its physical origin in the context of a perturbative picture: for instance, when the virtual photon in a DIS process hits a red quark, the spectators must be collectively anti-red in order to form a color-neutral bound state, and thus attract the struck quark (Fig. 1). In DY, when an anti-red antiquark annihilates with a target quark, the target quark must be red in order to merge into a photon, which carries no color. Since the proton was colorless before the scattering, the spectators must be anti-red and thus repel the approaching antiquark.

The significant distortion of parton distributions in impact parameter space (2) provides a natural mechanism for a Sivers effect. In semi-inclusive DIS, when the virtual photon strikes a u quark in a $\perp$ polarized proton, the u quark distribution is enhanced on the left side of the target (for a proton with spin pointing up when viewed from the virtual photon perspective). Although in general the final state interaction (FSI) is very complicated, we expect it to be on average attractive thus translating a position space distortion to the left into a momentum space asymmetry to the right and vice versa (Fig. 2) [6]. Since this picture is very intuitive, a few words of

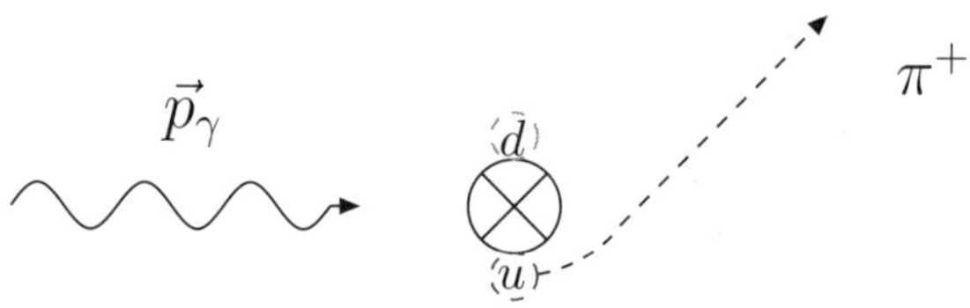

Fig. 2. The transverse distortion of the parton cloud for a proton that is polarized into the plane, in combination with attractive FSI, gives rise to a Sivers effect for u (d) quarks with a $\perp$ momentum that is on the average up (down).

caution are in order. First of all, such a reasoning is strictly valid only in mean field models for the FSI as well as in simple spectator models [7]. Furthermore, even in such mean field models there is no one-to-one correspondence between quark distributions in impact parameter space and unintegrated parton densities (e.g. Sivers function). While both are connected by a Wigner distribution [8], they are not Fourier transforms of each other. Nevertheless, since the primordial momentum distribution of the quarks (without FSI) must be symmetric we find a qualitative connection between the primordial position space asymmetry and the momentum space asymmetry (with FSI). Another issue concerns the x-dependence of the Sivers function. The x-dependence of the position space asymmetry is described by the GPD $E(x, 0, -\Delta_\perp^2)$. Therefore, within the above mechanism, the x dependence of the Sivers function should be related to the x dependence of $E(x, 0, -\Delta_\perp^2)$. However, the x dependence of E is not known yet and we only know the Pauli form factor $F_2 = \int dxE$. Nevertheless, if one makes the additional assumption that E does not fluctuate as a function of x then the contribution from each quark flavor q to the anomalous magnetic moment κ determines the sign of $E^q(x, 0, 0)$ and hence of the Sivers function. Making these assumptions, as well as the very plausible assumption that the FSI is on average attractive, one finds that $f_{1T}^{\perp u} < 0$, while $f_{1T}^{\perp d} > 0$. Both signs have been confirmed by a flavor analysis based on pions produced in a SIDIS experiment by the HERMES collaboration [9] and are consistent with a vanishing isoscalar Sivers function [10].

2. The Force on a Quark in SIDIS

The chirally even spin-dependent twist-3 parton distribution $g_2(x) = g_T(x) - g_1(x)$ is defined as

$$\int \frac{d\lambda}{2\pi} e^{i\lambda x} \langle PS|\bar{\psi}(0)\gamma^\mu\gamma_5\psi(\lambda n)|_{Q^2}|PS\rangle$$
$$= 2\left[g_1(x, Q^2)p^\mu(S\cdot n) + g_T(x, Q^2)S_\perp^\mu + M^2 g_3(x, Q^2)n^\mu(S\cdot n)\right].$$

neglecting m_q: $g_2(x) = g_2^{WW}(x) + \bar{g}_2(x)$, with $g_2^{WW}(x) = -g_1(x) + \int_x^1 \frac{dy}{y} g_1(y)$. $\bar{g}_2(x)$ involves quark-gluon correlations, e.g. [11,12]

$$\int dx x^2 \bar{g}_2(x) = \frac{d_2}{6} \tag{5}$$

with

$$g\left\langle P, S\left|\bar{q}(0)G^{+y}(0)\gamma^+ q(0)\right| P, S\right\rangle = MP^+P^+S^x d_2 \tag{6}$$

At low Q^2, g_2 has the physical interpretation of a spin polarizability, which is why the matrix elements (note that $\sqrt{2}G^{+y} = B^x - E^y$)

$$\chi_E 2M^2\vec{S} = \langle P, S|\, q^\dagger\vec{\alpha}\times g\vec{E}q\,|P, S\rangle \quad \chi_B 2M^2\vec{S} = \langle P, S|\, q^\dagger g\vec{B}q\,|P, S\rangle \tag{7}$$

are sometimes called spin polarizabilities or color electric and magnetic polarizabilities [13]. In the following we will discuss that at high Q^2 a better interpretation for these matrix elements is that of a 'force'.

As Qiu and Sterman have shown [14], the average transverse momentum of the ejected quark (here also averaged over the momentum fraction x

carried by the active quark) in a SIDIS experiment can be represented by the matrix element

$$\langle k_\perp^y \rangle = -\frac{1}{2P^+}\left\langle P,S\left|\bar{q}(0)\int_0^\infty dx^- G^{+y}(x^+=0,x^-)\gamma^+ q(0)\right|P,S\right\rangle \quad (8)$$

which has a simple physical interpretation: the average transverse momentum is obtained by integrating the transverse component of the color Lorentz force along the trajectory of the active quark — which is an almost light-like trajectory along the $-\hat{z}$ direction, with $z=-t$: The $\hat{y}$-component of the Lorentz force acting on a particle moving, with (nearly) the speed of light $\vec{v}=(0,0,-1)$, along the $-\hat{z}$ direction reads

$$g\sqrt{2}G^{y+} = g\left(E^y + B^x\right) = g\left[\vec{E}+\vec{v}\times\vec{B}\right]^y. \quad (9)$$

We now rewrite Eq. (8) as an integral over time

$$\langle k_\perp^y \rangle = -\frac{\sqrt{2}}{2P^+}\langle P,S|\,\bar{q}(0)\int_0^\infty dt G^{+y}(t,z=-t)\gamma^+ q(0)\,|P,S\rangle \quad (10)$$

in which the physical interpretation of $-\frac{\sqrt{2}}{2P^+}\langle P,S|\,\bar{q}(0)G^{+y}(t,z = -t)\gamma^+ q(0)\,|P,S\rangle$ as being the averaged force acting on the struck quark at time t after being struck by the virtual photon becomes more apparent.

In particular,

$$\begin{aligned} F^y(0) &\equiv -\frac{\sqrt{2}}{2P^+}\langle P,S|\,\bar{q}(0)G^{+y}(0)\gamma^+ q(0)\,|P,S\rangle \\ &= -\frac{1}{\sqrt{2}}MP^+S^x d_2 = -\frac{M^2}{2}d_2, - \end{aligned} \quad (11)$$

where the last equality holds only in the rest frame ($p^+ = \frac{1}{\sqrt{2}}M$) and for $S^x=1$, can be interpreted as the averaged transverse force acting on the active quark in the instant right after it has been struck by the virtual photon.

Lattice calculations of the twist-3 matrix element yield [15]

$$d_2^{(u)} = 0.010 \pm 0.012 \qquad\qquad d_2^{(d)} = -0.0056 \pm 0.0050 \quad (12)$$

renormalized at a scale of $Q^2 = 5$ GeV^2 for the smallest lattice spacing in Ref. [15]. Here the identity $M^2 \approx 5\text{GeV/fm}$ is useful to better visualize the magnitude of the force.

$$F_{(u)} = -25 \pm 30\text{MeV/fm} \qquad\qquad F_{(d)} = 14 \pm 13\text{MeV/fm}. \quad (13)$$

In the chromodynamic lensing picture, one would have expected that $F_{(u)}$ and $F_{(d)}$ are of about the same magnitude and with opposite sign. The same holds in the large N_C limit. A vanishing Sivers effect for an isoscalar target would be more consistent with equal and opposite average forces. However, since the error bars for d_2 include only statistical errors, the lattice result may not be inconsistent with $d_2^{(d)} \sim -d_2^{(u)}$.

The average transverse momentum from the Sivers effect is obtained by integrating the transverse force to infinity (along a light-like trajectory) $\langle k^y \rangle = \int_0^\infty dt F^y(t)$. This motivates us to define an 'effective range'

$$R_{eff} \equiv \frac{\langle k^y \rangle}{F^y(0)}. \tag{14}$$

Note that R_{eff} depends on how rapidly the correlations fall off along a light-like direction and it may thus be larger than the (spacelike) radius of a hadron. Of cource, unless the functional form of the integrand is known, R_{eff} cannot really tell us about the range of the FSI, but if the integrand does not oscillate

Fits of the Sivers function to SIDIS data yield 17 one finds about $|\langle k^y \rangle| \sim$ 100 MeV [17]. Together with the (average) value for $|d_2|$ from the littice this translates into an effective range R_{eff} of several fm. It would be interesting to compare R_{eff} for different quark flavors and as a function of Q^2, but this requires more precise values for d_2 as well as the Sivers function.

Note that a complementary approach to the effective range was chosen in Ref. 18, where the twist-3 matrix element appearing in Eq. (11) was, due to the lack of lattice QCD results, estimated using QCD sum rule techniques. Moreover, the 'range' was taken as a model *input* parameter to estimate the magnitude of the Sivers function.

A measurement of the twist-4 contribution f_2 to polarized DIS allows determination of the expectation value of different Lorentz/Dirac components of the quark-gluon correlator appearing in (6)

$$f_2 M^2 S^\mu = \frac{1}{2} \langle p, S | \bar{q} g \tilde{G}^{\mu\nu} \gamma_\nu q | p, S \rangle , \tag{15}$$

In combination with (6) this allows a decomposition of the force into electric and magnetic components using

$$F_E^y(0) = -\frac{M^2}{8} \chi_E \qquad\qquad F_B^y(0) = -\frac{M^2}{4} \chi_B \tag{16}$$

for a target nucleon polarized in the $+\hat{x}$ direction, where [13,16]

$$\chi_E = \frac{2}{3}(2d_2 + f_2) \qquad\qquad \chi_M = \frac{1}{3}(4d_2 - f_2). \tag{17}$$

A relation similar to (11) can be derived for the x^2 moment of the twist-3 scalar PDF $e(x)$. For its interaction dependent twist-3 part $\bar{e}(x)$ one finds for an unpolarized target [19]

$$4MP^+P^+e_2 = g\langle p|\,\bar{q}\sigma^{+i}G^{+i}q\,|P\rangle\,, \tag{18}$$

where $e_2 \equiv \int_0^1 dx x^2 \bar{e}(x)$. The matrix element on the r.h.s. of Eq. (18) can be related to the average transverse force acting on a transversely polarized quark in an unpolarized target right after being struck by the virtual photon. Indeed, for the average transverse momentum in the $+\hat{y}$ direction, for a quark polarized in the $+\hat{x}$ direction, one finds

$$\langle k^y\rangle = \frac{1}{4P^+}\int_0^\infty dx^- g\langle p|\,\bar{q}(0)\sigma^{+y}G^{+y}(x^-)q(0)\,|p\rangle\,. \tag{19}$$

A comparison with Eq. (18) shows that the average transverse force at $t=0$ (right after being struck) on a quark polarized in the $+\hat{x}$ direction reads

$$F^y(0) = \frac{1}{2\sqrt{2}p^+}g\langle p|\,\bar{q}\sigma^{+y}G^{+y}q\,|p\rangle = \frac{1}{\sqrt{2}}MP^+S^x e_2 = \frac{M^2}{2}e_2, \tag{20}$$

where the last identify holds only in the rest frame of the target nucleon and for $S^x = 1$.

The impact parameter distribution for quarks polarized in the $+\hat{x}$ direction was found to be shifted in the $+\hat{y}$ direction [20–22]. Applying the chromodynamic lensing model implies a force in the negative $-\hat{y}$ direction for these quarks and one thus expects $e_2 < 0$ for both u and d quarks. Magnitude: since $\kappa_\perp > \kappa$, expect odd force larger than even force and thus $|e_2| > |d_2|$.

It would be interesting to study not only whether the effective range is flavor dependent, but also whether there is a difference between the chirally even and odd cases. It would also be very interesting to learn more about the time dependence of the FSI by calculating matrix elements of $\bar{q}\gamma^+\left(\partial^+G^{+\perp}\right)q$, or even higher derivatives, in lattice QCD. Knowledge of not only the value of the integrand at the origin, but also its slope and curvature at that point, would be very useful for estimating the integral in Eq. (8).

Acknowledgments

I would like to thank the organizers of the Transversity 2008 workshop for the kind invitation and Daniel Boer and Yuji Koike for very helpful discussions. This work was supported in part by the DOE under grant number DE-FG03-95ER40965.

References

1. M.Burkardt, Phys. Rev. D **62**, 071503 (2000), Erratum-ibid. D **66**, 119903 (2002); M. Diehl, Eur. Phys. J. C **25**, 223 (2002); J.P. Ralston and B. Pire, Phys. Rev. D **66**, 111501 (2002).
2. M. Burkardt, Int. J. Mod. Phys. A **18**, 173 (2003).
3. D.W. Sivers, Phys. Rev. D **43**, 261 (1991).
4. A. Bacchetta et al., Phys. Rev. D **70**, 117504 (2004).
5. J.C. Collins, Phys. Lett. B **536**, 43 (2002).
6. M. Burkardt, Phys. Rev. D **66**, 114005 (2002); Phys. Rev. D **69**, 057501 (2004).
7. S.J. Brodsky, D.S. Hwang, and I. Schmidt, Nucl. Phys. B **642**, 344 (2002); M. Burkardt and D.S. Hwang, Phys. Rev. **D69**, 074032 (2004); L.P. Gamberg et al., Phys. Rev. D **67**, 071504 (2003); D. Boer, S.J. Brodsky, and D.S. Hwang, Phys. Rev. D **67**, 054003 (2003); A. Bacchetta et al., Phys. Lett **B578**, 109 (2004); M. Radici et al., hep-ph/0708.0232; L.P. Gamberg et al. hep-ph/0708.0324; D. Boer et al., Phys. Rev. D **67**, 054003 (2003); L.P. Gamberg et al., Phys. Rev. D **67** (2003) 071504.
8. A.V. Belitsky, X. Ji, and F. Yuan, Phys. Rev. D **69**, 074014 (2004).
9. A. Airapetian et al. (HERMES collaboration), Phys. Rev. Lett. **94**, 012002 (2005).
10. A. Martin (COMPASS collaboration), Czech. J. Phys. **56**, F33 (2006).
11. E. Shuryak and A.I. Vainshtein, Nucl. Phys. B **201**, 141 (1982).
12. R.L. Jaffe, Comm. Nucl. Part. Phys. **19**, 239 (1990).
13. B.W. Filippone and X. Ji, Adv. Nucl. Phys. **26**, 1 (2001).
14. J. Qiu and G. Sterman, Phys. Rev. Lett. **67**, 2264 (1991).
15. M. Göckeler et al., Phys. Rev. D **72**, 054507 (2005).
16. Z.-E. Mezziani et al., hep-ph/0404066.
17. M. Anselmino et al., hep-ph/0805.2677
18. A. Schäfer et al., Phys. Rev. D **47**, R1 (1993); E. Stein et al., hep-ph/9409212
19. Y. Koike and K. Tanaka, Phys. Rev. D **51**, 6125 (1995).
20. M. Diehl and P.Hägler, Eur. Phys. J. **C44**, 87 (2005).
21. M. Göckeler et al. (QCDSF collaboration), Phys. Rev. Lett. **98**, 222001 (2007); Ph. Hägler et al. (LHPC collaboration), hep-lat/0705.4295.
22. M. Burkardt and B. Hannafious, Phys. Lett. B **658**, 130 (2008).

CORRELATION FUNCTIONS IN HARD AND (SEMI-)INCLUSIVE PROCESSES

M. Schlegel*

*Theory Center, Jefferson Lab, 12000 Jefferson Avenue
Newport News, VA 23608, USA*

S. Meißner

*Institut für Theoretische Physik II, Ruhr-Universität Bochum
44780 Bochum, Germany*

A. Metz

*Department of Physics, Temple University
Philadelphia, PA 19122-6082, USA*

Possible relations between two *a priori* different classes of parton distributions, the Generalized Parton Distributions (GPDs) and the Transverse Momentum Dependent parton distributions (TMDs), are discussed in this note. Although these relations were proven to hold exactly only in simple models they imply an appealingly simple and intuitive explanation for single-spin asymmetries in semi-inclusive deep-inelastic scattering. In this context we perform a first classification of common mother functions of GPDs and TMDs, so-called Generalized Transverse Momentum Dependent parton distributions (GTMDs), investigate their GPD- and TMD-limits, and gain new insight into the nature of these relations.

Keywords: Wigner function, GTMD.

1. Introduction and Definitions

The non-perturbative substructure of hadrons is described in many hard inclusive and exclusive processes by parton distributions. The most prominent examples of parton distribution are the ordinary, collinear parton

*Notice: Authored by Jefferson Science Associates, LLC under U.S. DOE Contract No. DE-AC05-06OR23177. The U.S. Government retains a non-exclusive, paid-up, irrevocable, world-wide license to publish or reproduce this manuscript for U.S. Government purposes.

distributions (PDFs) depending only on a light-cone momentum fraction $x = k^+/P^+$. Here, k denotes the momentum of a parton inside a hadron moving with momentum P. In this note we use the usual light-cone coordinates $a^\pm = 1/\sqrt{2}(a^0 \pm a^3)$ and $\vec{a}_T = (a^1, a^2)$.

In contrast to ordinary PDFs, Generalized Parton Distributions (GPDs)[1] provide more information on the partonic substructure because of their definition by an off-diagonal matrix element of quark-quark operators. Such a matrix element reads for quarks with flavor q,

$$F^q_{ij}(x, \xi, \vec{\Delta}_T) = \int \frac{dz^-}{2\pi}\, e^{ixP^+z^-} \langle p',\, \lambda' |\, \bar{q}_j(-\tfrac{z}{2})\, [-\tfrac{z}{2}, \tfrac{z}{2}]\, q_i(\tfrac{z}{2})\, |p,\, \lambda\rangle \Big|_{z^+=z_T=0}, \tag{1}$$

where the usual GPD-variables $P = (p + p')/2$ and $\Delta = p' - p$ were used in conjunction with the skewness parameter $\xi = -\Delta^+/(2P^+)$. The helicities of incoming and outgoing hadrons are represented by λ and λ', while $[a, b]$ denotes a so-called gauge link operator which connects the two coordinates a and b by a straight Wilson line. Note that the Wilson line in (1) runs along a certain light-cone direction defined by a light-cone vector n. The non-perturbative object (1) plays an important role for hard exclusive processes and one is sensitive to it in Deeply-Virtual Compton Scattering (DVCS) or Meson Production (DVMP).[2–5] Eight GPDs parameterize the leading twist part of the matrix element (1), commonly referred to as H, E, $\tilde{H}$, $\tilde{E}$, H_T, E_T, $\tilde{H}_T$, $\tilde{E}_T$.[6]

In particular, GPDs provide information about the spatial distribution of partons in hadrons when being transformed into the impact parameter space.[7–10] The impact parameter b_T of a certain parton represents the transverse distance of this parton inside a hadron from the center of longitudinal momenta. The transverse momentum transfer $\vec{\Delta}_T$ and the impact parameter are considered as conjugate variables, and impact parameter GPDs can be defined by a Fourier-transformation of ordinary GPDs. For zero skewness, $\xi = 0$, the impact parameter representation of the nucleon states in (1) is similar to a mixed longitudinal momentum representation (defined by P^+) and a transverse coordinate representation (defined by the impact parameter). This allows one to write impact parameter GPDs as *diagonal* matrix elements of a quark-quark operator, provided by the following matrix element,

$$\begin{aligned}\mathcal{F}^q_{ij}(x, \vec{b}_T) &= \int \frac{d^2\Delta_T}{(2\pi)^2}\, e^{-i\vec{\Delta}_T\cdot\vec{b}_T}\, F^q_{ij}(x, 0, \vec{\Delta}_T) \\ &= \int \frac{dz^-}{2\pi}\, e^{ixP^+z^-} \langle P^+;\, \vec{0}_T |\, \bar{q}_j(z_1)\, [z_1\,,\, z_2]\, q_i(z_2)\, |P^+;\, \vec{0}_T\rangle\,, \end{aligned} \tag{2}$$

with $z_{1/2} = \mp\frac{1}{2}z^- n + b_T$. In such a way impact parameter GPDs can be interpreted as probability densities of partons distributed in a plane perpendicular to the hadron momentum.

Transverse Momentum Dependent parton distributions (TMDs) not only describe the longitudinal but also the transverse motion of partons inside hadrons. They appear in QCD-descriptions of hard semi-inclusive processes like semi-inclusive deep-inelastic scattering (SIDIS) or the Drell-Yan process (DY) (see, e.g., Refs.[11–13]). Like ordinary PDFs, TMDs are defined in terms of diagonal matrix elements. However, due to the additional dependence on the transverse parton momentum $\vec{k}_T$, the quark-quark operator is not only separated on the light-cone but also in the transverse direction. Such a matrix element reads

$$\Phi^q_{ij}(x, \vec{k}_T, S) = \int \frac{dz^- d^2 z_T}{(2\pi)^3}\, \mathrm{e}^{ik\cdot z} \langle P, S|\, \bar{q}_j(-\tfrac{z}{2})\, \mathcal{W}[-\tfrac{z}{2}, \tfrac{z}{2}]\, q_i(\tfrac{z}{2})\, |P, S\rangle \Big|_{z^+=0} . \tag{3}$$

The gauge link operator $\mathcal{W}$ ensures a color gauge-invariant definition of TMDs. It turns out that the Wilson line is process dependent. In SIDIS it describes pure final state interactions and pure initial state interactions for DY. The Wilson line for these two processes in a covariant gauge reads[14–16]

$$\mathcal{W}[-\tfrac{z}{2}, \tfrac{z}{2}|\,\eta] = [-(\tfrac{z^-}{2}, 0, \tfrac{\vec{z}_T}{2})\,,\, (\eta\,\infty, 0, -\tfrac{\vec{z}_T}{2})] \times [(\eta\,\infty, 0, \tfrac{\vec{z}_T}{2})\,,\, (\tfrac{z^-}{2}, 0, \tfrac{\vec{z}_T}{2})] . \tag{4}$$

The parameter η distinguishes between future-pointing Wilson lines ($\eta = 1$) for SIDIS and past-pointing Wilson lines ($\eta = -1$) for DY. It is defined as $\eta = \mathrm{sign}(n^0)$ (cf. discussion below (1)). Again, the leading twist part of the matrix element (3) can be parameterized by eight TMDs, the unpolarized distribution f_1, the Sivers function[17,18] $f_{1T}^{\perp}$, the helicities g_{1L} and g_{1T}, and the chirally-odd distributions, transversity h_1, pretzelosity $h_{1T}^{\perp}$, $h_{1L}^{\perp}$, and the Boer-Mulders function[19] $h_1^{\perp}$. The Sivers and Boer-Mulders functions are (naive) time-reversal odd (T-odd) functions which would vanish under time-reversal if the gauge link operator (4) were neglected. This feature leads to the famous sign change of these functions in SIDIS and DY.[15]

2. Relations between GPDs and TMDs

Although GPDs and TMDs belong to *a priori* different classes of parton distributions, non-trivial relations have been suggested and discussed in the literature.[20–26] Of particular interest is a relation between the Sivers function $f_{1T}^{\perp}$ and the GPD E. It can be understood in the following way:

Let's define the average transverse momentum of unpolarized quarks in a transversely polarized nucleon $\langle k_T^i \rangle_T$ and express it in terms of the first k_T-moment of the Sivers function $f_{1T}^{\perp(1)}(x)$,

$$\langle k_T^i \rangle_T = \int d^2k_T\, k_T^i\, \tfrac{1}{2}\Big[\mathrm{Tr}[\gamma^+\Phi(S_T)] - \mathrm{Tr}[\gamma^+\Phi(-S_T)]\Big] \propto \epsilon_T^{ij} S_T^j\, f_{1T}^{\perp(1)}(x). \tag{5}$$

On the other hand one can insert the matrix element Φ of Eq. (3) into (5). Then, by integrating over k_T one can derive a so-called soft-gluonic pole matrix element.[16] Transforming this result into the impact parameter space yields

$$\langle k_T^i \rangle_T = \int d^2b_T \int \frac{dz^-}{2\pi}\, \mathrm{e}^{iP^+z^-} \times \\ \langle P^+, \vec{0}_T, S_T |\, \bar{q}(z_1)\gamma^+[z_1; z_2]\, I^i(z_2)q(z_2)\, | P^+, \vec{0}_T; S_T \rangle. \tag{6}$$

This equation almost looks like a b_T-integral over the GPD E in the impact parameter space. However, Eq. (6) contains an additional operator I^i which is defined by the gluonic field-strength tensor,

$$I^i(z^-) = \int dy^-\, [z^-, y^-]\, gF^{+i}(y^-)\, [y^-, z^-]. \tag{7}$$

This operator enters the matrix element (6) because of the initial / final state interactions necessary for the Sivers function to exist. A relation between the Sivers function and the GPD E in the impact parameter representation exists if one treats the operator I^i semi-classically as a c-number and separates it from the impact parameter GPD E in the following way,

$$\epsilon_T^{ij} S_T^j\, f_{1T}^{\perp(1)}(x) \propto \int d^2b_T \mathcal{I}^i(x, \vec{b}_T)(\epsilon_T^{rs} b_T^r S_T^s) \frac{\partial}{\partial \vec{b}_T^2} \mathcal{E}(x, \vec{b}_T^2). \tag{8}$$

The function $\mathcal{I}$ refers to the operator I in (7) treated semi-classically. It is called "lensing function".[21] Although (8) is of course an approximation, this relation provides an appealingly simple explanation of the Sivers asymmetry in terms of two physical effects, initial / final state interactions described by the lensing function and a spatial distortion of the distribution of partons in the transverse plane induced by orbital angular momentum and described by the impact parameter GPD $\mathcal{E}$. In an analogous way one can also derive a relation between the Boer-Mulders function $h_1^\perp$ and a combination of chirally-odd GPDs $\mathcal{E}_T + 2\tilde{\mathcal{H}}_T$ in the impact parameter space.

Interestingly, it was shown that the relations between the Sivers function and the GPD E as well as the Boer-Mulders function and chirally-odd

GPDs hold exactly in spectator models for lowest order diagrams.[22,26,27] A scalar diquark spectator was used to check the relations $f_{1T}^{\perp} \leftrightarrow E$ and $h_1^{\perp} \leftrightarrow (E_T + 2\tilde{H}_T)$ for a nucleon[22,26] while they turn out to be valid also for a gluon spectator in a QCD-inspired quark target model.[26] The second relation $h_1^{\perp} \leftrightarrow (E_T + 2\tilde{H}_T)$ was also checked for an antiquark spectator in a pion model.[27] Non-zero results for GPDs can be obtained already from spectator models at tree-level while it is necessary to consider one-loop diagrams for T-odd TMDs. Remarkably, one can extract unique but model-dependent lensing functions from such calculations. Moreover, non-trivial relations were also found for gluon distributions.[26] However, the fact that these relations can be verified in models has to be attributed to the simplicity of the models. It was already argued in Ref.[26] that the relations are likely to become invalid even in spectator models if higher-order diagrams are taken into account.

An interesting relation between the pretzelosity distribution $h_{1T}^{\perp}$ and the chirally-odd GPD $\tilde{H}_T$ was not only established in a spectator model[26] but also in a light-cone constituent quark model.[28] For such a relation initial/final state interactions are not relevant since $h_{1T}^{\perp}$ is a T-even TMD. Thus, for a relation of such a kind one does not need a lensing function.

3. Generalized Transverse Momentum Dependent Parton Distributions

New model-independent insight into the possible relations between TMDs and GPDs was gained from the observation that GPDs and TMDs can be considered as different limiting cases of a more general class of *mother functions* which were called Generalized Transverse Momentum Dependent parton distributions (GTMDs).[27,29,30] When being transformed into the impact parameter space GTMDs are related to so-called Wigner functions, quantum mechanical analogues of classical phase space distributions.[31]

GTMDs or, equivalently, k_T-dependent GPDs, have been mentioned in the literature in the context of subleading twist contributions to DVMP.[32,33] Gluon GTMDs have been suggested for diffractive vector meson[34] and Higgs[35] production.

A classification of GTMDs can be performed by considering a matrix element of the following form,

$$W_{ij}^{\lambda\lambda'}(x, \vec{k}_T; \xi, \vec{\Delta}_T) = \int \frac{dz d^2 z_T}{(2\pi)^3}\, e^{ik\cdot z} \langle p', \lambda' | \bar{q}_j(-\tfrac{z}{2})\, \mathcal{W}[-\tfrac{z}{2}, \tfrac{z}{2}]\, q_i(\tfrac{z}{2}) | p, \lambda \rangle \Big|_{z^+=0}. \tag{9}$$

It is clear that this matrix element (9) reduces to (1) upon integration over k_T while it reduces to (3) for $\Delta \to 0$. The Wilson line in (9) is chosen here to be the same as in (4) in order to recover TMDs in SIDIS and DY.

GTMDs can be extracted from (9) by tracing the matrix element with certain Dirac matices. A parameterization of these traces is particularly easy for a scalar hadron like a pion.[27] In such a way one obtains four leading twist GTMDs for a pion. Since we are interested in investigating non-trivial relations between GPDs and T-odd TMDs (there is only a Boer-Mulders function for a pion) we restrict ourselves to presenting only the parameterization of (9) in terms of GTMDs describing transversely polarized quarks,

$$\frac{1}{2}\mathrm{Tr}\left[i\sigma^{i+}\gamma_5 W^\pi\right] = -\frac{i\epsilon_T^{ij}k_T^j}{M}\mathcal{H}_1^{\perp,\pi} - \frac{i\epsilon_T^{ij}\Delta_T^j}{M}\mathcal{H}_1^{\Delta,\pi}, \tag{10}$$

where the chirally-odd GTMDs $\mathcal{H}_1^{\perp,\pi}$ and $\mathcal{H}_1^{\Delta,\pi}$ are understood to depend on the variables x, ξ, $\vec{k}_T^2$, $\vec{\Delta}_T^2$ and $\vec{k}_T \cdot \vec{\Delta}_T$. The decomposition of the matrix element (9) into GTMDs is only restricted by parity. In contrast to GPDs and TMDs, GTMDs are complex-valued functions. Time-reversal and hermiticity determine the following dependence of GTMDs on the parameter η which distinguishes between initial and final state interactions and separates the real and imaginary parts of GTMDs,

$$\mathcal{X}(\eta) = \Re[\mathcal{X}] + i\eta\Im[\mathcal{X}]. \tag{11}$$

The two limiting cases for the chirally-odd GTMDs are easy to perform. The pion Boer-Mulders function can be expressed in terms of the GTMD $\mathcal{H}_1^{\perp,\pi}$,

$$h_1^\perp(x,\vec{k}_T^2) = \eta\Im[\mathcal{H}_1^{\perp,\pi}](x,0;\vec{k}_T^2,0,0), \tag{12}$$

while the chirally-odd GPD H_1^π[27] reads,

$$H_1^\pi(x,\xi,t) = \int d^2k_T\left[\Re[\mathcal{H}_1^{\Delta,\pi}] + \frac{\vec{k}_T\cdot\vec{\Delta}_T}{\vec{\Delta}_T^2}\Re[\mathcal{H}_1^{\perp,\pi}]\right](x,\xi;\vec{k}_T^2,\vec{\Delta}_T^2,\vec{k}_T\cdot\vec{\Delta}_T). \tag{13}$$

A certain relation between the Boer-Mulders function and the GPD H_1^π is given by the fact that the GTMD $\mathcal{H}_1^{\perp,\pi}$ contributes to both functions. It is not surprising that the imaginary part appears in (12) due to the T-odd character of $h_1^\perp$ while only real parts of GTMDs are expected to contribute to GPDs. However, since H_1^π also gets a contribution from the other chirally-odd GTMD $\mathcal{H}_1^{\Delta,\pi}$ this analysis indicates that there is no *exact* relation between the Boer-Mulders function and H_1^π.

In order to investigate a connection between the Sivers function and the GPD E a (transversely) polarized hadron such as a nucleon instead of a pion is needed. The decomposition of the matrix element (9) into GTMDs is more complicated for a nucleon. For brevity we only consider unpolarized quarks, and we obtain four leading twist GTMDs,

$$\mathrm{Tr}[\gamma^+ W] = \frac{1}{P^+}\bar{u}(p',\lambda')\Big[\tfrac{P^+}{M}\mathcal{F}_1 + \tfrac{i\sigma^{i+}k_T^i}{M}\mathcal{F}_1^\perp + \tfrac{i\sigma^{i+}\Delta_T^i}{M}\mathcal{F}_1^\Delta + \tfrac{P^+}{M}\tfrac{i\sigma^{ij}k_T^i\Delta_T^j}{M^2}\mathcal{F}_1^{\perp\Delta}\Big]u(p,\lambda). \quad (14)$$

Again, the limits can be investigated very easily. The Sivers function reads in terms of GTMDs,

$$f_{1T}^\perp(x,\vec{k}_T^2) = -\eta\Im[\mathcal{F}_1^\perp](x,0;\vec{k}_T^2,0,0), \quad (15)$$

and is given by the imaginary part of the GTMD $\mathcal{F}_1^\perp$. A combination of the GPDs H and E is given by

$$\tfrac{1}{2}(H+E)(x,\xi,t) = \int d^2k_T\,\Big[\Re[\mathcal{F}_1^\Delta] + \frac{\vec{k}_T\cdot\vec{\Delta}_T}{\vec{\Delta}_T^2}\Re[\mathcal{F}_1^\perp]\Big](x,\xi;\vec{k}_T^2,\vec{\Delta}_T^2,\vec{k}_T\cdot\vec{\Delta}_T). \quad (16)$$

As in the pion case we observe a certain relation between the Sivers function and the GPD E in the sense that the GTMD $\mathcal{F}_1^\perp$ can be found in both limits (15) and (16). Due to contributions of other GTMDs to E again an exact relation between $f_{1T}^\perp$ and E is excluded. Considering the discussion in Sec. 2 we would expect such a behavior. On the other hand, this analysis by no means prohibits approximate relations.

Despite the fact that an exact relation between the T-even pretzelosity distribution $h_{1T}^\perp$ and the chirally-odd GPD $\tilde{H}_T$ seems to hold in models (Sec. 2), we emphasize that a GTMD analysis for these functions rules out an exact model-independent relation.

4. Conclusion

In this note we discussed relations between TMDs and GPDs suggested in the literature. A classification of *mother functions* of TMDs and GPDs allowed us to express TMDs and GPDs in terms of GTMDs. We observed in particular a certain connection between the Sivers function and the GPD E as well as a connection between the Boer-Mulders function and a chirally-odd GPD H_1^π for a pion. However, exact model-independent non-trivial relations are ruled out by this analysis.

References

1. D. Mueller, D. Robaschik, B. Geyer, F. M. Dittes and J. Horejsi, *Fortschr. Phys.* **42**, p. 101 (1994).
2. K. Goeke, M. V. Polyakov and M. Vanderhaeghen, *Prog. Part. Nucl. Phys.* **47**, 401 (2001).
3. M. Diehl, *Phys. Rept.* **388**, 41 (2003).
4. A. V. Belitsky and A. V. Radyushkin, *Phys. Rept.* **418**, 1 (2005).
5. S. Boffi and B. Pasquini, *Riv. Nuovo Cim.* **30**, p. 387 (2007).
6. M. Diehl, *Eur. Phys. J.* **C19**, 485 (2001).
7. M. Burkardt, *Phys. Rev.* **D62**, p. 071503 (2000).
8. J. P. Ralston and B. Pire, *Phys. Rev.* **D66**, p. 111501 (2002).
9. M. Diehl, *Eur. Phys. J.* **C25**, 223 (2002).
10. M. Burkardt, *Int. J. Mod. Phys.* **A18**, 173 (2003).
11. P. J. Mulders and R. D. Tangerman, *Nucl. Phys.* **B461**, 197 (1996).
12. A. Bacchetta, M. Diehl, K. Goeke, A. Metz, P. Mulders and M. Schlegel, *JHEP* **02**, p. 093 (2007).
13. V. Barone, A. Drago and P. G. Ratcliffe, *Phys. Rept.* **359**, 1 (2002).
14. A. V. Belitsky, X. Ji and F. Yuan, *Nucl. Phys.* **B656**, 165 (2003).
15. J. C. Collins, *Phys. Lett.* **B536**, 43 (2002).
16. D. Boer, P. J. Mulders and F. Pijlman, *Nucl. Phys.* **B667**, 201 (2003).
17. D. W. Sivers, *Phys. Rev.* **D41**, p. 83 (1990).
18. D. W. Sivers, *Phys. Rev.* **D43**, 261 (1991).
19. D. Boer and P. J. Mulders, *Phys. Rev.* **D57**, 5780 (1998).
20. M. Burkardt, *Phys. Rev.* **D66**, p. 114005 (2002).
21. M. Burkardt, *Nucl. Phys.* **A735**, 185 (2004).
22. M. Burkardt and D. S. Hwang, *Phys. Rev.* **D69**, p. 074032 (2004).
23. M. Diehl and P. Hagler, *Eur. Phys. J.* **C44**, 87 (2005).
24. M. Burkardt, *Phys. Rev.* **D72**, p. 094020 (2005).
25. Z. Lu and I. Schmidt, *Phys. Rev.* **D75**, p. 073008 (2007).
26. S. Meissner, A. Metz and K. Goeke, *Phys. Rev.* **D76**, p. 034002 (2007).
27. S. Meissner, K. Goeke, A. Metz and M. Schlegel, *arXiv:0805.3165* (2008).
28. B. Pasquini, S. Cazzaniga and S. Boffi, *arXiv:0806.2298* (2008).
29. S. Meissner, A. Metz, M. Schlegel and K. Goeke, *arXiv:0710.5846* (2007).
30. S. Meissner, A. Metz and M. Schlegel, *arXiv:0807.1154* (2008).
31. A. V. Belitsky, X.-d. Ji and F. Yuan, *Phys. Rev.* **D69**, p. 074014 (2004).
32. M. Vanderhaeghen, P. A. M. Guichon and M. Guidal, *Phys. Rev.* **D60**, p. 094017 (1999).
33. M. Diehl and W. Kugler, *Eur. Phys. J.* **C52**, 933 (2007).
34. A. D. Martin, M. G. Ryskin and T. Teubner, *Phys. Rev.* **D62**, p. 014022 (2000).
35. V. A. Khoze, A. D. Martin and M. G. Ryskin, *Eur. Phys. J.* **C14**, 525 (2000).

TRANSVERSITY VIA EXCLUSIVE π–ELECTROPRODUCTION

G.R. Goldstein

Department of Physics and Astronomy, Tufts University, Medford, MA 02155, USA

S. Liuti and S. Ahmad

Physics Department, University of Virginia, Charlottesville, VA 22904, USA

Exclusive π^0 electroproduction from nucleons can be related to transversity, the tensor charge, and other quantities related to transversity. This process isolates C-parity odd and chiral odd combinations of t-channel exchange quantum numbers. In a hadronic picture the meson production amplitudes for intermediate energy and Q^2 are determined by C–odd Regge exchanges with final state interactions. The helicity structure for this C-odd process relates to the quark helicity flip, or chiral odd generalized parton distributions (GPDs).Various cross sections and asymmetries will be sensitive to the transversity.

Keywords: Transversity; Regge; GPD; Electroproduction; π^0 production.

1. Transversity questions

What is Transversity? It is a concept introduced in the 1970's by Goldstein and Moravcsik[1] as the appropriate spin quantization for two body scattering amplitudes, e.g. $f_{a,b;c,d}(s,t)$ that is most easily related to single spin asymmetry or azimuthal asymmetry observables. The transversity of a massive particle in its rest frame is its spin projection along an axis perpendicular to the two body scattering plane $(S \cdot p_{in} \times p_{out})$. In terms of helicities, transversity for spin $\frac{1}{2}$ particles is $|\pm\frac{1}{2}\rangle_T = \{|+\frac{1}{2}\rangle \pm (i)|-\frac{1}{2}\rangle\}/\sqrt{2}$.

The tensor charge is the first moment or the norm of the parton transversity distribution, $h_1(x)$. It is defined as the transversely polarized nucleon matrix element of local quark field operators,

$$\langle P, S_T|\bar{\psi}\sigma^{\mu\nu}\gamma_5\frac{\lambda^a}{2}\psi|P, S_T\rangle = 2\delta q^a(\mu^2)(P^\mu S_T^\nu - P^\nu S_T^\mu).$$

Like other charges, it is the integral of a distribution $(\delta q^a(x) - \delta\bar{q}^a(x))$, where $\delta q^a(x) = h_1^a(x)$ is the **transversity distribution**. It is essentially the probability to find a transversity $+\frac{1}{2}$ quark in a nucleon of transversity $+\frac{1}{2}$. Unlike the longitudinal distribution $g_1(x)$, $h_1(x)$ receives no contributions from gluons.

An important question is how the tensor charge can be determined, theoretically and experimentally? Some predictions and fits from various processes give: ($\delta u = 1.26, \delta d = -0.17$) from QCD Sum rules (He and Ji,[2]); ($\delta(u-d) = 1.090.02$) from Lattice (QCDSF, M. Gockeler *et al.*,[3]); ($\delta u = 0.48 \pm 0.09, \delta d = -0.62 \pm 0.30$) from phenomenological analysis (Anselmino *et al.*,[4]); ($\delta u = 0.58 \pm 0.20, \delta d = -0.11 \pm 0.20$) from axial vector dominance (Gamberg and Goldstein,[5]).

The Gamberg and Goldstein theoretical model is based on axial vector dominance by the $b_1(1235)$ and $h_1(1170)$–$h_1'(1380)$, with $J^{PC} = 1^{+-}$, that couple to the tensor Dirac matrix $\sigma^{\mu\nu}\gamma_5$. The Dirac matrix has C-parity minus, which is a crucial fact. Because the axial vector couplings involve an additional angular momentum, to obtain the tensor structure a transverse momentum enters the coupling - the pure pole term decouples at zero momentum transfer. The interpretation that was adopted was that the coupling involves the quark constituents and thus does not vanish at zero momentum transfer. The average transverse momentum thus gives non-zero results. The quarks can be explicitly represented in the interaction with the axial vectors via the GPDs, particularly in the ERBL region, which involves $q + \bar{q}$ exchange. Hence exclusive processes should be considered, where the GPDs provide a description of hard scattering from the constituents. Secondly, the pole at the axial mass can be extrapolated to the $t = 0$ limit using the Regge pole approach, which naturally allows extrapolation from physical poles to the physical scattering region. Hence there is an interplay between a partonic and hadronic mechanism. The exclusive photoproduction and electroproduction of π^0 or η mesons from nucleons have C-parity odd in the t-channel and hence can accommodate the appropriate axial vector exchanges.

2. Hadronic and partonic models

We will consider the exclusive reactions, $e + p \to e' + \pi^0 + p'$ and the related η and neutron target processes.[6] The relevant subprocess is $\gamma^* + p \to \pi^0 + p'$. The t-channel exchange picture involves the same C-parity odd, chiral odd states that include the 1^{+-} b_1 and h_1 mesons ($q + \bar{q}$; $S = 0, L = 1$ mesons) and the vector mesons, the 1^{--} ρ^0 and ω (($q + \bar{q}$; $S = 1, L = 0$ mesons).

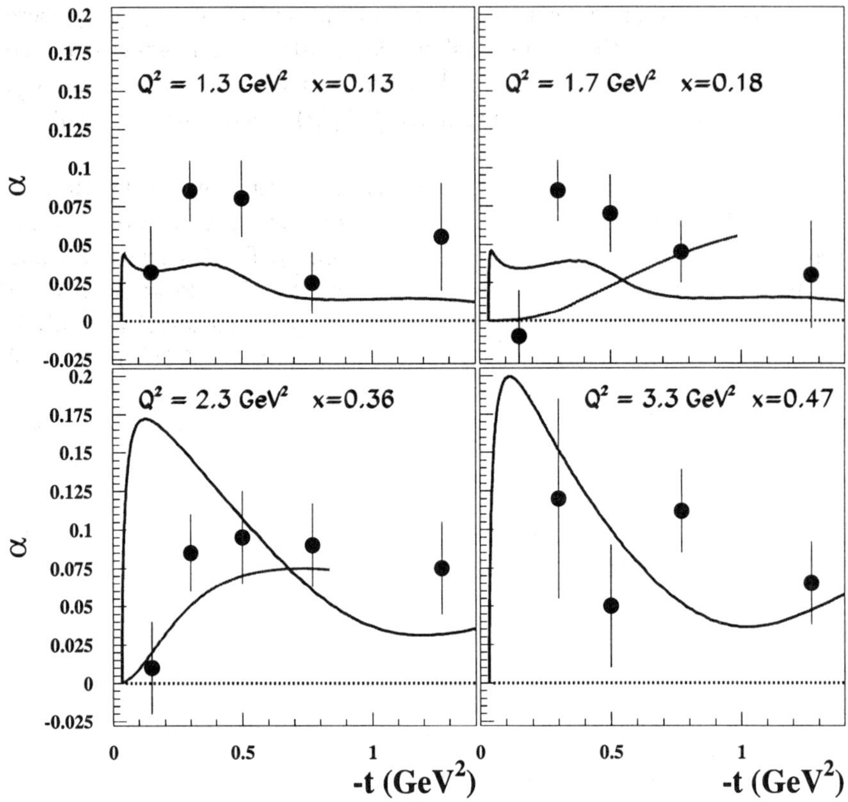

Fig. 1. Beam asymmetry in Regge and GPD pictures. Data from JLab.[12]

These axial vector mesons couple to the nucleon via the Dirac tensor $\sigma^{\mu\nu}\gamma_5$, while the vector mesons couple via γ^μ and/or $\sigma^{\mu\nu}$. Because of the C-parity there is no $\gamma^\mu\gamma^5$ coupling. This is quite significant in the GPD perspective - only chiral odd GPDs are involved, contrary to the accepted formulation.[7] While ref.(7) indicates that C-parity odd exchanges of 3 gluons, like the "odderon", are allowed, the authors relate the process to chiral even GPDs that can involve 1^{++} exchange quantum numbers. This can be the case for charged pseudoscalar production, where there is not a C-parity eigenstate in the t-channel, but not for the neutral case, which has *definite odd C-parity.*

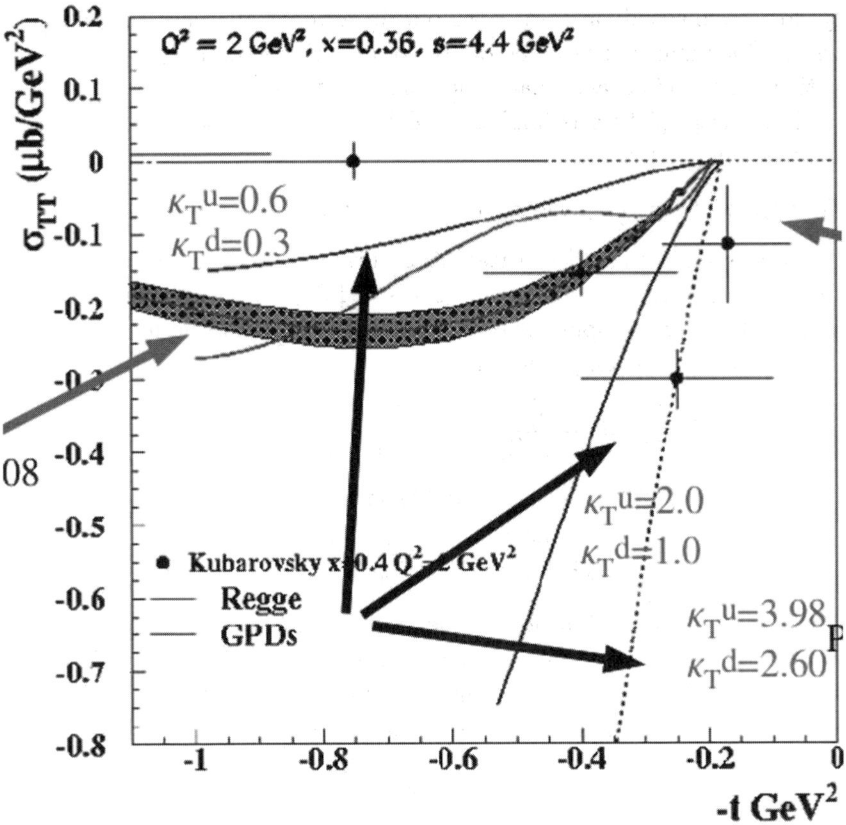

Fig. 2. $d\sigma_{TT}/dt$ Regge (wavy line) and GPD pictures (with 3 sets of κ_T pairs). See Eq. 10. Data preliminary.

3. Regge cut model

A successful Regge cut model was developed to fit photoproduction data many years ago.[8] That model essentially involves as input the vector and axial vector meson trajectories that factorize into couplings to the on-shell $\gamma + \pi^0$ vertex and the nucleon vertex. The cuts or absorptive corrections destroy that factorization, but fill in the small t and $t \approx -0.5$ GeV^2 amplitude zeroes. To connect to electroproduction, the upper vertex factor must acquire Q^2 dependence. This is accomplished by replacing the elementary, t-dependent couplings with Q^2 dependent transition form factors. In this Regge picture the factorization for the longitudinal virtual photon is not different from the transverse photon, except for the additional power of

Q^2 for the longitudinal case. With our form factor approach to the upper vertex (including Sudakov factors to soften the endpoint singularities) we anticipate a similar factorization for the transverse case.

The Regge picture is implemented by singling out the 6 independent helicity amplitudes.

$$f_1 = f_{1+,0+}, f_2 = f_{1+,0-}, f_3 = f_{1-,0+}, f_4 = f_{1-,0-}, f_5 = f_{0+,0-}, f_6 = f_{0+,0+}.$$

At large s and small $|t|$ the leading natural parity and unnatural parity Regge poles contribute to opposite sums and differences of pairs of helicity amplitudes. For example, for even signature exchanges, the natural parity ρ and ω, two of the amplitudes are

$$f_1 = f_4 = \frac{\beta_1^V}{\Gamma(\alpha_V(t))}\Delta\frac{1-e^{-i\pi\alpha_V(t)}}{sin(\pi\alpha_V(t))}\nu^{\alpha_V(t)}e^{-\Delta^2 c_V/2}$$

For the odd signature, unnatural parity b_1 and h_1, $f_1 = f_4 = 0$,

$$f_2 = +f_3 = \frac{\beta_1^A\Delta^2}{2M\Gamma(\alpha_A(t)+1)}\frac{1-e^{-i\pi\alpha_A(t)}}{sin(\pi\alpha_A(t))}\nu^{\alpha_A(t)}e^{-\Delta^2 c_A/2}.$$

The additional contributions to the longitudinal photon amplitudes are limited by parity, charge conjugation invariance and helicity conservation to the b_1 and h_1 only. To leading order in s they contribute only to f_5, which has the minimal Δ^1 dependence and is proportional to the same trajectories' contributions to f_2. The precise relation between these two amplitude contributions depends on Q^2 and can be related to the decay widths for the axial vectors using vector dominance of the photon. The vector meson trajectories do not couple to the helicity zero photon and pion vertex. These amplitudes are then altered by absorptive Regge-cut corrections.

4. Transversity, chiral odd GPDs and helicity

Now the crucial connection to the 8 GPDs that enter the partonic description of electroproduction is through the helicity decomposition,[9] where, for example, one of the chiral even helicity amplitudes is given by

$$A_{++,++}(X,\xi,t) = \frac{\sqrt{1-\xi^2}}{2}(H^q + \tilde{H}^q - \frac{\xi^2}{1-\xi^2}(E^q + \tilde{E}^q)),$$

while one of the chiral odd amplitudes is given by

$$A_{++,--}(X,\xi,t) = \sqrt{1-\xi^2}(H_T^q + \frac{t_0-t}{4M^2}\tilde{H}_T^q - \frac{\xi}{1-\xi^2}(\xi E_T^q + \tilde{E}_T^q)).$$

There are relations to PDFs, $H^q(X,0,0) = f_1^q(X)$, $\tilde{H}^q(X,0,0) = g_1^q(X)$, $H_T^q(X,0,0) = h_1^q(X)$. The first moments of these are the charge, the axial charge and the tensor charge, for each flavor q, respectively. Further, the first moments of $E(X,0,0)$ and $2\tilde{H}_T^q(X,0,0) + E_T^q(X,0,0)$ are the anomalous moments κ^q, κ_T^q, with the latter defined by Burkardt.[10]

Chiral even GPDs have been modeled in a thorough analysis[11] , based on diquark spectators and Regge behavior at small X, and consistent with constraints from PDFs, form factors and lattice calculations. That analysis is used to obtain chiral odd GPDs via a multiplicative factor that fits the phenomenological $h_1(x)$.[4] With that *ansatz* the observables can be determined in parallel with the Regge predictions.

The parameterization's form in the DGLAP region is

$$H(X,\zeta,t) = G(X,\zeta,t)R(X,\zeta,t),$$

where $R(X,\zeta,t)$ is a Regge motivated term that describes the low X and t behaviors, while the contribution of $G(X,\zeta,t)$ is obtained using a spectator model and is centered at intermediate/large values of X. Explicit expressions are given in Ahmad, *et al.*[11] The $\zeta = 0$ behavior is constrained by enforcing both the forward limit: $H^q(X,0,0) = q_{val}(X)$, where $q_{val}(X)$ is the valence quarks distribution, and the first moments that are related to the nucleon form factors. The $n = 1,2,3$ moments of the NS combinations: $H^{u-d} = H^u - H^d$, and $E^{u-d} = E^u - E^d$ are available from lattice QCD.[14] The parametrization was devised taking into account all of the above constraints.and gives an excellent description of recent Jefferson Lab data in the valence region.

The connection to the transversity GPDs is carried out similarly to Refs.[4] for the forward case by setting:

$$H_T^q(X,\zeta,t) = \delta q H^{q,val}(X,\zeta,t)$$

$$\overline{E}_T^q \equiv 2\widetilde{H}_T + E_T = \kappa_T^q H_T^q(X,\zeta,t)$$

where δq is the tensor charge, and κ_T^q is the tensor anomalous moment introduced, and connected to the transverse component of the total angular momentum by Burkardt.[10]

5. Exclusive π^0 electroproduction observables and results

The differential cross section for pion electroproduction off an unpolarized target is

$$\frac{d^4\sigma}{d\Omega dx d\phi dt} = \Gamma \left\{ \frac{d\sigma_T}{dt} + \epsilon_L \frac{d\sigma_L}{dt} + \epsilon \cos 2\phi \frac{d\sigma_{TT}}{dt} + \sqrt{2\epsilon_L(\epsilon+1)} \cos\phi \frac{d\sigma_{LT}}{dt} \right\}.$$

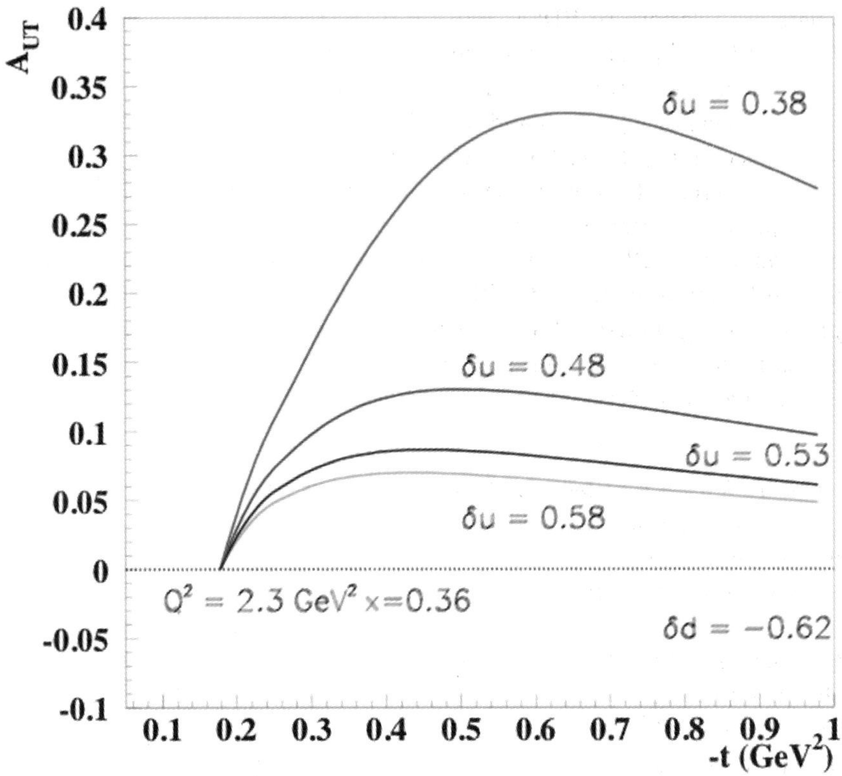

Fig. 3. Transverse spin asymmetry, A_{UT}, Eq.(11).

Each observable involves bilinear products of helicity amplitudes, or GPDs. For example, the cross section for the virtual photon linearly polarized out of the scattering plane minus that for the scattering plane is

$$\frac{d\sigma_{TT}}{dt} = \mathcal{N} \frac{1}{s \mid P_{CM} \mid^2} 2\Re e \left(f_1^* f_4 - f_2^* f_3 \right). \tag{10}$$

Another relevant observable is the target transverse polarization asymmetry

$$A_{UT} = \frac{2\Im m(f_1^* f_3 - f_4^* f_2)}{\frac{d\sigma_T}{dt}}. \tag{11}$$

Figure 1 shows the beam asymmetry calculated from both models along with recent data. It is remarkable that the order of magnitude is correct for the Regge model without any fitting to the electroproduction data.

The GPD results shown in the two middle Q^2 graphs are also reasonably close to data. Figure 2 shows some preliminary data and the variation of the GPD results for different values of the κ_T parameters. In Fig. 3 we show the sensitivity of A_{UT} to to the values of the u-quark and d-quark tensor charges. The values in the figure were taken by varying up to 20% the values of the tensor charge extracted from the global analysis,[4] *i.e.* $\delta u = 0.48$ and $\delta d = -0.62$, and fixing the transverse anomalous magnetic moment values to $\kappa_T^u = 0.6$ and $\kappa_T^d = 0.3$. This is the main result of this contribution: it summarizes our proposed method for a practical extraction of the tensor charge from π^o electroproduction experiments. Therefore our model can be used to constrain the range of transversity parameters allowed by the data.

Acknowledgments

This work is supported by the U.S. Department of Energy grants no. DE-FG02-92ER40702 (G.R.G.) and no. DE-FG02-01ER4120 (S.L). We thank L. Gamberg for initial conversations.

References

1. G.R. Goldstein and M.J. Moravcsik, Annals Phys. **98** 128 (1976).
2. H. He and X. Ji, Phys. Rev. **D52**, 2960 (1995).
3. M. Gockeler, *et al.*, arXiv:hep-lat/0710.2489.
4. M. Anselmino, *et al.*, Phys. Rev. **D75**, 054032 (2007).
5. L.P. Gamberg and G.R. Goldstein, Phys. Rev. Lett. **87** 242001 (2001).
6. S. Ahmad, G.R. Goldstein and S. Liuti, arXiv: hep-ph/0805.3568.
7. L. Mankiewicz, G. Piller and A. Radyushkin, Eur. Phys. Jour. **C10**, 307 (1999).
8. G. R. Goldstein and J. F. Owens, Phys. Rev. **D7** 865 (1973).
9. M. Diehl, Eur. Phys. Jour. **C19**, 485 (2001).
10. M. Burkardt, Phys. Rev. **D72**, 094020 (2005); *ibid* Phys. Lett. **B639** 462 (2006).
11. S. Ahmad, *et al.*, Phys. Rev. **D75**, 094003 (2007); *ibid*, arXiv: hep-ph/0708.0268.
12. R. De Masi, *et al.*, Phys. Rev. **C77**, 042201 (2008).
13. M. Diehl, Eur. Phys. Jour. C **19**, 485 (2001).
14. Ph. Hagler *et al.* [LHPC Collaborations], arXiv:0705.4295 [hep-lat].

ESTIMATE OF THE SIVERS ASYMMETRY AT INTERMEDIATE ENERGIES WITH RESCATTERING EXTRACTED FROM EXCLUSIVE PROCESSES

A. Bianconi

Dip. Chimica e Fisica per l'ingegneria e i materiali,
Università di Brescia, Via Valotti 9, Brescia 25100, Italy.
INFN, Gruppo collegato di Brescia.

Here an estimate for the Sivers asymmetry in semi-inclusive proton-proton scattering in *intermediate* energy regime (fixed target, 10-50 GeV beam energy) is presented, by a model where initial state interactions are extracted from the phenomenology of elastic proton-proton scattering.

1. General framework

This presentation is a summary of a longer work available in.[1]

I consider a quark in a hadron with $S_y = \pm\, 1/2$, where the scattering plane is formed by the directions z (longitudinal) and x (transverse). The quark 4-spinor is decomposed as

$$\Psi \equiv \psi_+ \Psi_{T+} + \psi_- \Psi_{T-} \tag{1}$$

on a S_y eigenstate basis. In the UR limit these two components completely describe a free quark state, and we may work on the "vector"

$$\vec{\psi} \equiv (\psi_+, \psi_-). \tag{2}$$

In particular, we exploit the UR relation

$$\bar{\Psi}\gamma_+\Psi \approx |\psi_+|^2 + |\psi_-|^2. \tag{3}$$

This is used to calculate the distribution for unpolarized quark in a polarized hadron with $\hat{S} = \pm\hat{y}$:

$$q(x,\vec{k}_T) \equiv q_U(x,k_T) - 2\frac{\vec{S}\wedge\vec{k}_T}{M} q_S(x,k_T) = q_U(x,k_T) \pm \frac{k_x}{M} q_S(x,k_T) \tag{4}$$

where q_S is the Sivers analyzing power.[2] On the theoretical side, $q(x, \vec{k}_T)$ may be defined as

$$q(x, \vec{k}_T) = Tr[\gamma_+ C(x, \vec{k}_T)], \tag{5}$$

$$C(x, \vec{k}_T) = \int dx d^2b \, e^{-ix\xi + i\vec{k}_T \cdot \vec{b}} < P|\Psi(0)\bar{\Psi}(0_+, \xi, \vec{b})|P > \tag{6}$$

$$\propto \sum_n |\psi_n(x, \vec{k}_T)|^2. \tag{7}$$

where the quark operators are affected by ISI (initial state interactions). In this work, the strategy will not be a direct calculation of $C(x, \vec{k}_T)$ from the definition eq. 6, but a calculation of a restricted number of amplitudes $\psi_n(x, k_T)$ in eq. 7. These are later squared and summed to obtain the correlator. I work in spacetime representation:

$$\vec{\psi}(x, \vec{k}_T) \equiv \int d\xi d^2b e^{-ix\xi} e^{i\vec{k}_T \cdot \vec{b}} \vec{\psi}(\xi, \vec{b}), \quad \vec{\psi}(\xi, \vec{b}) \equiv \begin{pmatrix} \psi_+(\xi, \vec{b}) \\ \psi_-(\xi, \vec{b}) \end{pmatrix} \tag{8}$$

In absence of ISI, the quark bound state has the very simple form

$$\vec{\psi}(\xi, \vec{b})_{PW} \equiv \phi(\xi)\phi'(b) \begin{pmatrix} 1 \\ 0 \end{pmatrix} \tag{9}$$

where PW (plane wave) means "no ISI", and $\phi(\xi)$ and $\phi'(b)$ have Gaussian shapes. So, the quark has no orbital angular momentum (with respect to the hadron c.m.) and the same spin as the hadron.

Proton-proton elastic scattering at 10-30 GeV beam energy is characterized by a strong normal analyzing power.[3–7] This is unfavoured by helicity-conservation in PQCD,[8] so several models exist for it,[9–23] but none affirmed as a "standard" model for this. Here I use a simple nuclear-style interpolating model, to transfer information from elastic to inclusive processes. To extract ISI from elastic proton-proton scattering, I need two assumptions:

Assumption 1): a set of independent quark-quark interactions is responsible for hadron-hadron scattering at $E =$ 10-60 GeV, transferred momenta $Q_T =$ 1-3 GeV/c.

Assumption 2): the same interactions produce initial state distortions of the wavefunction of a projectile quark passing through or near a target hadron in hadron-hadron semi-inclusive processes, at the same energy.

As a consequence of assumption (1), working in a Glauber-like style[24] the elastic scattering amplitude between a free quark and a bound quark in a hadron (fig. 1, top left) can be written as

$$T_{el,quark}(k_o,q) = f(k_o)\int d^2b e^{i\vec{q}\cdot\vec{b}}\vec{\chi}_f^+\left\{exp\left[i\hat{P}\int_{-\infty}^{\infty}\hat{T}(\vec{b},\xi)\right] - 1\right\}\vec{\chi}_o, \tag{10}$$

$$\hat{T}(\vec{b},\xi) \equiv \rho(\vec{b},\xi)\hat{T}'(k_o). \tag{11}$$

These relations describe an average of the quark-quark scattering amplitude over the target structure. In eqs. 10 and 11: (i) The coefficient $f(k_o)$ is a convention-dependent kinematic factor. (ii) $\vec{k}_o\hat{z} \mp \vec{q}$ is the initial/final momentum of the free quark, $\vec{q}$ is the transferred momentum, ξ is a longitudinal or a light-cone coordinate, $\hat{P}$ indicates path-ordered integration along a constant$-\vec{b}$ light-cone path. (iii) The 2-spinors $\vec{\chi}_o$, $\vec{\chi}_f$ assign the initial/final spin of the projectile quark. These asymptotic spinors are defined *not* to contain the spacetime dependence of the quark wavefunctions. In other words, the initial/final wavefunctions for the (free) projectile quark have the form $\vec{\chi}_{i,f}e^{i(k_o\xi\pm\frac{\vec{q}}{2}\cdot\vec{b})}$. (these will have to be generalized later, when we bound the projectile quark to a projectile hadron). (iv) $\hat{T}(\vec{b},\xi)$ is a 2x2 matrix in the normal spin space. $\hat{T}'(k_o)$ is the 2x2 matrix of *forward* scattering amplitudes between a projectile quark with momentum k_o and a target quark at rest. $\hat{T}'$ depends on k_o only. The composition of the target proton in terms of individual quarks is contained in the single-particle density function $\rho(\vec{b},\xi)$.

If the projectile quark belongs to a projectile hadron, eq. 10 must be specialized to include the projectile hadron structure (fig. 1, bottom left):

$$T_{EL}(p,q) = \int d\xi\int d^2b$$

$$\vec{\psi}^+(\xi,\vec{b})e^{i\vec{q}\cdot\vec{b}}\left\{exp\left[i\hat{P}\int_{-\infty}^{\infty}\hat{T}(\xi,\vec{b})d\xi\right] - 1\right\}\vec{\psi}(\xi,\vec{b}) \tag{12}$$

This describes the scattering between a projectile proton and a target hadron, for $q << p$, and will be used to fit the $\hat{T}$ operator starting from data on angular distribution and normal analyzing power in elastic proton-proton scattering.

The other specialization of eq. 10 is to give the ISI-distorted quark distribution amplitude (fig. 1, top right, where ISI are summarized by a gluon-shaped line):

$$\vec{\psi}_n(x,\vec{k}_T) = \int d\xi\int d^2b e^{-ix\xi+i\vec{k}_T\vec{b}}exp\left[i\hat{P}\int_{-\infty}^{\xi}\hat{T}(\xi,\vec{b})d\xi\right]\vec{\psi}_{n,PW}(\xi,\vec{b}) \tag{13}$$

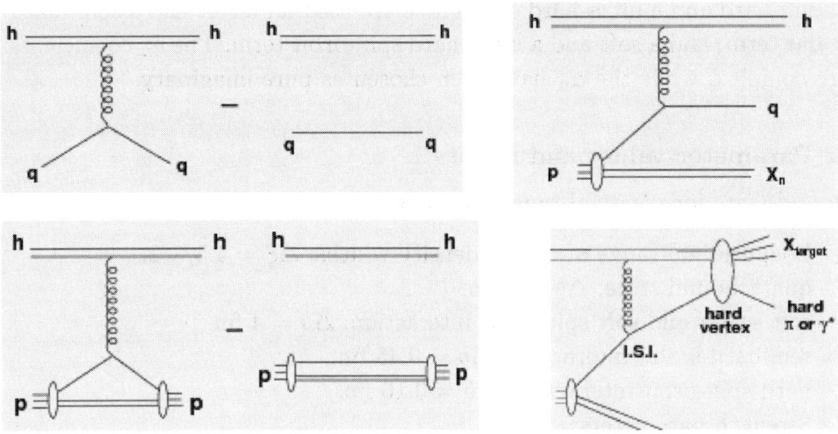

Fig. 1. Top left: quark-hadron scattering (resummed) amplitude, with subtraction of the no-scattering term. Bottom left: proton-proton corresponding scattering amplitude. Top right: ISI-affected quark distribution amplitude (including the no-ISI term). Bottom right: embedding the ISI-affected distribution in the full hard process.

where $\vec{\psi}_{n,PW}(\xi,\vec{b})$ is the amplitude for extracting a light-cone quark in position $(z_-, z_+, \vec{z}_T) = (P_+\xi, 0, \vec{b})$ while leaving the spectator in the state $|n>$, in absence of ISI. In fig. 1, bottom right, ISI are embedded in the full hard process. The gluon-shaped ISI also include the no-rescattering term.

Data on unpolarized and single-polarized *pp* scattering constraint a 2x2 matrix of the form:

$$\hat{T} \; = \; \sum \hat{T}_{scalar} \; + \; \sum \; \hat{T}_{spin-orbit}. \tag{14}$$

For the presented fits, two scalar ant two spin-orbit terms will be needed. In a base where S_y is diagonal, they have the form:

$$\hat{T}_{scalar,n} \; \equiv \; \delta_n \begin{pmatrix} 1 & 0 \\ 0 & 1 \end{pmatrix} \rho(\xi)\rho'_n(b). \tag{15}$$

$$\hat{T}_{spin-orbit,n} \; \equiv \; \alpha_n \begin{pmatrix} b_x & ib_y \\ -ib_y & -b_x \end{pmatrix} \rho(\xi)\rho'_n(b). \tag{16}$$

All the density functions have been chosen with gaussian form, normalized to $\rho_i(0) = 1$.

The longitudinal density $\rho(\xi)$ has been chosen as the same for all terms and equal to the squared quark distribution amplitude $|\phi(\xi)|^2$ introduced in eq. 9. Three transverse density gaussian functions $\rho'_n(b)$ are used: a soft, a semi-hard and a ultra-hard one. These are used for a soft and a semi-hard scalar term, and a soft and a ultra-hard spin-orbit term. The δ_n coefficients are complex, while the α_n have been chosen as pure imaginary.

2. Parameter values and results

Bound state longitudinal range: $\Delta\xi = 4.5$.

Longitudinal range of all the density widths: $\Delta\xi = 4.5/\sqrt{2}$.
quark bound state: $\Delta b = 1$ fm.
soft scalar and soft spin-orbit interaction: $\Delta b = 4$ fm.
semihard scalar interaction: $\Delta b = 0.45$ fm.
hard spin-orbit interaction: $\Delta b = 0.16$ fm.
Strength parameters:
soft scalar interaction: $\delta = 0.07$ (i-0.2).
semihard scalar interaction: $\delta = 0.01$ (1-i).
soft spin-orbit interaction: $\alpha = 0.0004$ i.
hard spin-orbit interaction: $\alpha = 0.0001$ i.

The longitudinal bound state $\phi(\xi)$ is constrained by comparison of $q(x, K_T = 0)$ with MRST[25] u-quark distribution (fig. 2). The phase of the strength of the soft scalar ISI term is fixed by the $\rho-$parameter in forward scattering at 20-50 GeV.[26] The corresponding width is fixed jointly with the transverse bound state $\phi'(b)$ by the forward peak in fig. 3. The spin-orbit terms are constrained by the fit in fig. 4, while the semi-hard scalar term by both the large$-q_T$ part of the data in fig. 3 and by fig. 4. Indeed, the fit in fig. 4 depends on the interference between the semihard scalar and both spin-orbit terms.

When the extracted T operator is used to calculate the Sivers asymmetry from eq. 13, we get the result shown in fig. 5, for $x = 0.3$. For other x values in the valence region, similar curves are obtained.

Although not in detail, the found curve has some *qualitative* features in common with most recent models and parameterizations for the Sivers function,[28–32] and with data at intermediate energies.[33] The main difference is the trend by my model to push asymmetries away from small k_T. At first sight, it does not agree with higher energy data from Compass.[34] This could be explained by the fact that the elastic scattering phenomenology fitted here refers to beam energies $<<$ 100 GeV. It could also be explained by the fact that the curve of fig. 5 needs to be convoluted with a k_T-dependent distribution/fragmentation partner. The k_T-width of this partner increases at increasing energy, spreading and lowering the resulting asymmetry. The

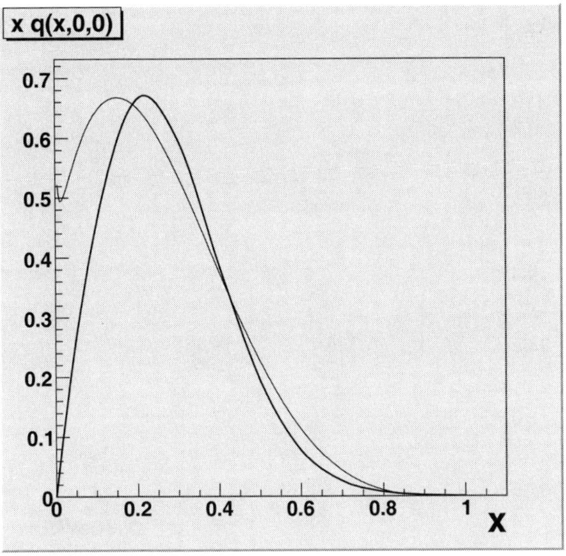

Fig. 2. Unpolarized quark distribution: MRST[25] collinear distribution $xu(x)$ at $Q^2 = 16\ \mathrm{GeV}^2$ and model $xq(x, k_T = 0)$. The model curve is the one that tends to zero at small x.

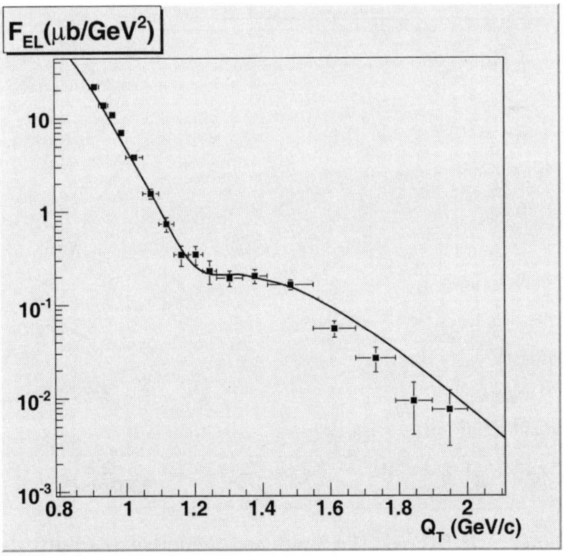

Fig. 3. Fit of elastic pp scattering data. $F_{EL} \equiv d\sigma/dt$. Data come from ref.[27]

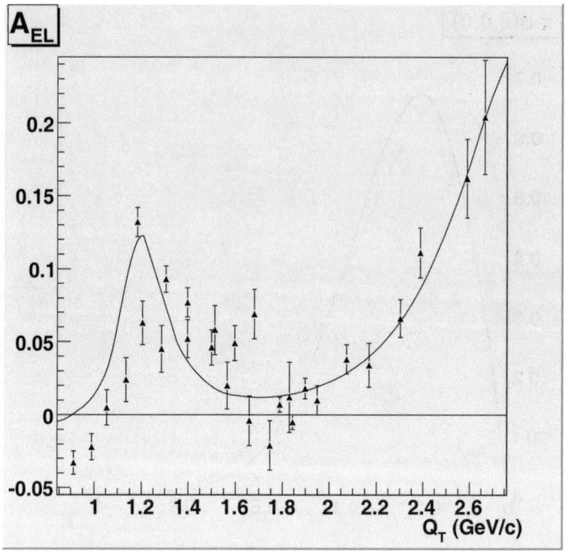

Fig. 4. Fit of the analyzing power in elastic pp data. $A_{EL} \equiv [F_{EL}(k_x) - F_{EL}(-k_x)]/[F_{EL}(k_x) + F_{EL}(-k_x)]$. Data come from refs.[3–7]

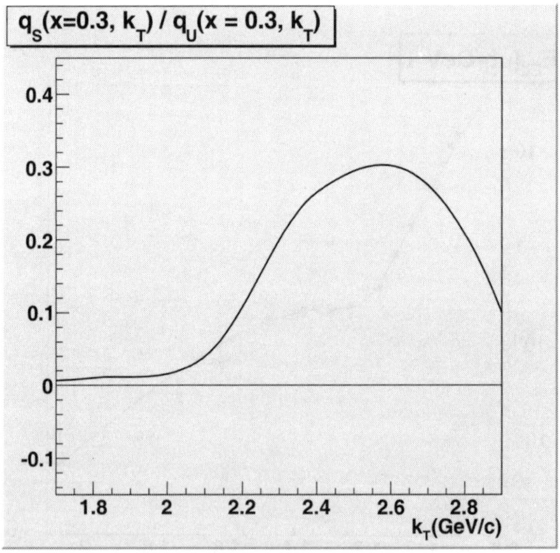

Fig. 5. Calculated ratio between the Sivers and the unpolarized distribution functions $q_S(x, k_T)$ and $q_U(x, k_T)$, (see eq. 4 in the text), for $x = 0.3$. For $k_T < 1.8$ GeV/c the curve does not overcome 0.02.

lowering effect may be marked, because of the rather narrow k_T-width of the curve in fig. 5.

References

1. A. Bianconi, arXiv:0806.0943.
2. D. Sivers, Phys. Rev. **D41**, 83 (1990); Phys. Rev. **D43**, 261 (1991).
3. D. G. Crabb et al., Phys. Rev. Lett. **65**, 3241 (1990).
4. P. R. Cameron et al., Phys. Rev. **D 32**, 3070 (1985).
5. D. C. Peaslee et al., Phys. Rev. Lett. **51**, 2359 (1983).
6. P. H. Hansen et al., Phys. Rev. Lett. **50**, 802 (1983).
7. J. Antille et al., Nucl. Phys. **B 135** 1 (1981).
8. G. R. Goldstein and M. J. Moravcsik, Ann. of Phys. **142** 219 (1982).
9. S. A. Gottlieb, D. Sivers and G. H. Thomas, Phys. Rev. **D 20**, 202 (1979).
10. G. R. Farrar, Phys. Rev. Lett. **56**, 1643 (1986).
11. S. J. Brodsky, C. E. Carlsson, and H. J. Lipkin Phys. Rev. **D 20**, 2278 (1979).
12. C. E. Bourrely, J. Soffer, and T. T. Wu, Phys. Rev. **D 19**, 3249 (1979).
13. P. A. Kazaks and D. L. Tucker, Phys. Rev. **D 37**, 222 (1988).
14. S. G. Brodsky and G. G. de Teramond, Phys. Rev. Lett. **60**, 1924 (1988).
15. M. Anselmino, Z. Phys. **C 13**, 63 (1982).
16. H. J. Lipkin, Nature **324**, 14 (1986); H. J. Lipkin, Phys. Lett. **B 181**, 164 (1987).
17. A. W. Hendry, Phys. Rev. **D 10**, 2300 (1974).
18. J. P. Ralston and B. Pire, Phys. Rev. Lett. **57**, 2330 (1986).
19. Y. Tomozawa, Phys. Rev. **D 36**, 2854 (1987).
20. G. F. Wolters, Phys. Rev. Lett. **45**, 776 (1980).
21. L. Durand and F. Halzen Nucl. Phys. **B 104**, 317 (1976).
22. M. Sakamoto and S. Wakaizumi, Progr. Theor. Phys. **62**, 1293 (1979).
23. C. Avilez, G. Cocho, and M. Moreno, Phys. Rev.**D 24**, 634 (1981).
24. R. J. Glauber, in "Lectures in Theoretical Physics", edited by W. Brittain and L. G. Dunham (Interscience Publ., New York, 1959) Vol. 1.
25. A. D. Martin, R. G. Roberts, W. S. Stirling, R. S. Thorne, Eur. Phys. J. **C 28** 455.
26. G. Bellettini et al., Phys. Lett. **14**, 164 (1965).
27. Asad et al., Nuclear Physics **B 255**, 273 (1985).
28. W. Vogelsang and F. Yuan, Phys. Rev. **D72**, 054028 (2005).
29. J. C. Collins, A. V. Efremov, K. Goeke, M. Grosse Perdekamp, S. Menzel, B. Meredith, A. Metz, and P. Schweitzer, Phys. Rev. **D73** 094023 (2006).
30. A. Bianconi and M. Radici, Phys. Rev. **D 73** (2006) 034018.
31. M. Anselmino, M. Boglione, U. D'Alesio, A. Kotzinian, S. Melis, F. Murgia, A. Prokudin, and C. Turk, arXiv:0805.2677.
32. A. Bacchetta, F. Forte and M. Radici, these proceedings.
33. L. Pappalardo (Hermes collaboration), these proceedings.
34. S. Levorato (Compass collaboration), these proceedings.

EXCLUSIVELY PRODUCED ρ^0 ASYMMETRIES ON THE DEUTERON AND FUTURE GPD MEASUREMENTS AT COMPASS

C. Schill
(On behalf of the COMPASS Collaboration)
Physikalisches Institut der Albert-Ludwigs-Universität Freiburg,
Hermann-Herder Str.3, D-79104 Freiburg, Germany.

Generalized parton distributions (GPDs) provide a new and powerful framework for a complete description of the nucleon structure. They can provide a three-dimensional picture of how the quarks and gluons form a nucleon. GPDs can be probed experimentally in hard exclusive meson production or deeply virtual Compton scattering (DVCS). The COMPASS experiment at CERN is a unique place to study these reactions. At COMPASS, a high energy polarized positive or negative muon beam is scattered off a polarized or unpolarized fixed target. First results for exclusive ρ^0 meson production are shown. The transverse target spin asymmetry for exclusively produced ρ^0 on a transversely polarized deuteron target has been measured. Prospects for future measurements of DVCS and exclusive meson production at COMPASS will be shown. The experiment will use the existing COMPASS spectrometer with a new target, a new recoil detector and extended calorimetry. Simulations for different models and a test of the recoil detector have been performed.

Keywords: nucleon structure, Generalized Parton Distributions, COMPASS, recoil detector

1. Introduction

Generalized parton distributions (GPDs) have been introduced about 10 years ago[1–3] and provide a comprehensive description of the nucleon. They can describe elastic form factors and polarized and unpolarized parton distributions at the same time. In addition, they may provide a handle on the orbital angular momentum of the quarks in the nucleon via the Ji sum rule.[3] GPDs may be accessed experimentally by measuring cross-sections and asymmetries in deeply virtual Compton scattering and exclusive meson production processes at COMPASS.

GPDs describe the quantum-mechanical amplitude for "taking out" a parton (quark or gluon) from a fast moving nucleon and "putting it back" with a different momentum, leading to a small momentum transfer to the nucleon. The GPDs are functions of three parameters, the longitudinal quark momentum fraction x, the longitudinal momentum transfer to the nucleon $\xi = x_{bj}/(2 - x_{bj})$ and the squared momentum transfer to the target nucleon t. The latter is the Fourier conjugate of the transverse impact parameter r of the scattering.

At leading twist four GPDs are necessary to describe the quark structure of the proton. The GPDs H and $\tilde{H}$ conserve the helicity of the proton and are generalizations of the ordinary parton distributions measured in DIS. In the forward limit $\xi = 0$ and $t = 0$ the unpolarized GPD H reproduces the unpolarized quark distribution $q(x)$. The polarized GPD $\tilde{H}$ gives the polarized quark distribution $\Delta q(x)$.

The GPDs E and $\tilde{E}$ allow a proton helicity flip and imply a transfer of orbital angular momentum to the nucleon. This is only possible for a non-zero transverse momentum transfer t, which is new with respect to ordinary parton distribution functions.

2. The COMPASS experiment

The COMPASS experiment is a fixed target experiment at the SPS accelerator at CERN. It scatters a longitudinally polarized muon beam on a polarized solid state target. From 2002 to 2004, COMPASS took data with a muon beam of 160 GeV and a longitudinally or transversely polarized ^6LiD target. For about 20% of the beam time, the target nucleons have been polarized transversely with respect to the beam polarization. The data presented here for the exclusive ρ^0 production have been taken with a transverse polarization of the target of about 50%.

The scattered muon and produced hadrons are detected in a two-stage spectrometer,[4] which provides high momentum resolution and excellent particle identification.

3. Exclusive ρ^0 production

One observable sensitive to the GPDs E and H is the transverse target spin asymmetry in the exclusive ρ^0 vector meson production. The reaction studied is

$$\mu N \rightarrow \mu \rho^0 N,$$

where N is a quasi-free nucleon in the polarized deuteron target. The reaction can be described as virtual photoproduction of a ρ^0: $\gamma^* N \to \rho^0 N$. To interpret the results in terms of GPDs, the following conditions have to be fulfilled: The virtual photon has to be longitudinally polarized, since only for longitudinal photons factorization between the hard reaction and the soft GPD part has been shown. The negative four-momentum transfer squared of the virtual photon Q^2 has to be large[5] and the momentum transfer squared t between initial and final nucleon has to be small.

At COMPASS, the outgoing ρ^0 and the scattered muon is detected, but the recoil proton is not. The ρ^0 is reconstructed from the invariant mass of the two pions by applying an invariant mass cut of $|M_{\pi^+\pi^-} - M_\rho| < 0.3$ GeV. The exclusivity is insured by a cut on the missing energy of $|E_{miss}| < 2.5$ GeV. To further minimize the non-exclusive background, a cut on the transverse momentum of the ρ^0 of $p_T^2 < 0.5$ (GeV/c)^2 has been applied. For a reliable definition of the azimuthal angle of the produced ρ^0 with respect to the lepton scattering plane, a lower limit has been imposed on the transverse momentum $p_T^2 > 0.01$ (GeV/c)^2.

Deep-inelastic scattering events were selected by a cut on the negative four-momentum transfer $Q^2 > 1$ (GeV/c)^2 and on the invariant mass of the hadronic system $W > 5.0$ GeV/c. The energy fraction of the virtual photon with respect to the beam momentum is limited in the range $0.1 < y < 0.9$. After all cuts $270k$ exclusively produced ρ^0 mesons have been obtained in total.

The azimuthal target spin asymmetry has been evaluated with respect to the angle $(\phi_h - \phi_S)$ (Fig. 1), where ϕ_h is the azimuthal angle of the produced ρ^0 meson with respect to the lepton scattering plane and ϕ_S the azimuthal angle of the target spin vector.

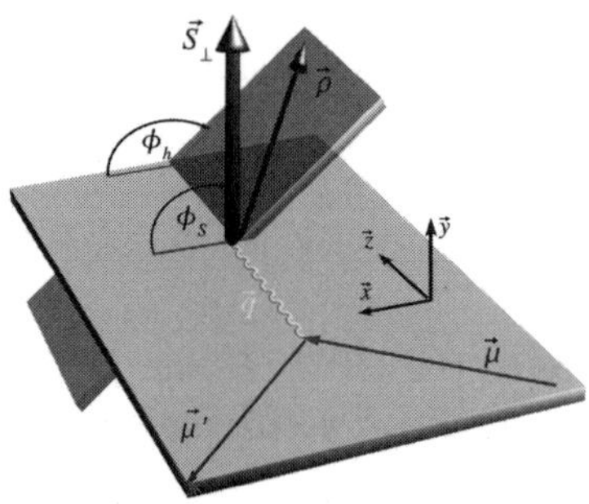

Fig. 1. Definition of ϕ_h and ϕ_S for exclusive ρ^0 production.

The transversely polarized target consists of two target cells with opposite polarization direction. Their polarization is reversed about once per week. To minimize systematic effects, the asymmetry is extracted from the counting rates in both cells and both target configurations with the double ratio method described in.[6] The extracted raw asymmetry ϵ is normalized to the target polarization $\langle P \rangle$ and the dilution factor $f = 0.38$ of the target:

$$A_{UT} = \frac{\epsilon}{f \cdot \langle P \rangle} \tag{1}$$

4. Results for exclusive ρ^0 production

The transverse target spin asymmetry A_{UT} in exclusive ρ^0 production is plotted in Fig. 2 as a function of x_{bj} and p_T. The measured asymmetries are small and within their statistical precision compatible with zero. Further analysis will allow to disentangle the contribution of the incoherent production on a quasi-free nucleon from the coherent production on the deuteron. Both mechanisms can be distinguished by the produced transverse momentum, since coherent production happens preferable at lower transverse momenta.

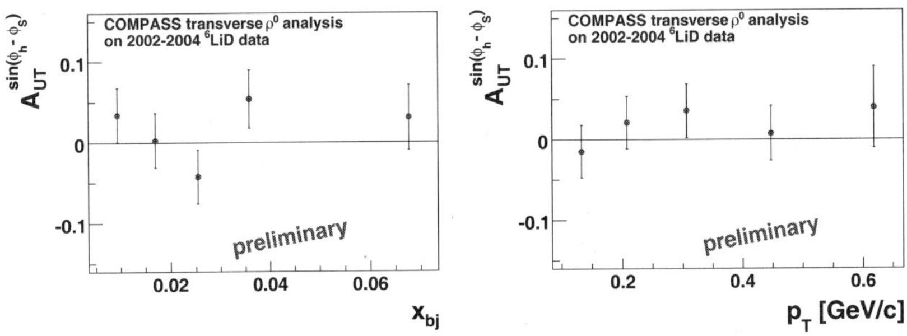

Fig. 2. Transverse ρ^0 target spin asymmetries as a function of x_{bj} (left) and p_T (right).

For a comparison to GPD model calculations, the asymmetry for longitudinal photons has to be extracted.[5] A method to extract the asymmetry for longitudinal photons was proposed in:[7] the contributions of longitudinal and transverse ρ^0 can be estimated from the angular distribution of the decay products. By assuming s-channel helicity conservation, we can determine the contribution from longitudinal photons.

In addition, in 2007 we took data with a transversely polarized proton target. Repeating the analysis on our proton data will provide new results on A_{UT} on the proton.

5. Future GPD measurements at COMPASS

The COMPASS experiment at CERN is a perfect place to study generalized parton distributions.[8] The GPDs will be accessed through deeply virtual Compton scattering and hard exclusive meson production. Due to the high energy of the muon beam available at COMPASS (between 100 and 190 GeV) the kinematic range covered by the proposed GPD program will be large enough to provide a bridge between the HERA collider experiments[9,10] at very small x_{bj} and the JLAB[11,12] and HERMES[13] experiments at large x_{bj} (Fig. 3).

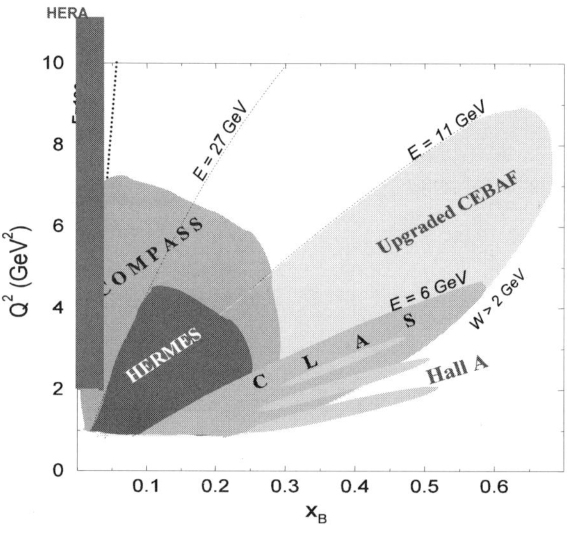

Fig. 3. Kinematic coverage for various DVCS experiments.

The measurements of generalized parton distributions at COMPASS will be performed with an unpolarized liquid hydrogen or deuterium target surrounded by a recoil detector.[8] The CERN beam-line can provide both positive and negative muons with opposite polarization of 80%. For beam charge asymmetry measurements the beam charge can be changed once a day. The goal of the experiment is to measure the cross section as well as spin and charge asymmetries as a function of x_{bj}, Q^2 and t. At 190 GeV, the DVCS process is dominant over the competing Bethe-Heitler process and the cross section can be measured. At 100 GeV, the possibility to use muon beams of opposite charge and polarization will allow to measure the charge and spin asymmetries arising from the interference between DVCS and the Bethe-Heitler process. Measurements of the cross section difference of oppositely charged and polarized muons give access to the real part of the DVCS amplitude, while the cross section sum provides information about the imaginary part of the DVCS amplitude.[14,15] These observables have great sensitivity to GPD models.[16] Hard exclusive meson production cross sections will also be measured at the same time providing different constraints on GPDs.

6. Simulations

Simulations of the beam charge asymmetry for COMPASS have been performed using three different model assumptions for the DVCS cross section. Two models are based on double distribution parameterizations of GPDs. Model 1 assumes $H(x, \xi, t) \propto q(x)F(t)$,[17] while model 2 includes a correlation between x and t: $H(x, \xi, t) \propto q(x)e^{tb_\perp^2}$ with a parameter $b_\perp$ describing the transverse extension of the partons. This ansatz reproduces the chiral quark soliton model.[18] The model 3 uses a dual parametrization, where the x, ξ dependence is separated from the ξ, t dependence. These calculations have been performed by.[16]

With a luminosity of $1.3 \cdot 10^{32}$ cm^{-2}s^{-1} and an global efficiency of 25% and a running of 150 days the statistical precision shown in Fig. 4 can be achieved. It is possible to split the data into 6 bins in Q^2 from 1.5 (GeV/c)^2 to 7.5 (GeV/c)^2 and in 3 bins in x_{bj} from 0.03 to 0.27. A separation between the different models can be achieved as shown in Fig. 4.

7. Experimental realization

The GPD experiment will use the existing COMPASS spectrometer with several new detectors added. The interaction will take place in a newly designed 2.5 m long liquid hydrogen or deuterium target. It will be surrounded by a recoil detector for the detection of the recoiling particles and to insure exclusivity of the reaction. It will use the time-of-flight technique to determine the recoiling proton momentum with a few percent precision.

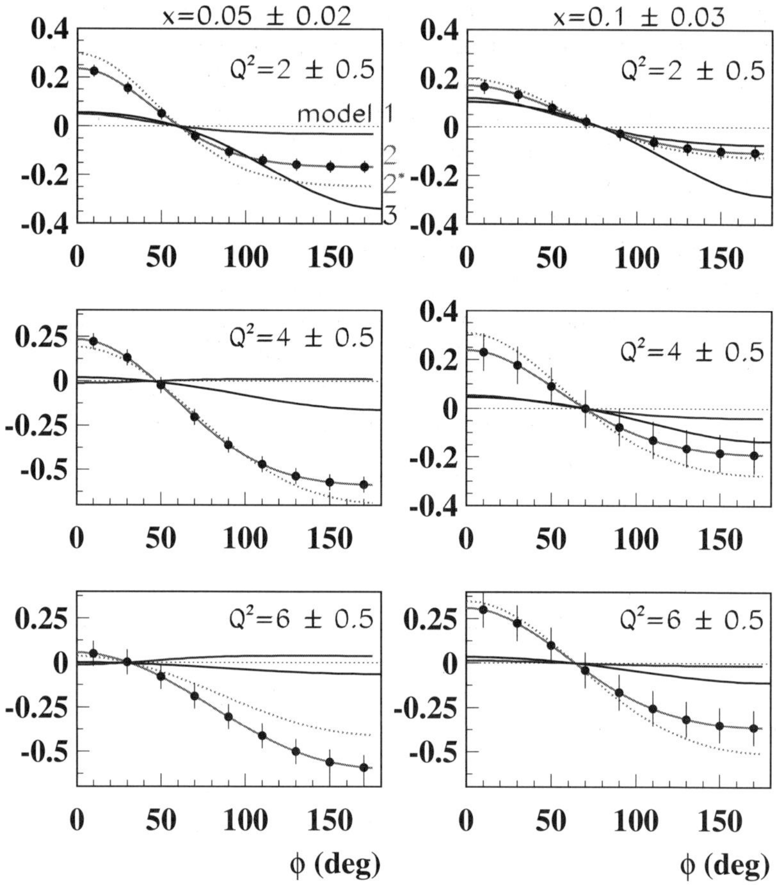

Fig. 4. Projected error bars for a measurement of the beam charge DVCS asymmetry as an example for these simulations in two domains of x_{bj} (0.05 on left and 0.10 on right) and 3 bins of Q^2 (2, 4, 6 $(\mathrm{GeV/c})^2$) from top to bottom. Projected error bars are compared to 3 different GPD models described in the text: model 1 (blue line), model 2 (line with points and error bars), model 3 (black line).

The recoil detector will consist of two concentric barrels of scintillators read at both sides. To allow for 10 bins in t, a minimum time resolution of 300 ps is needed. A full size prototype has been tested in the COMPASS experimental area in fall 2006 and achieved a time resolution of 310 ps.

In case of DVCS, the photon will be detected in the existing forward calorimeters, an additional wide angle calorimeter covering lab angles up to 20 degrees will be added to improve the acceptance at high x_{bj} and to reject the π^0 background.

8. Outlook

A big advantage of studying GPDs at the COMPASS experiment is that its kinematic domain covers both the valence and the sea quark region in terms of x_{bj}. Currently a proposal for GPD measurements at COMPASS is being prepared. In 2008-2010 the recoil detector, the liquid hydrogen target and the large angle calorimeter are being constructed. Data taking could start with the full set-up as early as 2010. Therefore first results are expected well before possible future projects at the JLAB 12 GeV upgrade and the FAIR project at the GSI.

References

1. D. Mueller *et al.*, *Fortsch. Phys.* **42**, 101 (1994).
2. A.V. Radyushkin, *Phys. Lett. B 385* 333 (1996).
3. X. Ji, *Phys. Rev. Lett.* **78**, 610 (1997).
4. COMPASS Collaboration, P. Abbon *et al.*, *Nucl. Instr. Meth.* **A 577**, 455 (2007).
5. COMPASS Collaboration, M. Alekseev *et al.*, *Eur. Phys. J.* **C 52**, 255 (2007).
6. COMPASS Collaboration, E.S. Ageev *et al. Nucl. Phys.* **B 765**, 31 (2007).
7. M. Diehl and S. Sapeta, *Eur. Phys. J.* **C 41**, 515 (2005).
8. COMPASS Collaboration, "Outline for generalized parton distribution measurements with COMPASS at CERN", *CERN report*, CERN-SPSC-2005-007, Jan. 2005.
9. H1 Collaboration, A. Aktas *et al.*, *Eur. Phys. J.* **C 46**, 585 (2006).
10. ZEUS Collaboration, S. Chekanov *et al.*, *Nucl. Phys.* **B 695**, 3 (2004).
11. CLAS Collaboration, S. Chen *et al.*, *Phys. Rev. Lett.* **97**, 072002 (2006).
12. JLab Hall A Collaboration, C. Munoz Camacho *et al.*, *Phys. Rev. Lett.* **97**, 262002 (2006).
13. HERMES Collaboration, A. Airapetian *et al.*, *Phys. Rev.* **D 75**, 011103 (2007).
14. M. Diehl *et al.*, *Phys. Lett.* **B 411**, 193 (1997).
15. A.V. Belitsky, D. Müller and A. Kirchner, *Nucl. Phys.* **B 629**, 323 (2002).
16. V. Guzey and T Teckentrup, *Phys. Rev.* **D 74**, 054027 (2006).
17. M. Vanderhaegen *et al.*, *Phys. Rev.* **D 60**, 094017 (1999).
18. K. Goeke *et al.*, *Prog. Part. Nucl. Phys.* **47**, 401 (2001).

TRANSVERSITY AND TRANSVERSE-MOMENTUM-DEPENDENT DISTRIBUTION MEASUREMENTS FROM PHENIX AND BRAHMS

C. Aidala
(On behalf of the PHENIX and BRAHMS Collaborations)
Physics Department, University of Massachusetts,
Amherst, MA 01003, U.S.A.

A variety of measurements performed utilizing transversely polarized proton-proton collisions at the Relativistic Heavy Ion Collider (RHIC) are now available. Recent results from the PHENIX and BRAHMS experiments are presented and discussed.

Keywords: transverse spin; proton structure.

1. Introduction

Surprisingly large transverse single-spin asymmetries (SSA's) were initially observed in hadronic collisions at the ZGS at Argonne National Laboratory[1,2] as well as the PS at CERN[3] in the late 1970's and early 1980's, for center-of-mass energies between 5 and 10 GeV. Subsequent experimental data over a wide range of energies indicate that similar effects exist for particle production at center-of-mass energies of tens of GeV and even as high as 200 GeV. The higher-energy data allow perturbative QCD (pQCD) to be used in attempting to interpret the results. While much remains to be learned, exciting progress has been made in the understanding of the observed transverse single-spin asymmetries at high energies, and the wealth of experimental data now becoming available is helping to drive the field forward.

The complex task of understanding measurements performed in hadronic collisions is now being facilitated as relevant quantities are being studied in semi-inclusive deep-inelastic scattering (SIDIS) and e^+e^- annihilation. A definitively non-zero Collins fragmentation function (FF) measured in e^+e^- annihilation was recently published by the BELLE experiment.[4] The availability of the Collins FF makes it possible to extract the transversity distribution from asymmetry measurements in SIDIS and proton-proton collisions, and first extractions of transversity have been published,[5] utilizing the available SIDIS data on the proton from HERMES[6,7] and the deuteron from COMPASS.[8] First extractions of the transverse-momentum-dependent Sivers distribution function from these SIDIS data have also been released.[9,10] As constraints start to be provided on transversity and the various transverse-momentum-dependent distribution and fragmentation functions, more can in turn be learned from hadronic collision data.

2. Using pQCD to describe polarization-averaged cross sections at RHIC

Polarization-averaged cross sections for pion production in proton-proton collisions at 200 GeV at both midrapidity as well as the forward region have been measured at RHIC and found to be well described by next-to-leading-order (NLO) pQCD.[11–13] Pion cross section measurements have also been performed at RHIC for $\sqrt{s} = 62.4$ GeV at mid- and forward rapidities[14,15] and are shown in Fig. 1 compared to pQCD calculations. On the left one can see the cross section for midrapidity neutral pions as a function of transverse momentum (p_T), compared to NLO as well as next-to-leading-log (NLL) pQCD calculations performed using p_T as the choice of renormalization and factorization scale. The bottom panels indicate the differences obtained in the calculations if scale choices are instead $p_T/2$ or $2p_T$. On the right side of Fig. 1 are shown the cross sections versus p_T for positive and negative pions measured at rapidities of 2.7 and 3.3, compared to NLO pQCD calculations. While the description of the data is not as successful as at midrapidity, the agreement is still reasonably good.

3. The PHENIX and BRAHMS experiments at RHIC

3.1. *PHENIX*

PHENIX is one of two large, ongoing experiments studying proton spin structure at RHIC. The PHENIX detector[16] consists of two multipurpose spectrometer arms at midrapidity ($|\eta| < 0.35$) covering a total of π in azimuth and two larger-rapidity spectrometer arms covering $1.2 < |eta| < 2.4$ and $\Delta\phi = 2\pi$ dedicated to the measurement of muons. The midrapidity spectrometers are instrumented to track and identify charged particles as well as detect electromagnetic probes. In 2006-07 two large-rapidity electro-

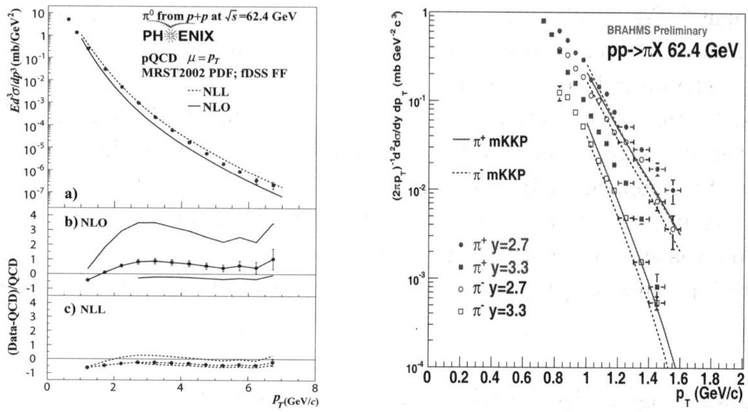

Fig. 1. Invariant cross section for midrapidity ($|\eta| < 0.35$) neutral pion (left) and forward charged pion (right) production at $\sqrt{s} = 62.4$ GeV, compared to perturbative QCD calculations.

magnetic calorimeters covering $3.1 < |\eta| < 3.7$, the muon piston calorimeters (MPC), were added. The PHENIX experiment was designed to detect rare probes, with fast data acquisition (greater than 5 kHz in proton-proton running) and sophisticated triggering capabilities.

The stable direction of the spin of the proton beams as they circulate in the RHIC ring is vertical. Spin rotator magnets immediately outside the PHENIX interaction region allow for the choice between transverse and longitudinal polarization of the colliding beams, independent of other RHIC experiments.

3.2. *BRAHMS*

The BRAHMS experiment, which finished taking data in 2006, consists of two movable spectrometer arms for the measurement of identified charged hadrons over a wide range of rapidity and transverse momentum.[17] The Forward Spectrometer (FS), which can be rotated from 2.3. to 15. degrees relative to the beam line, consists of 4 dipole magnets with a bending power of up to 9.2 Tm. There are five tracking stations, and a segmented time-of-flight wall as well as a ring-imaging Cherenkov detector are used for particle identification. The Mid-Rapidity Spectrometer (MRS), which can be rotated from 34. to 90. degrees relative to the beam line, is a single-dipole-magnet spectrometer with a solid angle of approximately 5 msr and a magnetic bending power up to 1.2 Tm. The MRS contains two time projection chambers followed by two highly segmented scintillator time-of-flight walls.

With no spin rotator magnets outside the BRAHMS interaction region, all proton-proton collisions at BRAHMS are transversely polarized in the vertical direction.

4. Results

A number of results are now available from transversely polarized data taken by the BRAHMS and PHENIX experiments at center-of-mass energies of 200 and 62.4 GeV. The transverse single-spin asymmetries discussed below are all left-right asymmetries, which can be calculated by

$$A_N^{\text{Left}} = \frac{1}{P}\frac{N^{\uparrow} - RN^{\downarrow}}{N^{\uparrow} + RN^{\downarrow}}$$

where A_N^{Left} indicates the asymmetry calculated to the left of the polarized beam, P is the beam polarization, $N^{\uparrow}$ ($N^{\downarrow}$) is the particle yield from bunches polarized up (down), and $R = \frac{L^{\uparrow}}{L^{\downarrow}}$ is the relative luminosity between up- and down-polarized bunches. Both beams at RHIC are polarized; in the calculation of single-spin asymmetries, the polarization of one beam is considered while averaging over the polarization states of the other.

In the early 1990's large transverse single-spin asymmetries in forward

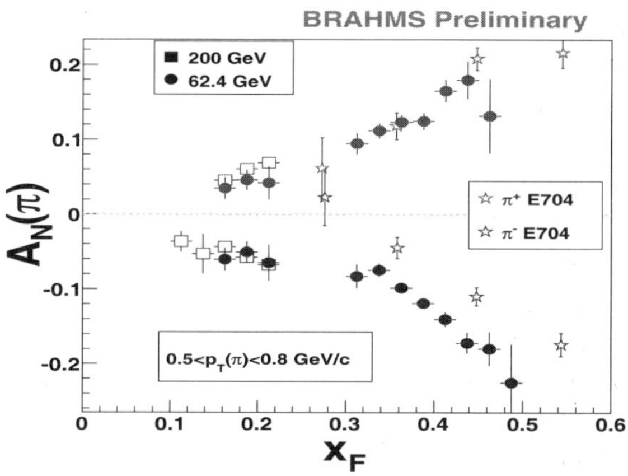

Fig. 2. Top: Charged pion asymmetries measured at 200 and 62.4 GeV by the BRAHMS experiment and at 19.4 GeV by the E704 experiment, shown for overlapping kinematic ranges (see text).

pion production were observed by the E704 experiment at Fermilab at a center-of-mass energy of 19.4 GeV.[18,19] The asymmetries of the charged pion species were found to be of similar magnitude with opposite sign, while the neutral pions exhibited a positive asymmetry, smaller in magnitude. The magnitude of the asymmetries increased with increasing Feynman-x ($x_F = 2p_L/\sqrt{s}$). Transverse SSA's in pion production at RHIC were initially measured by the STAR experiment at $\sqrt{s} = 200$ GeV.[20] The asymmetry in forward neutral pion production was found to be positive and increasing with x_F, as at E704. Subsequent measurements of forward charged pions by BRAHMS at 200 GeV revealed the same pattern as observed at 19.4 GeV: positive and negative pions exhibit near mirror symmetry in their behavior versus x_F, with a larger magnitude than the neutral pions. In 2006, data at the intermediate energy of $\sqrt{s} = 62.4$ GeV was taken at RHIC, and measurements of charged pions by BRAHMS[21] and neutral pions by PHENIX[22] again demonstrated the same pattern. In Fig. 2, the asymmetries for charged pions at all three energies are shown in overlapping kinematic ranges. The data at 62.4 and 200 GeV are for a transverse momentum (p_T) selection of 0.5-0.8 GeV/c; the 19.4 GeV data is for $p_T > 0.7$ GeV/c. The positive pion points overlap at all three energies, suggesting some kind of scaling behavior. The negative pion asymmetry at the lowest energy, however, appears to have a somewhat smaller magnitude than at the two higher energies. In any case, the similarities among the three energies are striking and suggest a common origin.

The behavior of the pion asymmetries has also been studied in bins of p_T and rapidity, as shown for example in Fig. 3 for charged and neutral pions at 62.4 GeV. The magnitude of the charged pion asymmetries appears to increase with increasing transverse momentum up to a p_T of approximately 1.0 GeV/c; however, the neutral pion data imply an opposite trend. Additional high-statistics data allowing detailed study of the asymmetries as a function of various kinematic variables will be important to clarify the trends currently observed as well as to pin down theoretical models which can correctly describe the behavior of the asymmetries in terms of multiple kinematic variables. At present, several effects are believed to be potential contributors to the observed asymmetries, with existing descriptions based on twist-three, Sivers, and Collins mechanisms.[23–25] Contrary to a prior claim,[26] the magnitude of the pion asymmetries observed at at RHIC at 200 GeV can in fact be explained by the Collins mechanism.[27,28]

The BRAHMS experiment has also measured transverse SSA's in kaon production at 62.4 and 200 GeV, and as the pions, they exhibit similar be-

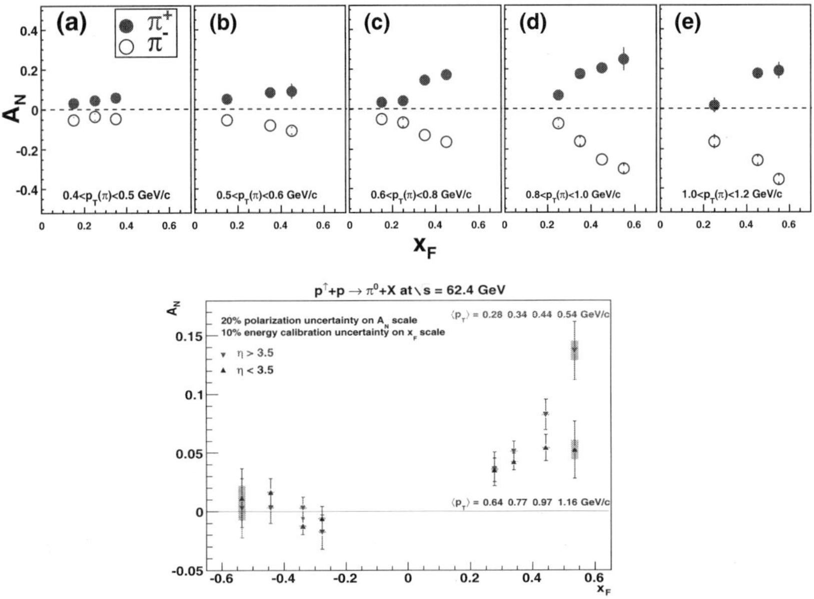

Fig. 3. Transverse single-spin asymmetry of charged and neutral pions for $\sqrt{s}$ = 62.4 GeV in bins of transverse momentum, measured by BRAHMS[21] (top), and for two different pseudorapidity ranges, measured by PHENIX[22] (bottom).

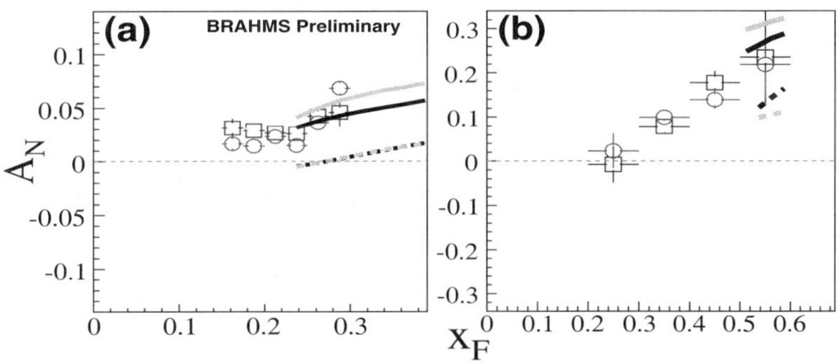

Fig. 4. Transverse SSA in kaon production at 200 GeV (left) and 62.4 GeV[21] (right), measured by BRAHMS. Positive (negative) kaons are indicated by open circles (squares). Solid (dashed) curves are for positive (negative) kaon initial-state twist-three calculations with (black) and without (grey) sea-quark contributions.

havior at the two energies. As can be seen in Fig. 4, the asymmetry of both charges is positive and of a similar magnitude. As shown in the figure, current theoretical calculations underpredict the negative kaon asymmetries.

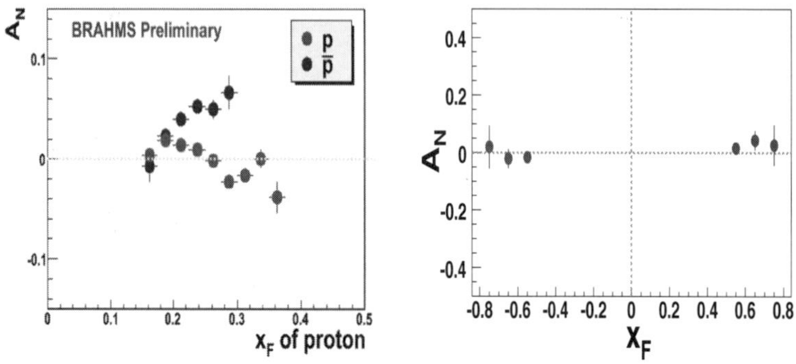

Fig. 5. Transverse SSA in proton and antiproton production at 200 GeV (left) and 62.4 GeV[21] (right), measured by BRAHMS.

In addition to pions and kaons, BRAHMS has also measured proton asymmetries, as shown in Fig. 5. While the proton asymmetry is consistent with zero for both energies, the antiprotons at 200 GeV demonstrate a clear positive asymmetry. Unfortunately the data for a measurement of antiprotons at 62.4 GeV are not available. These surprising results remain completely unexplained.

In contrast to the forward pion and kaon asymmetries, a measurement of neutral pions and charged hadrons at midrapidity for $1 < p_T < 5$ GeV/c was performed by PHENIX and found to be consistent with zero within a few percent.[29] Improvement of a factor of approximately 200 compared to the results published is anticipated in measurements at PHENIX of the transverse SSA of midrapidity neutral pions based on data taken in 2006 and 2008. The published result is sensitive mainly to gluon-gluon scattering and has been used to constrain the gluon Sivers function.[30] The more recent data are expected to provide at low p_T an even stronger constraint on the gluon Sivers function and at higher p_T increased sensitivity to quark effects.

It was proposed several years ago that the transverse SSA in D meson production would be an excellent channel to probe the gluon Sivers function.[31] A first measurement has now been made of the transverse SSA in the production of single muons from heavy flavor decays, shown versus x_F in Fig. 6. Results are also available as a function of transverse momentum. The significant systematic uncertainties indicated in the figure are due mainly to the estimation of the background contribution and should be improved in the future by the planned forward silicon detector upgrade to PHENIX. A direct comparison between the calculations of Anselmino *et al.*,[31] done for D mesons, and the present results, which are for single muons resulting

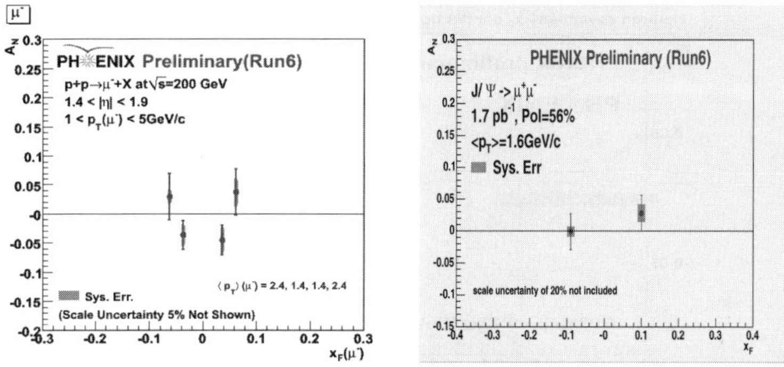

Fig. 6. Transverse SSA measured by PHENIX for $\sqrt{s} = 200$ GeV in single muons from heavy flavor decays (left) and J/Ψ production (right).

from the decay of both charm and bottom mesons, is not currently possible. However, it is worth noting that the magnitude of the observed muon asymmetries at positive x_F is significantly smaller than the calculation of 10-15% asymmetries in a similar x_F and p_T range for D mesons. We thus hope that with some further work in translating between D mesons and their decay muons it will be possible to use these data to place a stronger constraint on the gluon Sivers function than presently available.

Looking further at what we can learn from transverse single-spin asymmetries in heavy flavor measurements, recent work indicates that the transverse SSA in J/Ψ production may be sensitive to the J/Ψ production mechanism.[32] A non-zero asymmetry is predicted in $p+p$ collisions in the color-singlet model; a zero asymmetry in the color-octet model. This is in contrast to J/Ψ production in semi-inclusive DIS, in which a non-zero transverse single-spin asymmetry is predicted in the color-octet model but zero asymmetry in the color-singlet model. Preliminary results on the J/Ψ transverse SSA in $p+p$ collisions can be seen in Fig. 6. Early estimates suggest that an asymmetry of the order of a few percent at positive x_F, consistent with the data, may be possible due to *indirect* J/Ψ production coming from feed-down from the χ_c, assuming a gluon Sivers function of approximately $0.5x(1-x)xG(x)$.[33] However, more detailed theoretical work will be needed to clarify what can be learned from the current experimental result; improvement in the experimental measurement is expected once analysis of the 2008 transverse data set from PHENIX has been completed.

A large, negative asymmetry in the production of very forward neutrons at $\sqrt{s} = 200$ GeV has been observed by the PHENIX experiment.[34] This asymmetry has been found to be enhanced by requiring coincidence with

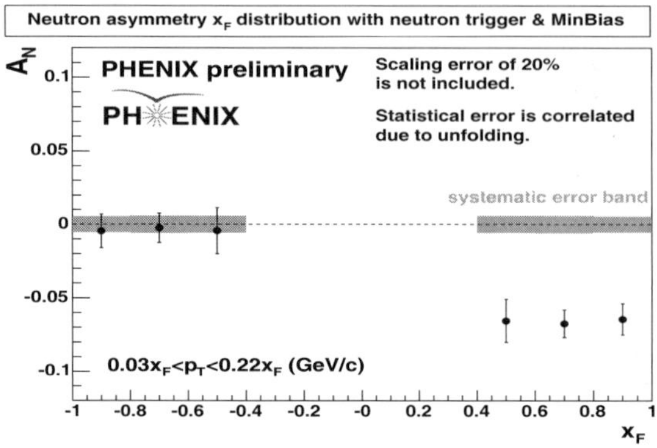

Fig. 7. Transverse SSA of forward neutrons produced in coincidence with forward charged particles for $\sqrt(s) = 200$ GeV, measured by PHENIX.

forward charged particles in the pseudorapidity range of $3.0 < |\eta| < 3.9$ (Fig. 7). As can be seen in the figure, the forward asymmetry exhibits no dependence on x_F, while the asymmetry in the backward direction is consistent with zero. A large negative transverse SSA in the production of forward neutrons at $\sqrt{s} = 200$ GeV has also been observed by a small experiment located at RHIC Interaction Point 12,[35] and similar asymmetries have been observed by PHENIX for center-of-mass energies of 62.4 as well as 410 GeV. Unlike other results presented above, the kinematics of the neutron measurement are not in a regime described by perturbative processes. The large measured asymmetries suggest interference between a spin-flip amplitude, such as one-pion exchange, which dominates lower-energy neutron production, and non-flip amplitudes, but a clear understanding of the large asymmetries observed has yet to be developed.

5. Conclusions

A wealth of transverse single-spin asymmetry measurements is now available from the PHENIX and BRAHMS experiments at RHIC, the majority of which are in a kinematic region in which the polarization-averaged cross sections are well described by pQCD. For inclusive pion measurements, enough statistics are available to allow first studies of the asymmetry dependence on p_T and rapidity in addition to x_F. The behavior of the SSA's as a function of various kinematic variables is essential to constrain phenomenological models of the underlying physics generating the asymmetries. With first extractions of the Collins fragmentation function as well as the transversity and Sivers distributions now available thanks to measurements in e^+e^- and SIDIS, an increasing amount can be learned from proton-proton collision data. The long future of the RHIC spin program promises a myriad of further transverse-spin results, and hadronic data will play a crucial role in our understanding of transversity and transverse-momentum-dependent distributions as the field comes to maturity.

References

1. R. D. Klem *et al.*, *Phys. Rev. Lett.* **36**, 929 (1976).
2. W. H. Dragoset *et al.*, *Phys. Rev.* **D18**, 3939 (1978).
3. J. Antille *et al.*, *Phys. Lett.* **B94**, p. 523 (1980).
4. K. Abe *et al.*, *Phys. Rev. Lett.* **96**, p. 232002 (2006).
5. M. Anselmino *et al.*, *Phys. Rev.* **D75**, p. 054032 (2007).
6. A. Airapetian *et al.*, *Phys. Rev. Lett.* **84**, 4047 (2000).
7. A. Airapetian *et al.*, *Phys. Rev. Lett.* **94**, p. 012002 (2005).
8. V. Y. Alexakhin *et al.*, *Phys. Rev. Lett.* **94**, p. 202002 (2005).
9. M. Anselmino *et al.*, *arXiv:0805.2677 [hep-ph]* (2008).
10. S. Arnold, A. V. Efremov, K. Goeke, M. Schlegel and P. Schweitzer (2008).
11. A. Adare *et al.*, *Phys. Rev.* **D76**, p. 051106 (2007).
12. J. Adams *et al.*, *Phys. Rev. Lett.* **97**, p. 152302 (2006).
13. I. Arsene *et al.*, *Phys. Rev. Lett.* **98**, p. 252001 (2007).
14. A. Adare *et al.*, *arXiv:0801.4555 [nucl-ex]* (2008).
15. F. Videbaek, *arXiv:0801.1696 [nucl-ex]* (2008).
16. K. Adcox *et al.*, *Nucl. Instrum. Meth.* **A499**, 469 (2003).
17. M. Adamczyk *et al.*, *Nucl. Instrum. Meth.* **A499**, 437 (2003).
18. D. L. Adams *et al.*, *Phys. Lett.* **B264**, 462 (1991).
19. D. L. Adams *et al.*, *Phys. Lett.* **B261**, 201 (1991).
20. J. Adams *et al.*, *Phys. Rev. Lett.* **92**, p. 171801 (2004).
21. I. Arsene *et al.*, *arXiv:0801.1078 [nucl-ex]* (2008).
22. M. Chiu, *AIP Conf. Proc.* **915**, 539 (2007).
23. J.-w. Qiu and G. Sterman, *Phys. Rev.* **D59**, p. 014004 (1999).
24. D. W. Sivers, *Phys. Rev.* **D41**, p. 83 (1990).
25. J. C. Collins, *Nucl. Phys.* **B396**, 161 (1993).
26. M. Anselmino, M. Boglione, U. D'Alesio, E. Leader and F. Murgia, *Phys. Rev.* **D71**, p. 014002 (2005).
27. F. Yuan, *Phys. Lett.* **B666**, 44 (2008).
28. F. Murgia, these proceedings.
29. S. S. Adler *et al.*, *Phys. Rev. Lett.* **95**, p. 202001 (2005).
30. M. Anselmino, U. D'Alesio, S. Melis and F. Murgia, *Phys. Rev.* **D74**, p. 094011 (2006).
31. M. Anselmino, M. Boglione, U. D'Alesio, E. Leader and F. Murgia, *Phys. Rev.* **D70**, p. 074025 (2004).
32. F. Yuan, *Phys. Rev.* **D78**, p. 014024 (2008).
33. F. Yuan, private communication.
34. M. Togawa, *AIP Conf. Proc.* **915**, 689 (2007).
35. A. Bazilevsky *et al.*, *Phys. Lett.* **B650**, 325 (2007).

SIVERS AND COLLINS EFFECTS IN POLARIZED *pp* SCATTERING PROCESSES

M. Anselmino[1], M. Boglione[1], U. D'Alesio[2,3],
E. Leader[4], S. Melis[1], F. Murgia[3],

1 – Dipartimento di Fisica Teorica, Università di Torino and INFN, Sezione di Torino, Via P. Giuria 1, I-10125 Torino, Italy

2 – Dipartimento di Fisica, Università di Cagliari, Cittadella Universitaria di Monserrato, I-09042 Monserrato (CA), Italy

3 – INFN, Sezione di Cagliari, C.P. 170, I-09042 Monserrato (CA), Italy

4 – Imperial College London, Prince Consort Road, London SW7 2BW, U.K.

We summarize the present phenomenology of Sivers and Collins effects for transverse single spin asymmetries in polarized proton–proton collisions within the framework of the generalized parton model (GPM). We will discuss a reassessment of the Collins effect and some preliminary predictions for SSA's in $p^{\uparrow}p \to \pi, K + X$ processes at RHIC obtained using updated information from SIDIS data and a new set of meson fragmentation functions.

Keywords: Single spin asymmetries, *pp* collisions, Sivers and Collins effects

1. Introduction

Transverse single spin asymmetries (SSA) for inclusive single-particle production in polarized high-energy proton–proton collisions have historically motivated a huge theoretical and experimental effort to understand the role of spin and intrinsic parton motion in hadronic processes (see Ref. [1] for an up-to-date review). At present, other asymmetries are experimentally at reach for meson production in semi-inclusive deeply inelastic scattering (SIDIS), in the Drell-Yan (DY) process, or in inclusive two-particle production in proton–proton collisions, which are considered safer from a theoretical point of view. Many leading-twist azimuthal asymmetries can be studied for such processes, in a context where factorization has been proven and some universality issues for the soft, transverse momentum dependent (TMD) functions involved, like the Sivers distribution and the Collins fragmentation function (FF), have been clarified. The interplay among different theoretical models (the collinear twist-three formalism [2] and TMD approaches) has also been analyzed, at least for some kinematical regimes. SSA's in inclusive single-particle production in hadronic collisions look more involved. Being twist-three effects, several competing mechanisms may contribute. Factorization holds in the collinear twist-three approach [2]; however, in phenomenological applications, leading order (LO) unpolarized cross sections are used, which often underestimate experimental data. On the other hand, factorization has never been proven for TMD approaches, and has to be assumed. Other open points concern the universality properties of the TMD functions.

Nevertheless, the study of SSA's in $p^\uparrow p \to h + X$ processes is still very important and useful. Together with the well-known fixed target results of the Fermilab-E704 collaboration, an increasing amount of challenging high-energy experimental data are becoming available from RHIC experiments, with enlarged statistics and kinematical coverage. At this stage, a combined phenomenological analysis of these results with SIDIS and e^+e^- data can give crucial information on the universality properties of the Sivers and Collins functions and on possible factorization-breaking terms in hadronic collisions in the kinematical regimes experimentally accessible.

2. The generalized parton model

In our phenomenological analysis we adopt an LO pQCD approach including spin and intrinsic motion ($\boldsymbol{k}_\perp$) effects and all leading-twist polarized TMD distribution and fragmentation functions, which keep their simple partonic interpretation [3]. The complete $\boldsymbol{k}_\perp$ kinematics is included also in the partonic cross sections. As already mentioned, for this approach factorization has not been proven yet, and has to be considered as a reasonable assumption deserving careful phenomenological tests. The same is true for the (assumed) universality of the TMD functions involved.

In the context of the generalized parton model and the helicity formalism the polarized differential cross section for the inclusive process $A(S_A)\,B(S_B) \to h + X$, where A, B, are two spin-1/2 hadrons (*e.g.* two protons) and h an unpolarized particle (*e.g.* a meson or a photon), reads schematically (see Ref. [3] for details):

$$d\sigma^{A(S_A)\,B(S_B)\to h+X} \propto \tag{1}$$

$$\sum_{i=a,b,c}\sum_{\lambda_i,\lambda'_i}\int dx_i\, d\boldsymbol{k}_{\perp i}\, \rho^{a/A}_{\lambda_a,\lambda'_a}\, \hat{f}_{a/A,S_A}\, \rho^{b/B}_{\lambda_b,\lambda'_b}\, \hat{f}_{b/B,S_B}\, M_{\{\lambda_i\}} M^*_{\{\lambda'_i\}}\, D^h_{\lambda_c,\lambda'_c}\,.$$

When summing over all partons and their helicities, all allowed combinations of TMD PDF's⊗FF's⊗polarized partonic cross sections appear [3]. By taking appropriate sums and differences of these cross sections (with S_A or S_B changing sign) all spin asymmetries can be written down. As an example, the transverse SSA, A_N, is given by $A_N = (d\sigma^\uparrow - d\sigma^\downarrow)/(d\sigma^\uparrow + d\sigma^\downarrow)$, where $d\sigma^{\uparrow,\downarrow} \equiv d\sigma^{A^{\uparrow,\downarrow}\,B\to h+X}$. Notice that for the single-particle production case considered here, keeping at least the dominant $\boldsymbol{k}_\perp$ dependence in the partonic cross sections is crucial in order to get a non-vanishing SSA (this is not the case for double spin asymmetries). This also explains why, although involving leading-twist TMD PDF's and FF's, the global effect is higher-twist in $k_\perp/p_{hT}$. Again, this is not the case for inclusive double-particle production, where an observed low-energy scale is present (*e.g.* the unbalanced two-particle total transverse momentum).

3. Collins effect in *pp* collisions: a reassessment

We recently realized that a sign mistake was present in our calculation of the SSA's in hadronic collisions in the $qg \to qg$ channel contribution to the Collins mechanism (see also Ref. [4]). This induced strong cancelations among the dominant partonic channels at large x_F, leading, together with non trivial effects due to azimuthal phase integrations, to a sizeable suppression of the Collins effect and a potential dominance of the Sivers contribution [5]. After correction this conclusion gets substantial change. Although the role of azimuthal phases is confirmed, for E704 and RHIC kinematics the Collins effect is not strongly suppressed anymore. Therefore, it can play a substantial role and mix with the Sivers mechanism.

Let us however stress that the corrected sign mistake *plays no role* in several phenomenological analyses recently performed by our group (see Ref. [1] and refs. therein). In particular, there is no change for: 1) The Sivers asymmetry in $p^\uparrow p \to h + X$ and in the Drell-Yan process, and for the Λ transverse polarization in unpolarized hadronic collisions; 2) The azimuthal Sivers and Collins asymmetries in SIDIS processes (no gluonic channel is involved at LO); 3) The gluon Sivers effect in $p^\uparrow p \to D + X$ at RHIC energies. Furthermore, there are practically no changes for the bounds on the gluon Sivers function recently derived from the RHIC-PHENIX $p^\uparrow p \to \pi^0 + X$ data at mid-rapidity.

In the rest of this section we will present a selection of the most significant variations in the phenomenological analyses which *are* affected by this correction (an extended and comprehensive paper is in progress). The old results were obtained with parameterizations for the TMD PDF's and FF's which are not at present the most updated ones. In order to avoid confusion, and to clearly show the effect of the correction, we will present results using *exactly* the same choices adopted in the original calculations. An analysis of the Collins effect in *pp* collisions with the most up-to-date parameterizations is also in progress. Some preliminary results will be discussed in the next section. In Ref. [3] we have considered the maximized contribution of the Collins mechanism to the transverse SSA for the process $p^\uparrow p \to \pi + X$ in the kinematical configurations of the E704 and STAR experiments. In

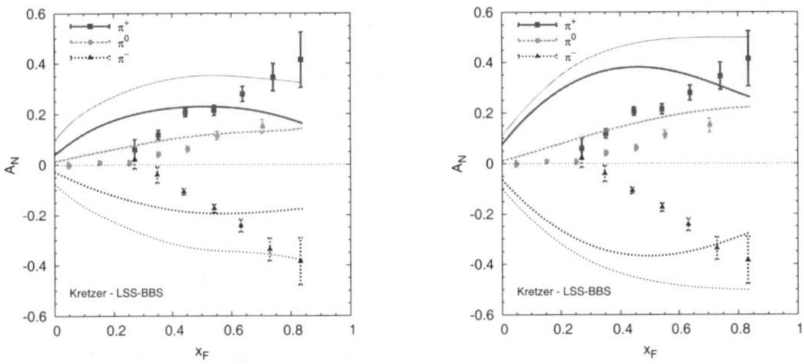

Fig. 1. Collins contribution (maximized effect) to pion SSA's for E704 kinematics, $\sqrt{s} = 20$ GeV, before (left) and after (right) correction. Thin lines are obtained setting to zero all azimuthal phases. The PDF/FF sets are: LSS-BBS/Kretzer.

Fig. 1 we compare the results for the E704 kinematics obtained before and after correction of the Collins contribution. Although the significant role of azimuthal phases in depleting the contribution at medium-large x_F is confirmed, the suppression of the Collins effect at large x_F is significantly reduced. Notice, however, that the plots show the maximized potential effect, with all partonic contributions saturated and added with the same sign. Similarly, in Fig. 2 we compare results, before (Ref. [3]) and after correction, for the Sivers and Collins contributions to the π^0 SSA at the RHIC-STAR kinematics. Different maximized contributions are shown separately (quark Sivers, gluon Sivers and transversity$\otimes$Collins terms). Again,

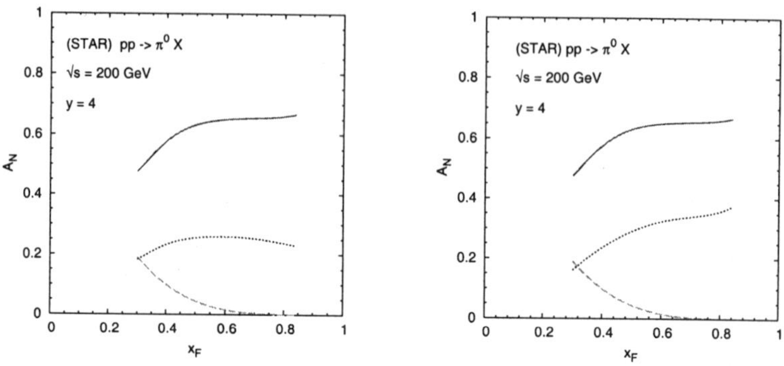

Fig. 2. Maximized contributions to $A_N(p^\uparrow p \to \pi^0 + X)$ for STAR kinematics, $\sqrt{s} = 200$ GeV, before (left) and after (right) correction: quark Sivers cont. (solid line); gluon Sivers cont. (dashed line); transversity⊗Collins cont. (dotted line). The PDF/FF sets are: MRST01/KKP.

after correction the suppression of the maximized potential Collins contribution to the SSA is sizeably reduced at large x_F (Sivers terms are clearly unaffected). In a recent paper [6], some of us have shown how predictions in very good agreement with RHIC data for pion SSA's can be obtained by using the central values of the parameterizations for the Sivers [7] and transversity distributions and the Collins function [8] then available from fits to the SIDIS and e^+e^- data. While the agreement with STAR data for $p^\uparrow p \to \pi^0 + X$ is still good (see left plot in Fig. 4), Fig. 3 shows how for BRAHMS measurements our corrected predictions overestimate the data, somehow weakening the conclusions reached in Ref. [6].

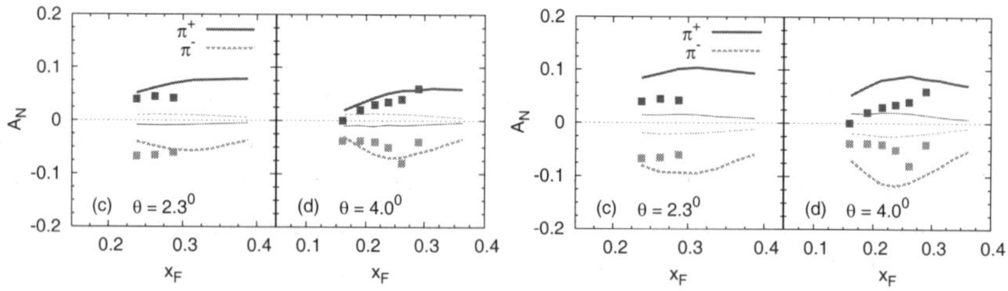

Fig. 3. Predictions for charged pion SSA's at RHIC-BRAHMS kinematics, $\sqrt{s} = 200$ GeV, before (left) and after (right) correction. The parameterizations for the Sivers, transversity and Collins functions are from Refs. [7,8]. Thin lines: Collins contribution alone; thick lines: Sivers + Collins contributions. The PDF/FF sets are: MRST01/Kretzer.

In conclusion, there is indeed a significant reduction of the Collins mechanism due to azimuthal phase factors. After correction, however, this effect alone is not sufficient to rule out sizeable Collins contributions to meson SSA's in *pp* collisions, at least for the E704 and RHIC kinematics. Therefore, and in contrast with Ref. [5], we cannot exclude a more complicated (but even more interesting) situation in which both Sivers and Collins effects may potentially give significant and comparable contributions to single-particle SSA's in hadronic collisions.

4. Sivers and Collins effects in $p^{\uparrow} p \to h + X$: a preliminary update

The analysis of the previous section was performed exactly in the same conditions as in the original papers. In the meantime both the amount and the precision of experimental data have significantly increased. Moreover, a new set of meson fragmentation functions, the DSS set, has become available [9] which was obtained by including in the fit unpolarized cross sections and multiplicities for meson production in *pp* collisions and in SIDIS. This set has a much larger s quark component for kaon FF's and a much larger LO gluon component than previous sets, for both pions and kaons.

The available SIDIS experimental data probe a (relatively) limited region of the Bjorken variable, x_B ($x_B \leq 0.3 - 0.4$). The parameterizations of the TMD distributions involved are therefore almost unconstrained at large light-cone momentum fractions. This implies large uncertainties in the predictions for SSA's in *pp* collisions at large positive x_F. The new high-precision Belle data put instead more stringent constraints on the Collins FF. The latest HERMES data on kaon SSA's are challenging, since positive kaons have on the average larger Sivers asymmetries than positive pions. This is difficult to understand in the simplified valence-like scenario adopted in the previous fits. Therefore, we have recently performed [10] a new fitting procedure for quark and, for the first time, sea-quark Sivers functions, using the DSS set and optimizing some parameters in order to: a) maximize the sensitivity in the x_B region covered by the HERMES and COMPASS results; b) describe well the kaon data. A new preliminary parameterization of the transversity and Collins functions, using the new SIDIS and e^+e^- data is also available [11]. Details and features of the new fits and parameterizations have been already presented in this workshop (see the contributions of M. Boglione and A. Prokudin). In Figs. 4 and 5 we compare predictions for pion SSA's in *pp* collisions at RHIC, using the corrected Collins contribution and the central-value parameterizations of the Sivers, Collins and transversity functions as obtained respectively in the previous (left plots) and in the new (right plots) fits. The results for STAR kinematics look much worse with the new fits, which however seem to reconcile (after correction of the Collins contribution) our predictions with the BRAHMS results (see Fig. 3).

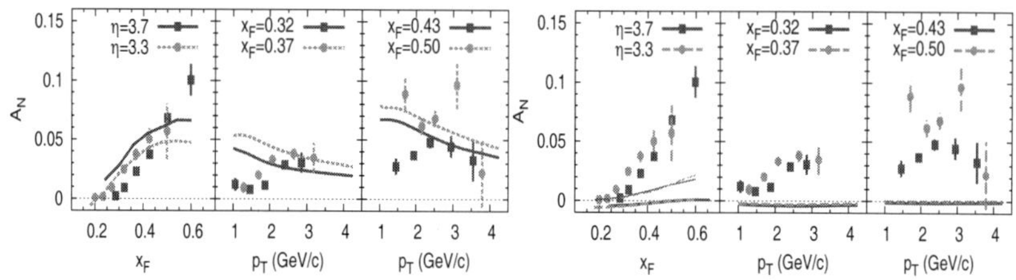

Fig. 4. Predictions for neutral pion SSA at RHIC-STAR kinematics, $\sqrt{s} = 200$ GeV, obtained with the previous [7,8] (left) and new [10,11] (right) parameterizations for the Sivers, transversity and Collins functions. Left: Kretzer FF set, Collins + Sivers contributions; right: DSS FF set, Sivers (thick lines) and Collins (thin lines, first panel only) contributions. PDF set: MRST01.

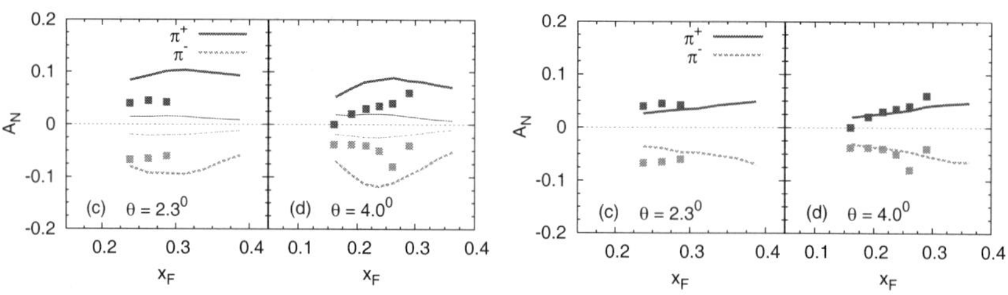

Fig. 5. Predictions for charged pion SSA's at RHIC-BRAHMS kinematics, $\sqrt{s} = 200$ GeV, obtained with the previous [7,8] (left) and new [10,11] (right) parameterizations for the Sivers, transversity and Collins functions. Left: Kretzer FF set, Collins (thin lines) and Collins + Sivers (thick lines) contributions; right: DSS FF set, Collins + Sivers contributions. PDF set: MRST01.

These preliminary results must be taken with great care. As already discussed, SIDIS results do not constrain the large x behaviour of the TMD distributions. As a consequence, predictions for SSA's at large x_B or x_F can be affected by huge uncertainties. They can also be strongly dependent on simplifying assumptions (necessary to reduce the number of free parameters in the fit) involving the parameters governing this large x behaviour. In the new fits [10,11], in the effort to probe the role of sea quarks in detail, we adopted the same large x behaviour for all Sivers distributions. This is at variance with the old valence-like fits [7,8], where the large x behaviour of the u and d quark Sivers functions were left totally independent. The results of Fig. 4 clearly show that in order to have a sizable Sivers SSA at large x_F, the u and d quark Sivers functions *cannot* have the same large x behaviour. Notice that this is not in contradiction with the results of the new fits, since present SIDIS data are basically unaffected by these details.

Concerning the Collins contribution, it is indeed bigger after correction. However, for neutral pion production large cancelations between the leading and non-leading Collins functions (which result with opposite signs from the fits) prevent it to be very large. Let us also remark that the huge LO gluon component of the DSS set gives a large contribution to the denominator of the SSA but almost no contribution to its numerator (dominated by quark initiated terms).

This discussion allows us to understand the preliminary results of Figs. 4 and 5. Clearly, in order to better investigate the possibility of simultaneously explain SIDIS and *pp* data on meson SSA's with a universal set of TMD functions, we need to perform, in the lack of experimental SIDIS data at large x_B, a more detailed analysis. This is in progress and will be presented elsewhere.

References

1. U. D'Alesio and F. Murgia, *Prog. Part. Nucl. Phys.* **61**, 394 (2008).
2. J.-w. Qiu and G. Sterman, *Phys. Rev. D* **59**, 014004 (1998).
3. M. Anselmino *et al.*, *Phys. Rev. D* **73**, 014020 (2006).
4. F. Yuan, *Phys. Lett. B* **666**, 44 (2008).
5. M. Anselmino, M. Boglione, U. D'Alesio, E. Leader and F. Murgia, *Phys. Rev. D* **71**, 014002 (2005).
6. M. Boglione, U. D'Alesio and F. Murgia, *Phys. Rev. D* **77**, 051502(R) (2008).
7. M. Anselmino *et al.*, *Phys. Rev. D* **72**, 094007 (2005).
8. M. Anselmino *et al.*, *Phys. Rev. D* **75**, 054032 (2007).
9. D. de Florian, R. Sassot and M. Stratmann, *Phys. Rev. D* **75**, 114010 (2007).
10. M. Anselmino *et al.*, arXiv:0805.2677 [hep-ph] (2008).
11. M. Anselmino *et al.*, arXiv:0807.0173 [hep-ph] (2008).

SIVERS FUNCTION IN CONSTITUENT QUARK MODELS

S. Scopetta

Dipartimento di Fisica, Università degli Studi di Perugia, and INFN, sezione di Perugia, via A. Pascoli, 06100 Perugia, Italy

A. Courtoy

Departament de Fisica Teòrica, Universitat de València and IFIC, CSIC 46100 Burjassot (València), Spain

F. Fratini

Dipartimento di Fisica, Università degli Studi di Perugia, via A. Pascoli, 06100 Perugia, Italy

V. Vento

Departament de Fisica Teòrica, Universitat de València and IFIC, CSIC 46100 Burjassot (València), Spain,
TH-Division, PH Department, CERN, CH-1211 Genève 23, Switzerland

A formalism to evaluate the Sivers function, developed for calculations in constituent quark models, is applied to the Isgur-Karl model. A non-vanishing Sivers asymmetry, with opposite signs for the u and d flavor, is found; the Burkardt sum rule is fulfilled up to 2%. Nuclear effects in the extraction of neutron single spin asymmetries in semi-inclusive deep inelastic scattering off ^{3}He are also evaluated. In the kinematics of JLab, it is found that the nuclear effects described by an Impulse Approximation approach are under control.

Keywords: DIS, transversity, neutron structure.

1. The Sivers function in Constituent Quark Models

The partonic structure of transversely polarized nucleons is still an open problem.[1] Semi-inclusive deep inelastic scattering (SIDIS) is one of the proposed processes to access the parton distributions (PDs) of transversely polarized hadrons. SIDIS of unpolarized electrons off a transversely polarized target shows "single spin asymmetries" (SSAs),[2] due to two physical mechanisms, whose contributions can be distinguished,[3–5] i.e. the Collins[2] and

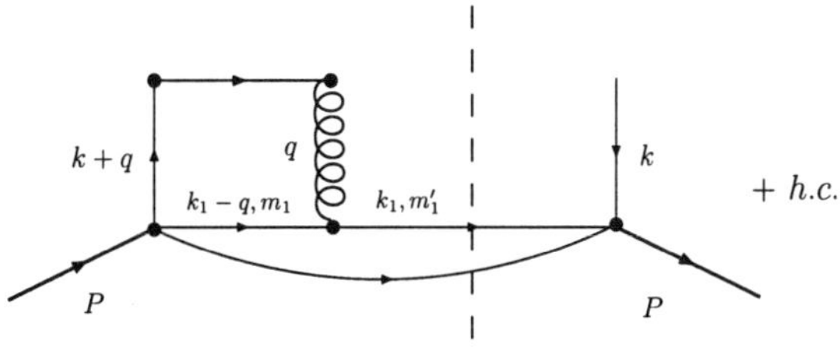

t

Fig. 1. The contributions to the Sivers function in the present approach.

the Sivers[6] mechanisms. The former is due to parton final state interactions (FSI) in the production of a hadron by a transversely polarized quark. The Sivers mechanism leads to a SSA which is the product of the unpolarized fragmentation function with the Sivers PD. The latter describes the number density of unpolarized quarks in a transversely polarized target: it is a time-reversal odd, Transverse Momentum Dependent (TMD) PD. From the existence of leading-twist Final State Interactions (FSI),[7,8] a non-vanishing Sivers function has been explained as generated by the gauge link in the definition of TMDs,[9,10] whose contribution does not vanish in the light-cone gauge, as happens for the standard PD functions. Recently, the first data of SIDIS off transversely polarized targets have been published, for the proton[11] and the deuteron.[12] It has been found that, while the Sivers effect is sizable for the proton, it becomes negligible for the deuteron, so that apparently the neutron contribution cancels the proton one, showing a strong flavor dependence of the mechanism. Different parameterizations of the available SIDIS data have been published,[13–15] still with large error bars. Since a calculation from first principles in QCD is not yet possible, several model evaluations have been performed, e.g. in a quark-diquark model;[7,9,16] in the MIT bag model;[17] in a light-cone model;[18] in a nuclear framework, relevant to proton-proton collisions.[19] We here describe a Constituent Quark Model (CQM) calculation of the Sivers function.[20] CQM calculations of PDs are based on a two steps procedure.[21] First, the matrix element of the proper operator is evaluated using the wave functions of the model; then, a low momentum scale, μ_0^2, is ascribed to the model calculation and QCD evolution is used to evolve the observable calculated in this low energy scale to the scale of DIS experiments. Such procedure has proven successful in describing the gross features of PDs[22] and GPDs,[23] by using different CQMs, e.g. the Isgur-Karl (IK) model.[24] Besides the fact that it successfully reproduces the low-energy properties of the nucleon, the IK model contains the one-gluon-exchange (OGE) mechanism.[25] In the present calculation, with respect to calculations of PDs and GPDs, the leading twist contribution to the FSI has to be taken into account. The main approximations have been: i) only the valence quark sector is investigated; ii) the leading twist FSI are taken into account at leading, OGE, order, which is natural in the IK model; iii) the resulting interaction has been obtained through a non-relativistic (NR) reduction of the relevant operator, according to the philosophy of constituent quark models,[25] leading to a potential V_{NR}. The Sivers function for a proton polarized along the y axis and for the quark of flavor $\mathcal{Q}$, $f_{1T}^{\perp\mathcal{Q}}(x, k_T)$, takes the form (cf. Fig. 1 for the labels of the momenta and helicities):

$$
\begin{aligned}
& f_{1T}^{\perp\mathcal{Q}}(x, k_T) \\
& = \Im\left\{ -ig^2 \frac{M^2}{k_x} \int d\vec{k}_1 d\vec{k}_3 \frac{d^2\vec{q}_T}{(2\pi)^2} \delta(k_3^+ - xP^+) \delta(\vec{k}_{3T} + \vec{q}_T - \vec{k}_T) \mathcal{M}^{\mathcal{Q}} \right\} \quad (1)
\end{aligned}
$$

where g is the strong coupling constant, M the proton mass, and

$$
\begin{aligned}
\mathcal{M}^{u(d)} = & \sum_{m_1, m_1', m_3, m_3'} \Phi^\dagger_{sf, S_z=1} \left(\vec{k}_3, m_3; \vec{k}_1, m_1; \vec{P} - \vec{k}_3 - \vec{k}_1, m_n \right) \\
& \times \frac{1 \pm \tau_3(3)}{2} V_{NR}(\vec{k}_1, \vec{k}_3, \vec{q}) \\
& \times \Phi_{sf, S_z=-1} \left(\vec{k}_3 + \vec{q}, m_3'; \vec{k}_1 - \vec{q}, m_1'; \vec{P} - \vec{k}_3 - \vec{k}_1, m_n \right) . \quad (2)
\end{aligned}
$$

Using the spin-flavor wave function of the proton in momentum space, Φ_{sf}, corresponding to a given CQM, the Sivers function, Eq. (1), can be evaluated. From Eq. (2), one notices that the helicity conserving part of the global interaction does not contribute to the Sivers function. Besides, in an extreme NR limit, it turns out to be identically zero: in our scheme, it is precisely the interference of the small and large components in the four-spinors of the free quark states which leads to a non-vanishing Sivers function. This holds even from the component with $l = 0$ of the target wave function. While, in other approaches,[17] these interference terms arise due to the wave function, they are produced here by the interaction.

The above-described formalism is now applied to the IK model. The detailed procedure and the final expressions of the Sivers function in this model can be found in Ref.[20] To evaluate numerically Eq. (1), g (i.e. $\alpha_s(Q^2)$) has to be fixed. The prescription[21] is used to fix μ_0^2, according to the amount of momentum carried by the valence quarks in the model. Here, assuming that all the gluons and sea pairs in the proton are produced perturbatively

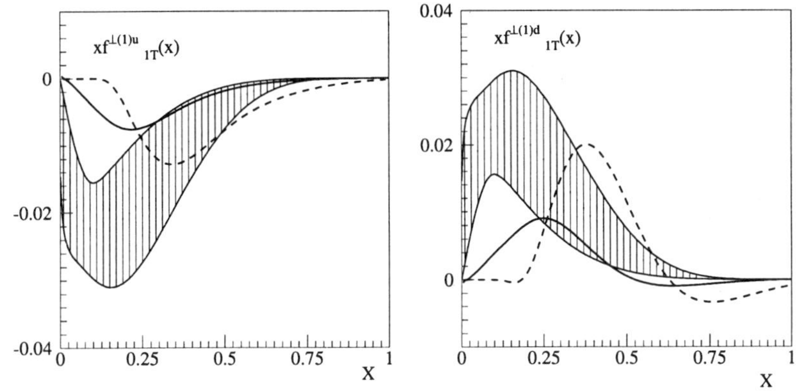

Fig. 2. Left (right): the quantity $f_{1T}^{\perp(1)u(d)}(x)$, Eq. (3). Dashed curve: IK at μ_0^2. Full curve: the evolved distribution at NLO. Patterned area: parameterization by[14] (see text).

according to NLO evolution equations, in order to have $\simeq 55\%$ of the momentum carried by the valence quarks at a scale of 0.34 GeV^2 one finds that $\mu_0^2 \simeq 0.1$ GeV^2 if $\Lambda_{QCD}^{NLO} \simeq 0.24$ GeV. This yields $\alpha_s(\mu_0^2)/(4\pi) \simeq 0.13$.[21] The results of the present approach for the first moments of the Sivers function, defined as

$$f_{1T}^{\perp(1)\mathcal{Q}}(x) = \int d^2\vec{k}_T \frac{k_T^2}{2M^2} f_{1T}^{\perp\mathcal{Q}}(x, k_T) \;, \tag{3}$$

are given by the dashed curves in Fig. 2. They are compared with a parameterization of the HERMES data, taken at $Q^2 = 2.5$ GeV^2 : The patterned area represents the $1-\sigma$ range of the best fit proposed in Ref.[14] The magnitude of the results is close to that of the data, although they have a different shape: the maximum (minimum) is predicted at larger values of x. Actually μ_0^2 is much lower, $Q^2 = 2.5$ GeV^2. A proper comparison requires QCD evolution of TMDPDs, what is, to large extent, unknown. We nevertheless perform a NLO evolution of the model results assuming, for $f_{1T}^{\perp(1)\mathcal{Q}}(x)$, the same anomalous dimensions of the unpolarized PDFs. From the final result (full curve in Fig. 2), one can see that the agreement with data improves dramatically and the trend is reasonably reproduced at least for $x \geq 0.2$. Although the performed evolution is not exact, the procedure highlights the necessity of evolving the model results to the experiment scale and it suggests that the present results could be consistent with data, still affected by large errors.

Properties of the Sivers function can be inferred from general principles.

The Burkardt Sum Rule (BSR)[26] states that, for a proton polarized in the positive y direction, $\sum_{Q=u,d} \langle k_x^Q \rangle = 0$ with

$$\langle k_x^Q \rangle = - \int_0^1 dx \int d\vec{k}_T \frac{k_x^2}{M} f_{1T}^{\perp Q}(x, k_T) \ , \tag{4}$$

and must be satisfied at any scale. Within our scheme, at the scale of the model, it is found $\langle k_x^u \rangle = 10.85$ MeV, $\langle k_x^d \rangle = -11.25$ MeV and, in order to have an estimate of the quality of the agreement of our results with the sum rule, we define the ratio $r = |\langle k_x^d \rangle + \langle k_x^u \rangle| / |\langle k_x^d \rangle - \langle k_x^u \rangle|$ obtaining $r \simeq 0.02$, so that we can say that our calculation fulfills the BSR to a precision of a few percent. One should notice that the agreement which is found is better than that found in other model calculations,[16,17] especially for what concerns the fulfillment of the Burkardt Sum Rule.

2. The Sivers function from neutron (^{3}He) targets

As explained in the previous section, the experimental scenario which arises from the analysis of SIDIS off transversely polarized proton and deuteron targets[11,12] is puzzling. The data show an unexpected flavor dependence in the azimuthal distribution of the produced pions. With the aim at extracting the neutron information to shed some light on the problem, a measurement of SIDIS off transversely polarized ^{3}He has been addressed,[27] and two experiments, aimed at measuring the azimuthal asymmetries in the production of leading $\pi^\pm$ from transversely polarized ^{3}He, are forth-coming at JLab.[28] Here, a realistic analysis of SIDIS off transversely polarized ^{3}He[29] is described. The expressions of the Collins and Sivers contributions to the azimuthal Single Spin Asymmetry (SSA) for the production of leading pions have been derived, in impulse approximation (IA), including the initial transverse momentum of the struck quark. The final equations are involved and they are not reported here. They can be found in.[29] The same quantities have been then evaluated in the kinematics of the JLab experiments. Wave functions[30] obtained within the AV18 interaction[31] have been used for a realistic description of the nuclear dynamics, using overlap integrals evaluated in Ref.,[32] and the nucleon structure has been described by parameterizations of data or model calculations.[13,33] The crucial issue of extracting the neutron information from ^{3}He data will be now discussed. As a matter of facts, a model independent procedure, based on the realistic evaluation of the proton and neutron effective polarizations in ^{3}He,[34] called respectively p_p and p_n in the following, is widely used in DIS to take into account effectively the momentum and energy distributions of the bound nucleons in

^{3}He. It is found that the same extraction technique can be applied also in the kinematics of the proposed experiments, although fragmentation functions, not only parton distributions, are involved, as it can be seen in Figs. 1 and 2. In these figures, the free neutron asymmetry used as a model in the calculation, given by a full line, is compared with two other quantities. One is:

$$\bar{A}_n^i \simeq \frac{1}{d_n} A_3^{exp,i} \,, \tag{5}$$

where i stands for "Collins" or "Sivers", $A_3^{exp,i}$ is the result of the full calculation, simulating data, and d_n is the neutron dilution factor. The latter quantity is defined as follows, for a neutron n (proton p) in ^{3}He:

$$d_{n(p)}(x,z) = \frac{\sum_q e_q^2 f^{q,n(p)}(x)\, D^{q,h}(z)}{\sum_{N=p,n} \sum_q e_q^2 f^{q,N}(x) D^{q,h}(z)} \tag{6}$$

and, depending on the standard parton distributions, $f^{q,N}(x)$, and fragmentation functions, $D^{q,h}(z)$, is experimentally known (see[29] for details). $\bar{A}_n^i$ is given by the dotted curve in the figures. The third curve, the dashed one, is given by

$$A_n^i \simeq \frac{1}{p_n d_n} \left(A_3^{exp,i} - 2 p_p d_p A_p^{exp,i} \right) \,, \tag{7}$$

i.e. ^{3}He is treated as a nucleus where the effects of its spin structure, of Fermi motion and binding, can be taken care of by parameterizing p_p and p_n. One should realize that Eq. (5) is the relation which should hold between the ^{3}He and the neutron SSAs if there were no nuclear effects, i.e. the ^{3}He nucleus were a system of free nucleons in a pure S wave. In fact, Eq. (5) can be obtained from Eq. (7) by imposing $p_n = 1$ and $p_p = 0$. It is clear from the figures that the difference between the full and dotted curves, showing the amount of nuclear effects, is sizable, being around 10 - 15 % for any experimentally relevant x and z, while the difference between the dashed and full curves reduces drastically to a few percent, showing that the extraction scheme Eq. (7) takes safely into account the spin structure of ^{3}He, Fermi motion and binding effects. This important result is due to the kinematics of the JLab experiments, which helps in two ways. First of all, to favor pions from current fragmentation, z has been chosen in the range $0.45 \le z \le 0.6$, which means that only high-energy pions are observed. Secondly, the pions are detected in a narrow cone around the direction of the momentum transfer. As it is explained in,[29] this makes nuclear effects in the fragmentation functions rather small. The leading nuclear effects

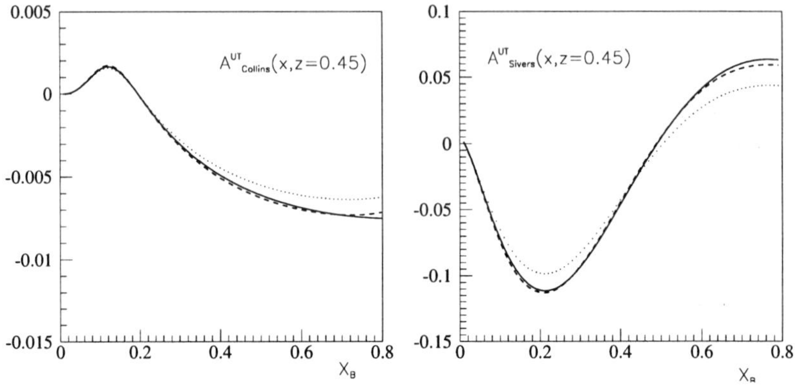

Fig. 3. Left (right) The model neutron Collins (Sivers) asymmetry for π^- production (full) in JLab kinematics, and the one extracted from the full calculation taking into account the p_p (dashed), or neglecting it (dotted). The results are shown for z=0.45 and $Q^2 = 2.2$ GeV^2, typical values in the kinematics of the JLab experiments.

are then the ones affecting the parton distributions, already found in DIS, and can be taken into account in the usual way, i.e., using Eq. (7) for the extraction of the neutron information. In the figures, one should not take the shape and size of the asymmetries seriously, being the obtained quantities strongly dependent on the models chosen for the unknown distributions.[33] One should instead consider the difference between the curves, a model independent feature which is the most relevant outcome of the present investigation. Eq. (7) is therefore a valuable tool for the experiments.[28] The evaluation of final state interactions effects and the inclusion of more realistic models of the nucleon structure are in progress.

Acknowledgments

This work is supported in part by the INFN-CICYT agreement, by the Generalitat Valenciana under the contract AINV06/118; by the Sixth Framework Program of the European Commission under the Contract No. 506078 (I3 Hadron Physics); by the MEC (Spain) (FPA 2007-65748-C02-0, AP2005-5331 and PR2007-0048).

References

1. V. Barone, A. Drago and P. Ratcliffe, Phys. Rept. **359** 1 (2002).
2. J. C. Collins, Nucl. Phys. B **396**, 161 (1993).

3. P. J. Mulders and R. D. Tangerman, Nucl. Phys. B **461** (1996) 197.
4. A. M. Kotzinian and P. J. Mulders, Phys. Lett. B **406** (1997) 373.
5. D. Boer and P. J. Mulders, Phys. Rev. D **57** (1998) 5780; A. Bacchetta *et al*, JHEP **0702**, 093 (2007).
6. D. W. Sivers, Phys. Rev. D **41**, 83 (1990).
7. S. J. Brodsky *et al*, Phys. Lett. B **530**, 99 (2002).
8. S. J. Brodsky *et al.*,Phys. Rev. D **65**, 114025 (2002).
9. J. C. Collins, Phys. Lett. B **536**, 43 (2002); X. d. Ji and F. Yuan, Phys. Lett. B **543**, 66 (2002); A. V. Belitsky *et al.*, Nucl. Phys. B **656**, 165 (2003).
10. A. Drago, Phys. Rev. D **71**, 057501 (2005).
11. A. Airapetian *et al.*, Phys. Rev. Lett. **94**, 012002 (2005).
12. V. Y. Alexakhin *et al.*, Phys. Rev. Lett. **94**, 202002 (2005).
13. M. Anselmino *et al.*, Phys. Rev. D **72**, 094007 (2005).
14. J. C. Collins *et al.*, Phys. Rev. D **73**, 014021 (2006).
15. W. Vogelsang and F. Yuan, Phys. Rev. D **72**, 054028 (2005).
16. L.P. Gamberg *et al.*, Phys. Rev. D **67**, 071504(R) (2003); A. Bacchetta *et al.*, Phys. Lett. B **578**, 109 (2004); A. Bacchetta *et al.*, arXiv:0807.0323 [hep-ph].
17. F. Yuan, Phys. Lett. B **575**, 45 (2003); I. O. Cherednikov *et al.*, Phys. Lett. B **642**, 39 (2006).
18. Z. Lu and B. Q. Ma, Nucl. Phys. A **741**, 200 (2004).
19. A. Bianconi, arXiv:hep-ph/0702186.
20. A. Courtoy, F. Fratini, S. Scopetta, V. Vento, Phys. Rev. D (2008), in press; arXiv:0801.4347 [hep-ph].
21. M. Traini *et al.*, Nucl. Phys. A **614**, 472 (1997).
22. S. Scopetta and V. Vento, Phys. Lett. B **424**, 25 (1998).
23. S. Scopetta and V. Vento, Eur. Phys. J. A **16**, 527 (2003); S. Boffi, B. Pasquini and M. Traini, Nucl. Phys. B **649**, 243 (2003).
24. N. Isgur and G. Karl, Phys. Rev. D **18**, 4187 (1978).
25. A. De Rujula *et al.* Phys. Rev. D **12**, 147 (1975).
26. M. Burkardt, Phys. Rev. D **69** (2004) 091501.
27. S. J. Brodsky and S. Gardner, Phys. Lett. B **643** (2006) 22.
28. E-06-010 Proposal to JLab-PAC29, J.-P. Chen and J.-C. Peng Spokespersons; E-06-011 Proposal to JLab-PAC29, E. Cisbani and H. Gao Spokespersons.
29. S. Scopetta, Phys. Rev. D **75**, 054005 (2007).
30. A. Kievsky *et al.*, Nucl. Phys. A **577**, 511 (1994).
31. R. B. Wiringa *et al.*, Phys. Rev. C **51** (1995) 38.
32. A. Kievsky *et al.*, Phys. Rev. C **56**, 64 (1997); E. Pace *et al.*, Phys. Rev. C **64**, 055203 (2001).
33. D. Amrath *et al.*, Phys. Rev. D **71**, 114018 (2005).
34. C. Ciofi degli Atti *et al.*, Phys. Rev. C **48**, 968 (1993).

SIVERS, BOER-MULDERS AND TRANSVERSITY IN DRELL-YAN PROCESSES

M. Anselmino[1], M. Boglione[1], U. D'Alesio[2,3],
S. Melis[1], F. Murgia[3], A. Prokudin[1,4]

1 - Dipartimento di Fisica Teorica, Università di Torino and INFN, Sezione di Torino, Via P. Giuria 1, I-10125 Torino, Italy

2 - Dipartimento di Fisica, Università di Cagliari, Cittadella Universitaria di Monserrato, I-09042 Monserrato (CA), Italy

3 - INFN, Sezione di Cagliari, C.P. 170, I-09042 Monserrato (CA), Italy

4 - Di.S.T.A., Università del Piemonte Orientale "A. Avogadro", Alessandria, Italy

Polarized and unpolarized Drell-Yan processes are crucial to investigate transverse momentum dependent distribution functions. Some of them are directly accessible even when integrating on the leptonic degree of freedom, others require the study of the angular decay distribution of the virtual photon. Exploiting our recent fit of SIDIS data on the Sivers functions we present predictions for Sivers effect at COMPASS-DY.

1. Introduction and Formalism

The Drell-Yan (DY) process , $AB \to l^+l^-X$, is the most suitable process to study the spin structure of hadrons. In fact, contrary to other processes, spin-dependent Parton Distribution Functions (PDF's) are not coupled to any (unknown) fragmentation functions. Transverse Momentum Dependent distribution functions (TMD's), like the Sivers, the Boer-Mulders and the Transversity functions can be directly accessed. Moreover, similarly to the Semi-inclusive Deep Inelastic Scattering(SIDIS), factorization with inclusion of transverse momentum has been proven.[1]

The effect of the intrinsic motion has been observed for the first time in Drell-Yan processes. Taking into account the intrinsic partonic momenta the final detected dilepton is no more collinear to the colliding hadrons. Let us consider two colliding hadrons A and B moving along the Z axis in A-B

c.m. frame. Neglecting masses, their four-momenta are:

$$p_A = \frac{\sqrt{s}}{2}\Big(1,0,0,1\Big), \qquad p_B = \frac{\sqrt{s}}{2}\Big(1,0,0,-1\Big), \tag{1}$$

where $s \equiv (p_A + p_B)^2$. Denoting with x_a and x_b the light-cone fractions of partons a and b respectively, the partonic momenta can be written as:

$$p_a = x_a \frac{\sqrt{s}}{2}\left(1 + \frac{k_{\perp a}^2}{x_a^2 s}, \frac{2\,\boldsymbol{k}_{\perp a}}{x_a\sqrt{s}}, 1 - \frac{k_{\perp a}^2}{x_a^2 s}\right), \tag{2}$$

$$p_b = x_b \frac{\sqrt{s}}{2}\left(1 + \frac{k_{\perp b}^2}{x_b^2 s}, \frac{2\,\boldsymbol{k}_{\perp b}}{x_b\sqrt{s}}, -1 + \frac{k_{\perp b}^2}{x_b^2 s}\right). \tag{3}$$

At Born level ($q\bar{q}$ direct annihilation) the virtual photon γ^* has momentum q which is the sum of the two partonic momenta p_a and p_b:

$$q = (p_a + p_b) = (q_0, \boldsymbol{q}_T, q_L) \tag{4}$$

$$M^2 = (p_a + p_b)^2 = q^2 \tag{5}$$

M is the virtual photon mass while $\boldsymbol{q}_T = q_T(\cos\phi_\gamma, \sin\phi_\gamma)$ is the virtual photon (that is the dilepton) transverse momentum.

From Eqs. (2,3,4), we obtain $\boldsymbol{q}_T = \boldsymbol{k}_{\perp a} + \boldsymbol{k}_{\perp b}$. Therefore in a DY process the observed q_T-spectrum gives direct information about the intrinsic transverse motion of partons. This picture is valid only when $q_T^2 \ll M^2$, otherwise the transverse q_T can be considered as generated by pQCD correction (gluon emission, etc.).

Partonic transverse motion not only modifies the DY kinematics, but, as we know, plays a role also in the distribution sector: TMD's contribute to both polarized and unpolarized cross sections. The unpolarized differential cross-sections for the Drell-Yan process can be written as:

$$\begin{aligned} d\sigma^{AB\to l^+l^-X} &= f_{a/A}(x_a, k_{\perp a}) \otimes f_{b/B}(x_b, k_{\perp b}) \otimes \frac{1}{2}\left(|M_2|^2 + |M_3|^2\right) \\ &+ \Delta f^a_{s_y/A}(x_a, \boldsymbol{k}_{\perp a}) \otimes \Delta f^b_{s_y/B}(x_b, \boldsymbol{k}_{\perp b}) \otimes \Re[M_3\, M_2^*] \end{aligned} \tag{6}$$

where $M_2 \equiv M_{-,+;-,+}$, $M_3 \equiv M_{-,+;+,-}$, and $M_{\lambda_c\lambda_d;\lambda_a,\lambda_b}$ are the helicity amplitudes for the elementary scattering process $ab \to cd$. The unpolarized cross section, Eq. (6), is given by the sum of two contributions. The first one is present also in the collinear approximation and it is due to the convolution of two unpolarized PDF's. The second one vanishes in collinear approximation and is proportional to the product of two Boer-Mulders functions: $\Delta f^a_{s_y/A}(x_a, \boldsymbol{k}_{\perp a})$ and $\Delta f^b_{s_y/B}(x_b, \boldsymbol{k}_{\perp b})$. This second term, as we will see, can be responsible for the large $\cos(2\phi)$ asymmetry observed in unpolarized DY scattering.[2,3] If one of the hadron, e.g. A, is transversely polarized then:

$$d\sigma^{A^\uparrow B\to l^+l^-X} = d\sigma^{AB\to l^+l^-} + \frac{1}{2}\Delta\sigma^\uparrow\,, \tag{7}$$

$$\Delta\sigma^\uparrow \equiv d\sigma^\uparrow - d\sigma^\downarrow = d\sigma^{A^\uparrow B\to l^+l^-X} - d\sigma^{A^\downarrow B\to l^+l^-X}\,, \tag{8}$$

and $\Delta\sigma^\uparrow$ can be written as:

$$\begin{aligned}\Delta\sigma^\uparrow &= \Delta^N f_{a/A^\uparrow}(x_a,\boldsymbol{k}_{\perp a})\otimes f_{b/B}(x_b,k_{\perp b})\otimes\frac{1}{2}\left(|M_2|^2+|M_3|^2\right)\\ &+\Delta^- f^a_{s_y/\uparrow}(x_a,\boldsymbol{k}_{\perp a})\otimes\Delta f^b_{s_y/B}(x_b,\boldsymbol{k}_{\perp b})\otimes 2\,\Re[M_3\,M_2^*]\\ &-\Delta f^a_{s_x/\uparrow}(x_a,\boldsymbol{k}_{\perp a})\otimes\Delta f^b_{s_y/B}(x_b,\boldsymbol{k}_{\perp b})\otimes 2\,\Im[M_3\,M_2^*]\,.\end{aligned} \tag{9}$$

The first term of Eq. (9) is the so called Sivers effect. The second term is the product of a Boer-Mulders function for the hadron B multiplied by the function $\Delta^- f^a_{s_y/\uparrow}(x_a,\boldsymbol{k}_{\perp a})$. The third term term is similar to previous one but here we find a Boer-Mulders function times the function $\Delta f^a_{s_x/\uparrow}(x_a,\boldsymbol{k}_{\perp a})$. The functions $\Delta^- f^a_{s_y/\uparrow}(x_a,\boldsymbol{k}_{\perp a})$ and $\Delta f^a_{s_x/\uparrow}(x_a,\boldsymbol{k}_{\perp a})$ are related to the transversity functions $h_1(x,\boldsymbol{k}_\perp)$ and $h_{1T}^\perp(x,\boldsymbol{k}_\perp)$, see Appendix C of Ref. 4 for details. In particular:

$$\Delta^- f^a_{s_y/\uparrow}(x_a,\boldsymbol{k}_{\perp a}) = \left[h_1(x_a,k_{\perp a}) - \frac{1}{2}\left(\frac{k_{\perp a}}{m_p}\right)^2 h_{1T}^\perp\right]\cos\phi_a\,, \tag{10}$$

$$\Delta f^a_{s_x/\uparrow}(x_a,\boldsymbol{k}_{\perp a}) = \left[h_1(x_a,k_{\perp a}) + \frac{1}{2}\left(\frac{k_{\perp a}}{m_p}\right)^2 h_{1T}^\perp\right]\sin\phi_a, \tag{11}$$

where ϕ_a is the parton a azimuthal angle.

2. Invariant Cross Sections

Starting from Eqs. (6, 9) we can calculate the invariant cross sections for the unpolarized and single transversely polarized DY processes. We neglects all terms of order greater then q_T/M. At this order $x_a \simeq (q_0+q_L)/\sqrt{s}$ and $x_b \simeq (q_0-q_L)/\sqrt{s}$ and the transverse momentum components, $\boldsymbol{k}_{\perp a}$ and $\boldsymbol{k}_{\perp b}$, kinematically decouple from x_a and x_b. In order to perform integrations in the transverse components we need simple parametrizations of the transverse momentum distributions in our TMD's. We assume a factorized gaussian distribution for the unpolarized PDF. The width of the gaussian is a parameter to be determined from data fit. Similarly for the other unknown TMD's, we assume that they are proportional to the unpolarized

PDF times another gaussian distribution (with width M_f different for each TMD's) and, if kinematically requested, powers of $k_\perp/M_f$:

$$\text{unpolarized PDF} \quad f_{a/A}(x_a, k_{\perp a}) = f_{a/A}(x_a) \frac{e^{-\frac{k_{\perp a}^2}{\langle k_{\perp a}^2 \rangle}}}{\pi \langle k_{\perp a}^2 \rangle},$$

$$\text{Boer-Mulders} \quad \Delta f_{a^\uparrow/A}(x_a, k_{\perp a}) = \Delta f_{a^\uparrow/A}(x_a) \frac{\sqrt{2e}\, k_{\perp a}}{M_{BM}} \frac{e^{-\frac{k_{\perp a}^2}{\langle k_{\perp a}^2 \rangle_{BM}}}}{\pi \langle k_{\perp a}^2 \rangle},$$

$$\text{Sivers} \quad \Delta^N f_{a/A^\uparrow}(x_a, \boldsymbol{k}_{\perp a}) = \Delta^N f_{a/A^\uparrow}(x_a) \frac{\sqrt{2e}\, k_{xa}}{M_S} \frac{e^{-\frac{k_{\perp a}^2}{\langle k_{\perp a}^2 \rangle_S}}}{\pi \langle k_{\perp a}^2 \rangle},$$

$$\text{Transversity} \quad h_1(x_a, k_{\perp a}) = h_1(x_a) \frac{e^{-\frac{k_{\perp a}^2}{\langle k_{\perp a}^2 \rangle}}}{\pi \langle k_{\perp a}^2 \rangle},$$

$$\frac{k_{\perp a}^2}{M_{h_{1T}^\perp}^2} h_{1T}^\perp(x_a, k_{\perp a}) = h_{1T}^\perp(x_a) \frac{e\, k_{\perp a}^2}{M_{h_{1T}^\perp}^2} \frac{e^{-\frac{k_{\perp a}^2}{\langle k_\perp^2 \rangle_{h_{1T}^\perp}}}}{\pi \langle k_\perp^2 \rangle}, \tag{12}$$

where: $\frac{1}{\langle k_\perp^2 \rangle_f} = \frac{1}{\langle k_\perp^2 \rangle} + \frac{1}{M_f^2}$ with $f = S,\ BM,\ h_{1T}^\perp$. Factorized parametrizations allow to perform an analytical integration at order $\mathcal{O}(q_T/M)$. From Eqs. (6,9,12) we obtain:

$$\frac{d\sigma^{AB \to l^+ l^-}}{d^4 q} = 2\, f_{a/A}(x_a)\, f_{b/B}(x_b) \frac{1}{\pi(\langle k_{\perp a}^2 \rangle + \langle k_{\perp b}^2 \rangle)} \frac{4\pi\alpha^2 e_q^2}{9M^2 s} e^{-\frac{q_T^2}{(\langle k_{\perp a}^2 \rangle + \langle k_{\perp b}^2 \rangle)}} \tag{13}$$

$$\frac{d\Delta\sigma^\uparrow}{d^4 q} = 2\sqrt{2e}\, \Delta^N f_{a/A^\uparrow}(x_a) f_{b/B}(x_b) \frac{q_T \sin(\phi_{S_A} - \phi_\gamma)}{M_{Siv}} \times \frac{4\langle k_\perp^2 \rangle_{Siv}^2\, \alpha^2 e_q^2}{9s\langle k_\perp^2 \rangle (\langle k_\perp^2 \rangle + \langle k_\perp^2 \rangle_{Siv})^2 M^2} e^{-\frac{q_T^2}{\langle k_\perp^2 \rangle + \langle k_\perp^2 \rangle_{Siv}}}. \tag{14}$$

As we can see, in Eq. (6) there are two contributions to the unpolarized cross section. However, after the integration over the leptonic momenta, the contribution from the Boer-Mulders function vanishes. Therefore in Eq. (14) only the contribution from the unpolarized PDF appears. From the simple ansatz of a gaussian distribution of the partonic transverse momenta it follows that the q_T-spectrum is also gaussian with $\langle q_T^2 \rangle = \langle k_{\perp a}^2 \rangle + \langle k_{\perp b}^2 \rangle$. Similarly, in the polarized case, Eq. (9), there are three contributions but only the Sivers effect is present in the invariant cross section, Eq. (14). The conclusion is that it is safe to look at the d^4q invariant cross section to study the Sivers effect; the study of the other contributions in both polarized and unpolarized cases, required the knowledge of the leptonic angular distribution.[5] Note that the Sivers effect, as expected, vanishes as $q_T \to 0$ and depends on $\sin(\phi_{S_A} - \phi_\gamma)$ where ϕ_{S_A} is the spin azimuthal angle in hadron c.m. frame.

3. Angular Distribution

The analysis of the properties of the hadronic tensor[6,7] for the unpolarized Drell-Yan process leads to the following structure:

$$\frac{d\sigma}{d\Omega' d^4q} = \frac{\alpha^2}{2(2\pi)^4 M^2\, s}\Big[W_T(1+\cos^2\theta') + W_L(1-\cos^2\theta') + W_\Delta \sin 2\theta' \cos\phi' + W_{\Delta\Delta}\sin^2\theta' \cos 2\phi'\Big]\,, \tag{15}$$

where primes denote that angles are in the dilepton rest frame. Starting from Eq. (6) we can derive the angular distribution of leptons in the dilepton rest frame. The dilepton rest frame is not uniquely defined. Our choice is to consider the so called Collins-Soper (CS) frame[7] (a useful invariant definition of this frame can be found in Ref. 8). In the CS frame we find:

$$\begin{aligned}\frac{d\sigma}{d^4q d\Omega_{cs}} = \frac{e_q^2\alpha^2}{12M^2\, s}\Bigg[&2\, f_{a/A}(x_a) f_{b/B}(x_b)\frac{e^{-\frac{q_T^2}{\langle k_{\perp a}^2\rangle + \langle k_{\perp b}^2\rangle}}}{\pi\langle k_{\perp a}^2\rangle + \langle k_{\perp b}^2\rangle}(1+\cos^2\theta_{cs})\\ &+\Delta f_{a^\uparrow/A}(x_a)\Delta f_{b^\uparrow/B}(x_b)\frac{q_T^2}{M_{BM}^a M_{BM}^b}\\ &\times\frac{2e\langle k_{\perp a}^2\rangle_{BM}^2\langle k_{\perp b}^2\rangle_{BM}^2 e^{-\frac{q_T^2}{\langle k_{\perp a}^2\rangle_{BM}+\langle k_{\perp b}^2\rangle_{BM}}}}{\langle k_{\perp a}^2\rangle\langle k_{\perp b}^2\rangle\left(\langle k_{\perp a}^2\rangle_{BM}+\langle k_{\perp b}^2\rangle_{BM}\right)^3}\\ &\cos(2\phi_{cs})\sin^2(\theta_{cs})\Bigg]\,.\end{aligned} \tag{16}$$

The Boer-Mulders effect is associated to the $\cos 2\phi$ asymmetry i.e. the $W_{\Delta\Delta}$ structure function in Eq. (15). After integration on Ω_{cs} this contribution vanishes. Similar expression can be found for the polarized case where the Sivers function plays the same role of the unpolarized PDF while the other two contributions play the role of the Boer-Mulders effect in the unpolarized case.

4. The Sivers Effect

At present, no set of data on the Sivers effect in DY processes is available. However many collaborations are planning to perform polarized DY experiments: namely the COMPASS collaboration at CERN, the PAX collaboration at GSI, PHENIX and STAR collaborations at RHIC and collaborations at J-PARC. Data on the transverse single spin asymmetry $A_{UT}^{\sin(\phi_h-\phi_S)}$ for polarized SIDIS processes, $\ell\, p\,(\boldsymbol{S}) \to \ell'\, h\, X$, collected by the HERMES[9] and COMPASS[10] collaborations allowed to perform a rather well constrained extraction of the Sivers function for u and d quarks, assuming a negligibly small Sivers sea.[11,12] Recently, much higher statistics data on the $A_{UT}^{\sin(\phi_h-\phi_S)}$ azimuthal asymmetries for SIDIS have become available.[13–15] The analysis of these new experimental sets of data leads[16] to a better understanding of the u and d Sivers functions at low-intermediate x and, a first insight into the sea contributions to the Sivers functions. In Ref. 16 the Sivers functions are parametrized as in Eqs. (12):

$$\begin{aligned}\Delta^N f_{a/A^\uparrow}(x_a, k_{\perp a}) &= \Delta^N f_{a/A^\uparrow}(x_a)\,\frac{\sqrt{2e}\,k_{\perp a}}{M_S}\,\frac{e^{-\frac{k^2_{\perp a}}{\langle k^2_{\perp a}\rangle_S}}}{\pi\langle k^2_{\perp a}\rangle} \\ &= 2\,\mathcal{N}_q(x)\,h(k_\perp)f_{q/p}(x)\,\frac{e^{-\frac{k^2_{\perp a}}{\langle k^2_{\perp a}\rangle}}}{\pi\langle k^2_{\perp a}\rangle}\end{aligned} \tag{17}$$

with

$$\mathcal{N}_q(x) = N_q\, x^{\alpha_q}(1-x)^{\beta_q}\,\frac{(\alpha_q+\beta_q)^{(\alpha_q+\beta_q)}}{\alpha_q^{\alpha_q}\,\beta_q^{\beta_q}}\,, \qquad h(k_\perp) = \sqrt{2e}\,\frac{k_\perp}{M_1}\,e^{-k_\perp^2/M_1^2}\,, \tag{18}$$

where $N_q \in [-1,1]$, α_q, β_q and $M_1 \equiv M_S$ (GeV/c) are free parameters to be determined by fitting the experimental data. Notice that $h(k_\perp) \le 1$ for any $k_\perp$ and $|\mathcal{N}_q(x)| \le 1$ for any x, therefore positivity bound for the Sivers function is automatically fulfilled. The unpolarized PDF $f_{q/p}(x, k_\perp)$ is parametrized as in Eq. (12) with $\langle k_\perp^2\rangle = 0.25$ (GeV/c)^2 fixed analyzing the Cahn effect in unpolarized SIDIS.[11] For the purposes of our fit, we used the unpolarized parton distribution functions $f_q(x, Q^2)$ as given in Ref. 17 (GRV98LO) and the fragmentation functions $D_q^h(z, Q^2)$ as given in Ref. 18 (DSS). We chose 11 free parameters for the Sivers functions ("broken sea" ansatz), summarized in Table 1. A peculiar property of the so called T-odd functions, like the Sivers function, is that they are process dependent. In particular it is predicted that the Sivers functions measured in SIDIS and DY processes have opposite sign.[19,20] An experimental verification of such prediction would be crucial to test factorization and non-universality of T-odd functions. Assuming that $\Delta^N f_{q/p^\uparrow}(x, k_\perp)_{DY} = -\Delta^N f_{q/p^\uparrow}(x, k_\perp)_{SIDIS}$, we use the Sivers functions determined through our fit to give predictions for the Sivers effect in DY processes. Here we present a prediction for the

Table 1. Best values of the free parameters for the "broken sea" ansatz.

$N_u = 0.35^{+0.08}_{-0.08}$	$N_d = -0.90^{+0.43}_{-0.10}$	$N_s = -0.24^{+0.62}_{-0.50}$
$N_{\bar{u}} = 0.04^{+0.22}_{-0.24}$	$N_{\bar{d}} = -0.40^{+0.33}_{-0.44}$	$N_{\bar{s}} = \ 1^{+0}_{-0.0001}$
$\alpha_u = 0.73^{+0.72}_{-0.58}$	$\alpha_d = 1.08^{+0.82}_{-0.65}$	$\alpha_{sea} = 0.79^{+0.56}_{-0.47}$
$\beta \ = 3.46^{+4.87}_{-2.90}$	$M_1^2 = 0.34^{+0.30}_{-0.16}$ (GeV/c)^2	

planned DY run at COMPASS ($\pi^{\pm}p \to \mu^+\mu^-$). For the unpolarized pion PDF we use the MRSS92 set.[21] The asymmetry $A_N = d\Delta\sigma^{\uparrow}/(2\,d\sigma)$ depends on $\sin(\phi_S - \phi_\gamma)$, Eq. (14), where ϕ_γ is the azimuthal angle of the virtual photon in hadron c.m. frame. Integration on ϕ_γ would lead to a vanishing asymmetry. In order to exploit data from all ranges of ϕ_γ we can take into account the change of sign in the different quadrants as in definition of $\langle A_N\rangle$, Eq. (19), or, similarly to what is done in SIDIS, construct a weighted asymmetry $A_N^{\sin(\phi_S-\phi_\gamma)}$, Eq. (20).

$$\langle A_N\rangle = \frac{4\int_0^{\pi/2}(d\sigma^{\uparrow} - d\sigma^{\downarrow})d\phi_\gamma}{\int_0^{2\pi}(d\sigma^{\uparrow} + d\sigma^{\downarrow})d\phi_\gamma}\,, \tag{19}$$

$$A_N^{\sin(\phi_S-\phi_\gamma)} = \frac{\int_0^{2\pi}(d\sigma^{\uparrow} - d\sigma^{\downarrow})\sin(\phi_S-\phi_\gamma)d\phi_\gamma}{\frac{1}{2}\int_0^{2\pi}(d\sigma^{\uparrow} + d\sigma^{\downarrow})d\phi_\gamma}\,, \tag{20}$$

$$\langle A_N\rangle = \frac{2}{\pi}A_N^{\sin(\phi_S-\phi_\gamma)}\,. \tag{21}$$

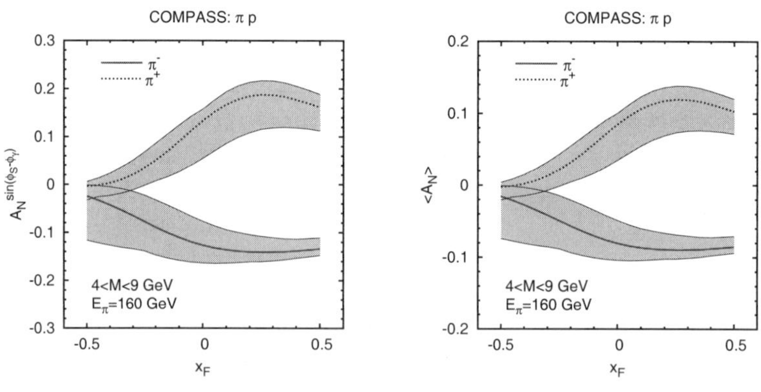

Fig. 1. The single spin asymmetries $A_N^{\sin(\phi_S-\phi_\gamma)}$ (left panel) and $\langle A_N\rangle$ (right panel) for the Drell-Yan process $\pi p^{\uparrow} \to \mu^+\mu- + X$ at COMPASS, as function of $x_F = x_a - x_b$, averaged over the invariant mass range $4 < M < 9$ and transverse momentum $0 < q_T < 1$ GeV/c, for a pion beam energy of 160 GeV/c.

From the definition, (20), it follows that $A_N^{\sin(\phi_S-\phi_\gamma)} \in [-1,1]$. We can see, in the left panel of Fig. 1, that in the positive x_F region the predicted asymmetry is sizable, 10-20%. Note that in this region $x_b < 0.4$, which is approximately the same region covered by SIDIS data.

References

1. J. C. Collins and A. Metz, *Phys. Rev. Lett.* **93**, p. 252001 (2004).
2. J. S. Conway *et al.*, *Phys. Rev.* **D39**, 92 (1989).
3. S. Falciano *et al.*, *Z. Phys.* **C31**, p. 513 (1986).
4. M. Anselmino *et al.*, *Phys. Rev.* **D73**, p. 014020 (2006).
5. D. Boer, *Phys. Rev.* **D60**, p. 014012 (1999).
6. C. S. Lam and W.-K. Tung, *Phys. Rev.* **D18**, p. 2447 (1978).
7. J. C. Collins and D. E. Soper, *Phys. Rev.* **D16**, p. 2219 (1977).
8. D. Boer and W. Vogelsang, *Phys. Rev.* **D74**, p. 014004 (2006).
9. A. Airapetian *et al.*, *Phys. Rev. Lett.* **94**, p. 012002 (2005).
10. V. Y. Alexakhin *et al.*, *Phys. Rev. Lett.* **94**, p. 202002 (2005).
11. M. Anselmino *et al.*, *Phys. Rev.* **D71**, p. 074006 (2005).
12. M. Anselmino *et al.*, *Phys. Rev.* **D72**, p. 094007 (2005).
13. M. Diefenthaler (2007), arXiv:0706.2242 [hep-ex].
14. A. Martin, *Czech. J. Phys.* **56**, F33 (2006).
15. M. Alekseev *et al.* (2008), arXiv:0802.2160 [hep-ex].
16. M. Anselmino *et al.* (2008), arXiv:0805.2677 [hep-ph].
17. M. Gluck, E. Reya and A. Vogt, *Eur. Phys. J.* **C5**, 461 (1998).
18. D. de Florian, R. Sassot and M. Stratmann, *Phys. Rev.* **D75**, p. 114010 (2007).
19. S. J. Brodsky, D. S. Hwang and I. Schmidt, *Nucl. Phys.* **B642**, 344 (2002).
20. J. C. Collins, *Phys. Lett.* **B536**, 43 (2002).
21. P. J. Sutton, A. D. Martin, R. G. Roberts and W. J. Stirling, *Phys. Rev.* **D45**, 2349 (1992).

TMDS AND DRELL-YAN EXPERIMENTS AT FERMILAB AND J-PARC

Jen-Chieh Peng

Department of Physics, University of Illinois at Urbana-Champaign
Urbana, IL 61801, U.S.A.

The roles of the Drell-Yan experiments in studying the Transverse-Momentum-Dependent (TMD) parton distributions are discussed. Recent results from the Fermilab E866 experiment on the angular distributions of Drell-Yan dimuons in $p+p$ and $p+d$ at 800 GeV/c are presented. These data are compared with the pion-induced Drell-Yan data, and with models which attribute the $\cos 2\phi$ azimuthal distribution to the presence of the transverse-momentum-dependent Boer-Mulders structure function $h_1^\perp$. Constraints on the magnitude of the sea-quark $h_1^\perp$ structure functions are obtained. Future prospects for studying the TMDs with Drell-Yan experiments at Fermilab and J-PARC are also discussed.

Keywords: Drell-Yan, azimuthal asymmetry, Boer-Mulders functions.

1. Introduction

While our current knowledge on the partonic substructure of nucleons is obtained largely from deep-inelastic scattering (DIS) experiments, the Drell-Yan process,[1] in which a charged lepton pair is produced in a hadron-hadron interaction via the electromagnetic $q\bar{q} \to l^+l^-$ process, also provides unique information on parton distributions. In particular, the Drell-Yan process has been used to determine the antiquark contents of nucleons and nuclei,[2] as well as the quark distributions of pions, kaons, and antiprotons.[3] Such information is difficult, if not impossible, to obtain from DIS experiments. As the Drell-Yan process can be well described by next-to-leading order QCD calculations,[4] a firm theoretical framework exists for utilizing the Drell-Yan process to extract the parton distributions.

The study of the transverse momentum dependent (TMD) parton distributions of the nucleon has received much attention in recent years as it provides new perspectives on the hadron structure and QCD.[5] These novel TMDs can be extracted from semi-inclusive deep-inelastic (SIDIS) scattering experiments. Recent measurements of the SIDIS by the HERMES[6] and

COMPASS[7] collaborations have shown clear evidence for the existence of the T-odd Sivers functions. These data also allow the first determination[8] of the magnitude and flavor structure of the Sivers functions and the nucleon transversity distributions.

The TMD and transversity parton distributions can also be probed in Drell-Yan experiments. As pointed out[9] long time ago, the double transverse spin asymmetry in polarized Drell-Yan, A_{TT}, is proportional to the product of transversity distributions, $h_1(x_q)h_1(x_{\bar{q}})$. The single transverse spin asymmetry, A_N, is sensitive to the Sivers function,[10] $f_{1T}^{\perp}(x)$ of the polarized proton (beam or target). Even unpolarized Drell-Yan experiments can be used to probe the TMD distribution function, since the cos2ϕ azimuthal angular dependence is proportional to the product of two Boer-Mulders functions,[11] $h_1^{\perp}(x_1)\bar{h}_1^{\perp}(x_2)$. A unique feature of the Drell-Yan process is that, unlike the SIDIS, no fragmentation functions are involved. Therefore, the Drell-Yan process provides an entirely independent technique for measuring the TMD functions. Furthermore, the proton-induced Drell-Yan process is sensitive to the sea-quark TMDs and can lead to flavor separation of TMDs when combined with the SIDIS data. Finally, the intriguing prediction[12] that the T-odd TMDs extracted from DIS will have a sign-change for the Drell-Yan process remains to be tested experimentally.

In this article, we first present the measurement of Drell-Yan azimuthal angular distributions in the Fermilab experiment E866. Implication of the data on the TMD Boer-Mulders function is discussed. Prospects for future Drell-Yan experiments for studying TMD and transversity distributions at Fermilab and J-PARC will also be discussed.

2. Fermilab dimuon experiments and TMDs

2.1. *Overview*

During the last two decades, a series of fixed-target dimuon production experiments (E772, E789, E866) have been carried out using 800 GeV/c proton beam at Fermilab. At 800 GeV/c, the dimuon data contain Drell-Yan continuum up to dimuon mass of $\sim$ 15 GeV as well as quarkonium (J/Ψ, Ψ', and Υ resonances) productions. The Drell-Yan process and quarkonium productions often provide complementary information, since Drell-Yan is an electromagnetic process via quark-antiquark annihilation while the quarkonium production is a strong interaction process dominated by gluon-gluon fusion at this beam energy.

The Fermilab dimuon experiments cover a broad range of physics topics. The Drell-Yan data have provided informations on the antiquark distributions in the nucleons[13–16] and nuclei.[17,18] These results showed the surprising results that the antiquark distributions in the nuclei are not enhanced,[17,18] contrary to the predictions of models which explain the EMC effect in term of nuclear enhancement of exchanged mesons. Moreover, the Drell-Yan cross section ratios $(p+d)/(p+p)$ clearly establish the flavor asymmetry of the $\bar{d}$ and $\bar{u}$ distributions in the proton, and they map out the x-dependence of this asymmetry.[14–16] A recent analysis of the Υ production cross section ratios for $(p+d)/(p+p)$ has shown that the gluon distributions in proton and neutron are very similar.[19] Pronounced nuclear dependences of quarkonium productions as a function of x_F and p_T have been observed for J/Ψ, Ψ', and Υ resonances .[20–23] The nuclear Drell-Yan cross sections also exhibit x_F as well as p_T dependences,[2,18,24,25] which are weaker than the quarkonium nuclear dependences but can provide information on the energy loss of quarks traversing the nucleus.[26]

Several review articles covering some of these results are available.[2,27,28] In the following, we will focus on the recent results from experiment E866 on the measurement of angular distributions of the Drell-Yan cross sections.

2.2. *Angular distributions of Drell-Yan and the Boer-Mulders function*

Despite the success of perturbative QCD in describing the Drell-Yan cross sections, it remains a challenge to understand the angular distributions of the Drell-Yan process. Assuming dominance of the single-photon process, a general expression for the Drell-Yan angular distribution is[29]

$$\frac{d\sigma}{d\Omega} \propto 1 + \lambda \cos^2\theta + \mu \sin 2\theta \cos\phi + \frac{\nu}{2} \sin^2\theta \cos 2\phi, \tag{1}$$

where θ and ϕ denote the polar and azimuthal angle, respectively, of the l^+ in the dilepton rest frame. In the "naive" Drell-Yan model, where the transverse momentum of the quark is ignored and no gluon emission is considered, $\lambda = 1$ and $\mu = \nu = 0$ are obtained. QCD effects[30] and non-zero intrinsic transverse momentum of the quarks[31] can both lead to $\lambda \neq 1$ and $\mu, \nu \neq 0$. However, λ and ν should still satisfy the relation $1 - \lambda = 2\nu$.[29] This so-called Lam-Tung relation, obtained as a consequence of the spin-1/2 nature of the quarks, is analogous to the Callan-Gross relation[32] in deep-inelastic scattering.

The first measurement of the Drell-Yan angular distribution was performed by the NA10 Collaboration for $\pi^- + W$ with the highest statistics at 194 GeV/c.[33,34] The $\cos 2\phi$ angular dependences showed a sizable ν, increasing with dimuon transverse momentum (p_T) and reaching a value of ≈ 0.3 at $p_T = 2.5$ GeV/c. The Fermilab E615 Collaboration subsequently performed a measurement of $\pi^- + W$ Drell-Yan production at 252 GeV/c with broad coverage in the decay angle θ.[35] The E615 data showed that the Lam-Tung relation, $2\nu = 1 - \lambda$, is clearly violated.

The NA10 and E615 results on the Drell-Yan angular distributions strongly suggest that new effects beyond conventional perturbative QCD are present. Brandenburg, Nachtmann and Mirke suggested that a factorization-breaking QCD vacuum may lead to a correlation between the transverse spin of the antiquark in the pion and that of the quark in the nucleon.[36] This would result in a non-zero $\cos 2\phi$ angular dependence consistent with the data. Several authors have also considered higher-twist effects from quark-antiquark binding in pions.[37,38] However, the model is strictly applicable only in the $x_\pi \to 1$ region, while the NA10 and E615 data exhibit nonperturbative effects over a much broader kinematic region.

More recently, Boer pointed out[39] that the $\cos 2\phi$ angular dependences observed in NA10 and E615 could be due to the k_T-dependent parton distribution function $h_1^\perp$. This so-called Boer-Mulders function[11] is an example of a novel type of k_T-dependent parton distribution function, and it characterizes the correlation of a quark's transverse spin and its transverse momentum, k_T, in an unpolarized nucleon. The Boer-Mulders function is the chiral-odd analog of the Sivers function and owes its existence to the presence of initial/final state interactions.[40] While the Sivers function is beginning to be quantitatively determined from the SIDIS experiments, very little is known about the Boer-Mulders function so far.

Several model calculations have been carried out for the Boer-Mulders functions. In the quark-diquark model, it was shown that the Boer-Mulders functions are identical to the Sivers functions when only the scalar diquark configuration is considered.[40,41] More recently, calculations taking into account both the scalar and the axial-vector diquark configurations found significant differences between the Sivers and Boer-Mulders functions.[42] In particular, the u and d quark Boer-Mulders functions are predicted to be both negative, while the Sivers function is negative for the u quark and positive for the d quark. Other calculations using the MIT bag model,[43] the relativistic constituent quark model,[44] the large-N_c model,[45] and the lattice QCD[46] also predicted negative signs for the u and d Boer-Mulders functions. Burkardt recently pointed out[47] that the origin of the negative signs for the Boer-Mulders functions lies in the phase of the p-wave lower component relative to the s-wave upper component of the quark wave function for the solution to the free Dirac equation. The model predictions for the same signs of the u and d Boer-Mulders functions remain to be tested experimentally.

2.3. *Results from Fermilab E866*

To shed additional light on the origins of the NA10 and E615 Drell-Yan angular distributions, we recently analyzed $p+p$ and $p+d$ Drell-Yan angular distribution data at 800 GeV/c from Fermilab E866. There had been no report on the azimuthal angular distributions for proton-induced Drell-Yan – all measurements were for polar angular distributions.[2,48,49] Proton-induced Drell-Yan data provide a test of theoretical models. For example, the $\cos 2\phi$ dependence is expected to be much reduced in proton-induced Drell-Yan if the underlying mechanism involves the Boer-Mulders functions. This is due to the expectation that the Boer-Mulders functions are small for the sea-quarks.[50] However, if the QCD vacuum effect[36] is the origin of the $\cos 2\phi$ angular dependence, then the azimuthal behavior of proton-induced Drell-Yan should be similar to that of pion-induced Drell-Yan. Finally, the validity of the Lam-Tung relation has never been tested for proton-induced Drell-Yan, and the present study provides a first test.

The Fermilab E866 experiment was performed using the upgraded Meson-East magnetic pair spectrometer. An 800 GeV/c primary proton beam with up to 2×10^{12} protons per 20-second beam spill was incident upon 50.8 cm long target flask containing either liquid hydrogen, liquid deuterium or vacuum. The detector system consisted of four tracking stations and a momentum analyzing magnet. From the momenta of the μ^+ and μ^-, kinematic variables of the dimuons ($x_F, m_{\mu\mu}, p_T$) were readily reconstructed. The muon angles θ and ϕ in the Collins-Soper frame[51] were also calculated. To remove the quarkonium background, only events with $4.5 < m_{\mu\mu} < 9$ GeV/c^2 or $m_{\mu\mu} > 10.7$ GeV/c^2 were analyzed. A total of ~54,000 $p+p$ and ~118,000 p+d Drell-Yan events covering the decay angular range $-0.5 < \cos\theta < 0.5$ and $-\pi < \phi < \pi$ remain.

The measurement of the $\cos 2\phi$ dependence for the $p+d$ Drell-Yan cross sections has already been reported.[52] As shown in Fig. 1, significantly smaller (but non-zero) $\cos 2\phi$ azimuthal angular dependence was observed in the $p+d$ reaction. While pion-induced Drell-Yan process is dominated by annihilation between a valence antiquark in the pion and a valence quark in the nucleon, proton-induced Drell-Yan process involves a valence quark annihilating with a sea antiquark in the nucleon. Therefore, the $p+d$ result suggests[52] that the Boer-Mulders functions for sea antiquarks are significantly smaller than those for valence quarks.

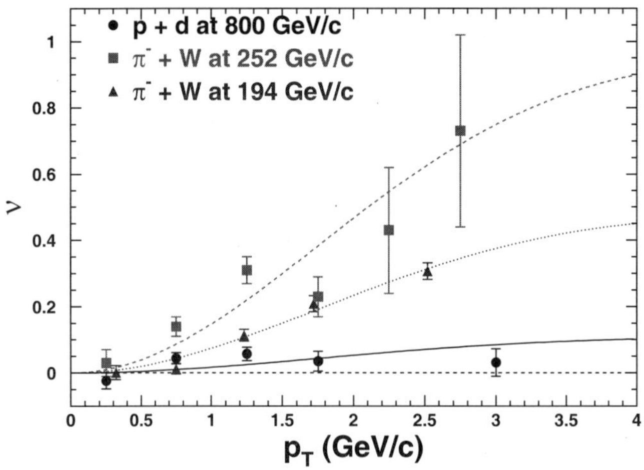

Fig. 1. Parameter ν vs. p_T in the Collins-Soper frame for three Drell-Yan measurements. Fits to the data using an empirical expression[39] are also shown.

A recent analysis[53] of the $p + d$ $\cos 2\phi$ data showed that the sea-quark Boer-Mulders functions are indeed smaller by a factor ~ 5 than the valence-quark Boer-Mulders functions. This analysis also indicated that the E866 $p+d$ data are consistent with the u and d Boer-Mulders functions having the same signs, as predicted by various models. However, the $p + d$ data alone can not provide an unambiguous determination of the flavor dependence of the Boer-Mulders functions. Moreover, it was recently pointed out[54,55] that QCD processes would lead to sizeable $\cos 2\phi$ effect which has not been taken into account in the extractions[39,53,56] of Boer-Mulders functions from the Drell-Yan data. The $p + p$ Drell-Yan data should provide further constraints on the flavor dependence of the Boer-Mulders functions.[56] It is also interesting to compare the $p+p$ and $p+d$ data with the prediction of QCD.

Figure 2 shows the angular distribution parameters λ, μ, and ν vs. p_T for the E866 $p + p$ and $p + d$ Drell-Yan data. Within statistics, the angular distributions of $p + p$ are consistent with those of $p + d$. Also shown in Fig. 2 is the quantity $2\nu - (1 - \lambda)$, which should vanish if the Lam-Tung relation is valid. Figure 2 shows that the Lam-Tung relation is indeed quite well satisfied within statistics. This is different from the observation of a significant violation of the Lam-Tung relation by the E615 collabortaion in the $\pi^- + W$ reaction at 252 GeV/c.[35]

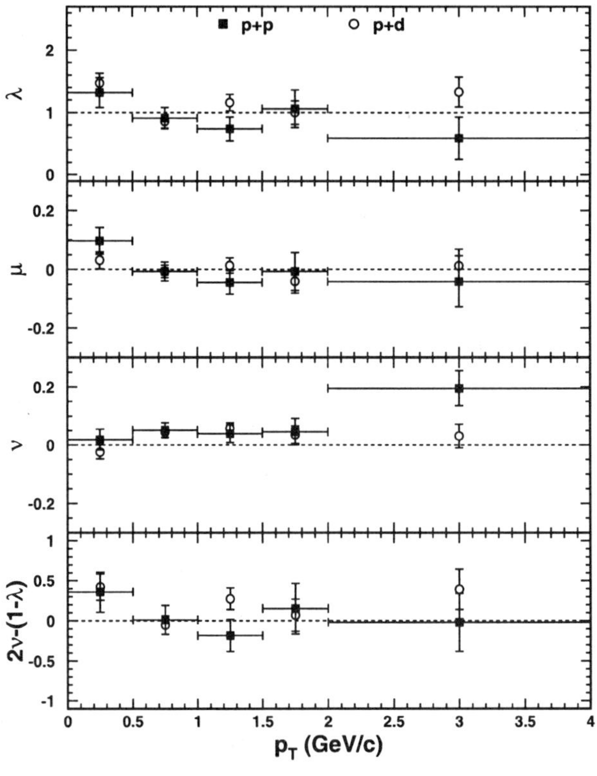

Fig. 2. Parameters λ, μ, ν and $2\nu - (1 - \lambda)$ vs. p_T in the Collins-Soper frame. Solid squares (open circles) are for E866 $p + p$ $(p + d)$ at 800 GeV/c.

Figure 3 shows the parameter ν vs. p_T for the $p + p$ and $p + d$ Drell-Yan data. The solid curves are calculations[53,56] for $p + p$ and $p + d$ using parametrizations based on a fit to the $p + d$ Drell-Yan data. The larger values of ν for $p + p$ compared to $p + d$ are in qualitative agreement with the prediction. However, the shape of the predicted p_T dependence is different from the data. This strongly suggests that there could be other mechanisms contributing to the $\cos 2\phi$ azimuthal angular dependence at large p_T region. In a recent paper,[55] the QCD contribution to the $\cos 2\phi$ azimuthal angular dependence is given as

$$\nu = \frac{Q_\perp^2/Q^2}{1+\frac{3}{2}Q_\perp^2/Q^2}, \tag{2}$$

where $Q_\perp$ is the dimuon transverse momentum. The predicted QCD contribution is shown as the dotted curve is Fig. 3. Although neither the prediction based on Boer-Mulders functions nor the QCD prediction could describe the entire range of p_T, it is evident taht the QCD contribution becomes important at high p_T and the Boer-Mulders functions contribute at lower p_T. An analysis combining both effects is required before a reliable extraction of the Boer-Mulders functions could be obtained.

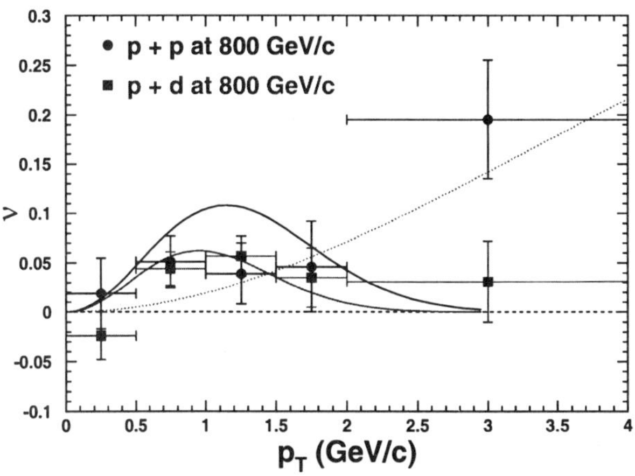

Fig. 3. Parameter ν vs. p_T for the $p+p$ and $p+d$ Drell-Yan data. The solid curves are calculations[53] for $p+p$ and $p+d$ using parametrizations based on a fit to the $p+d$ Drell-Yan data. The dotted curve is the contribition from QCD process (Eq. 2).

3. Future prospects at Frmilab and J-PARC

Future fixed-target dimuon experiments have been proposed at the 120 GeV Fermilab Main Injector (FMI) and the 50 GeV J-PARC facilities. The Fermilab proposal,[57] E906, has been approved and is expected to start data-taking around 2011. Two dimuon proposals (P04[58] and P24[59]) have also been submitted to the J-PARC for approval. The lower beam energies at FMI and J-PARC present opportunities for extending the Drell-Yan measurements to larger x ($x > 0.25$). For given values of x_1 and x_2, the Drell-Yan cross section is proportional to $1/s$, hence a gain of ~ 16 times in the Drell-Yan cross sections can be obtained at the J-PARC energy of 50 GeV. This would extend the measurement of Boer-Mulders functions to larger x. In addition, the nuclear dependence of the Boer-Mulders functions can also be measured.

The dimuon physics program at J-PARC is proposed to be carried out in several stages. Since 30 GeV proton beam will be available at the initial phase of J-PARC, the first measurements will focus on J/Ψ production at 30 GeV. This will be followed by measurements of Drell-Yan and quarkonium production at 50 GeV after the beam energy is upgraded to 50 GeV. Experiments using polarized target could already be performed with unpolarized beams. When polarized proton beam becomes available at J-PARC, a rich and unique program on spin physics could also be pursued at J-PARC using the dimuon spectrometer.

An important feature of J/Ψ production using 30 or 50 GeV proton beam is the dominance of the quark-antiquark annihilation subprocess. This is in striking contrast to J/Ψ production at 800 GeV (Fermilan E866) or at 120 GeV (Fermilab E906), where the gluon-gluon fusion is the dominant process. This suggests an exciting opportunity to use J/Ψ production at J-PARC as an alternative method to probe antiquark distribution.

With the possibility to accelerate polarized proton beams at J-PARC, the spin structure of the proton can also be investigated with the proposed dimuon experiments. In particular, polarzied Drell-Yan process with polarized beam and/or polarized target at J-PARC would allow a unique program on spin physics complementary to polarized DIS experiments and the RHIC-Spin program. Specific physics topics include the measurements of T-odd Boer-Mulders distribution function in unpolarized Drell-Yan, the extraction of T-odd Sivers distribution functions in singly transversely polarized Drell-Yan, the helicity distribution of antiqaurks in doubly longitudinally polarized Drell-Yan, and the transversity distribution in doubly transversely polarized Drell-Yan. It is worth noting that polarized Drell-Yan is one of the major physics program at the GSI Polarized Antiproton Experiment (PAX). The RHIC-Spin program will likely provide the first results on polarized Drell-Yan. However, the high luminosity and the broad kinematic coverage for the large-x region at J-PARC would allow some unique measurements to be performed in the J-PARC dimuon experiments.

Acknowledgments

I am very grateful to Lingyan Zhu, Paul Reimer, and my collaborators on the E772, E789, and E866 experiments at Fermilab. Illuminating comments from Matthias Burkardt, Feng Yuan, Bo-Qiang Ma, Werner Vogelsang, and Jianwei Qiu are greatly appreciated. I would also like to acknowledge the collaboration with Shin'ya Sawada and Yuji Goto on the J-PARC dimuon proposals.

References

1. S.D. Drell and T.M. Yan, Phys. Rev. Lett. **25**, 316 (1970); Ann. Phys. (NY) **66**, 578 (1971).
2. P.L. McGaughey, J.M. Moss, and J.C. Peng, Annu. Rev. Nucl. Part. Sci. **49**, 217 (1999).
3. I.R. Kenyon, Rep. Prog. Phys. **45**, 1261 (1982); K. Freudenreich, Int. J. Mod. Phys. **A5**, 3643 (1990).
4. W.J. Stirling and M.R. Whalley, J. Phys. **G19**, D1 (1993).
5. V. Barone, A. Drago, and P. G. Ratcliffe, Phys. Rep. **359**, 1 (2002).
6. HERMES Collaboration, A. Airapetian *et al.*, Phys. Rev. Lett. **94**, 012002 (2005); M. Diefenthaler, arXiv: 0706.2242.
7. COMPASS Collaboration, V. Yu. Alexakhin *et al.*, Phys. Rev. Lett. **94**, 202002 (2005); M. Alekseev *et al.*, arXiv: 0802.2160.
8. W. Vogelsang and F. Yuan, Phys. Rev. D **72**, 054028 (2005); M. Anselmino *et al.*, Phys. Rev. D **72**, 094007 (2005); M. Anselmino *et al.*, arXiv:0807.0166.
9. J.P. Ralston and D.E. Soper, Nucl. Phys. B **152**, 109 (1979).
10. D. Sivers, Phys. Rev. D **41**, 83 (1990).
11. D. Boer and P.J. Mulders, Phys. Rev. D **57**, 5780 (1998).
12. J.C. Collins, Phys. Lett. B **536**, 43 (2002).
13. E772 Collaboration, P.L. McGaughey *et al.*, Phys. Rev. Lett. **69**, 1726 (1992).
14. E866 Collaboration, E.A. Hawker *et al.*, Phys. Rev. Lett. **80**, 3715 (1998).
15. E866 Collaboration, J.C. Peng *et al.*, Phys. Rev. D **58**, 092004 (1998).
16. E866 Collaboration, R.S. Towell *et al.*, Phys. Rev. D **64**, 052002 (2001).
17. E772 Collaboration, D.M. Alde *et al.*, Phys. Rev. Lett. **64**, 2479 (1990).
18. E866 Collaboration, M.A. Vasiliev *et al.*, Phys. Rev. Lett. **83**, 2304 (1999).
19. E866 Collaboration, L.Y. Zhu *et al.*, Phys. Rev. Lett. **100**, 062301 (2008).
20. E772 Collaboration, D.M. Alde *et al.*, Phys. Rev. Lett. **66**, 133 (1991).
21. E772 Collaboration, D.M. Alde *et al.*, Phys. Rev. Lett. **66**, 2285 (1991).
22. E789 Collaboration, M.S. Kowitt *et al.*, Phys. Rev. Lett. **72**, 1318 (1994).
23. E789 Collaboration, M.J. Leitch *et al.*, Phys. Rev. D **52**, 4251 (1995).
24. E772 Collaboration, M.B. Johnson *et al.*, Phys. Rev. Lett. **86**, 4483 (2001).
25. M.B. Johnson *et al.*, Phys. Rev. C **65**, 025203 (2002).
26. G.T. Garvey and J.C. Peng, Phys. Rev. Lett. **90**, 092302 (2003).
27. G.T. Garvey and J.C. Peng, Prog. Part. Nucl. Phys. **47**, 203 (2001).
28. P.E. Reimer, J. Phys. G **34**, S107 (2007).
29. C.S. Lam and W.K. Tung, Phys. Rev. D **18**, 2447 (1978).

30. P. Chiappetta and M. LeBellac, Z. Phys. C **32**, 521 (1986).
31. J. Cleymans and M. Kuroda, Phys. Lett. B **105**, 68 (1981).
32. C.G. Callan and D. J. Gross, Phys. Rev. Lett. **22**, 156 (1969).
33. NA10 Collaboration, S. Falciano *et al.*, Z. Phys. C **31**, 513 (1986).
34. NA10 Collaboration, M. Guanziroli *et al.*, Z. Phys. C **37**, 545 (1988).
35. E615 Collaboration, J.S. Conway *et al.*, Phys. Rev. D **39**, 92 (1989); J.G. Heinrich *et al.*, Phys. Rev. D **44**, 1909 (1991).
36. A. Brandenburg, O. Nachtmann, and E. Mirkes, Z. Phys. C **60**, 697 (1993).
37. A. Brandenburg, S.J. Brodsky, V.V. Khoze, and D. Müller, Phys. Rev. Lett. **73**, 939 (1994).
38. K.J. Eskola, P. Hoyer, M. Väntinnen, and R. Vogt, Phys. Lett. B **333**, 526 (1994).
39. D. Boer, Phys. Rev. D **60**, 014012 (1999).
40. D. Boer, S.J. Brodsky, and D.S. Hwang, Phys. Rev. D **67**, 054003 (2003).
41. L.P. Gamberg, G.R. Goldstein, and K.A. Oganessyan, Phys. Rev. D **67**, 071504(R) (2003); A. Bacchetta, A. Schäfer, and J.-J. Yang, Phys. Lett. B **578**, 109 (2004).
42. L.P. Gamberg, G.R. Goldstein, and M. Schlegel, Phys. Rev. D **77**, 094016 (2008).
43. F. Yuan, Phys. Lett. B **575**, 45 (2003).
44. B. Pasquini, M. Pincetti, and S. Boffi, Phys. Rev. D **76**, 034020 (2007).
45. P.V. Pobylitsa, hep-ph/0301236.
46. M. Göckeler *et al.*, Phys. Rev. Lett. **98**, 222001 (2007).
47. M. Burkardt and B. Hannafious, Phys. Lett. B **658**, 130 (2008).
48. E866 Collaboration, T.H. Chang *et al.*, Phys. Rev. Lett. **91**, 211801 (2003).
49. E866 Collaboration, C.N. Brown *et al.*, Phys. Rev. Lett. **86**, 2529 (2001).
50. Z. Lu, B.-Q. Ma, and I. Schmidt, Phys. Rev. D **75**, 014026 (2007).
51. J.C. Collins and D.E. Soper, Phys. Rev. **D16**, 2219 (1977).
52. E866 Collaboration, L.Y. Zhu *et al.*, Phys. Rev. Lett. **99**, 082301 (2007).
53. B. Zhang, Z. Lu, B.-Q. Ma, and I. Schmidt, Phys. Rev. D **77**, 054011 (2008).
54. D. Boer and W. Vogelsang, Phys. Rev. D **74**, 014004 (2006).
55. E.L. Berger, J.-W. Qiu, and R.A. Rodriguez-Pedraza, Phys. Lett. B **656**, 74 (2007); Phys. Rev. D **76**, 074006 (2007).
56. B. Zhang, Z. Lu, B.-Q. Ma, and I. Schmidt, arXiv: 0807.0503.
57. http://www.phy.anl.gov/mep/drell-yan; D. Geesaman, P. Reimer, *et al.*, Fermilab Proposal P906 (1999).
58. http://j-parc.jp/NuclPart/pac_0606/pdf/p04-Peng.pdf; J. Peng, S. Sawada, *et al.*, J-PARC Proposal P04 (2006).
59. http://j-parc.jp/NuclPart/pac_0801/pdf/Goto.pdf; Y. Goto, H. Sato, *et al.*, J-PARC Proposal P24 (2007).

DOUBLE POLARISATION OBSERVABLES AT PAX

M. Nekipelov
(On behalf of the PAX Collaboration)
Institut für Kernphysik, Forschungszentrum Jülich,
Jülich, 52425, Germany

A polarised antiproton beam at the Facility for Antiproton and Ion Research, proposed by the PAX collaboration, will provide new physics possibilities, uniquely accessible at the planned High Energy Storage Ring at FAIR. Our proposal to realise an asymmetric collider, in which polarised antiprotons with momenta of about 3.5 GeV/c collide with polarised protons with momenta up to 15 GeV/c, is well suited to perform a direct measurement of the transversity distribution function h_1.

Keywords: antiproton polarisation, transversity.

1. Introduction

The polarised antiproton-proton interactions at the High Energy Storage Ring (HESR) at the future Facility for Antiproton and Ion Research (FAIR) will provide unique access to a number of new fundamental physics observables, which can be studied neither at other facilities nor at HESR without transverse polarisation of protons and antiprotons.

Unlike the more conventional unpolarised quark distribution $q(x, Q^2)$ and the helicity distribution $\Delta q(x, Q^2)$, the transversity $h_1^q(x, Q^2)$ can neither be accessed in inclusive deep-inelastic scattering of leptons off nucleons nor can it be reconstructed from the knowledge of $q(x, Q^2)$ and $\Delta q(x, Q^2)$. It may contribute to some single-spin observables, but always blurred by the coupling to other unknown functions. The transversity distribution is directly accessible in a unique way via the double transverse spin asymmetry A_{TT} in the Drell-Yan production of lepton pairs (see Fig. 1).

The theoretical expectations for A_{TT} in the Drell-Yan process with transversely polarised antiprotons interacting with a transversely polarised proton target at HESR are in the range 0.3–0.4.[1,2] With the expected beam polarisation achieved using a dedicated low-energy antiproton polariser ring

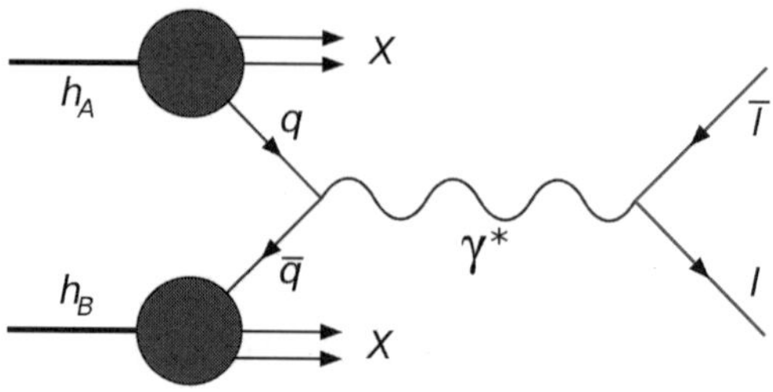

Fig. 1. Drell-Yan process: a quark from one hadron and an anti-quark from another hadron annihilate to create a pair of leptons.

(APR) of $P \approx 0.2$–0.3[a] and a luminosity of 10^{31}, the PAX experiment is uniquely suited for the definitive observation of $h_1^q(x, Q^2)$ of the proton for the valence quarks.

2. Beam polarisation via spin filtering

The possibility to test the nucleon structure via double spin asymmetries in polarised proton-antiproton reactions at the HESR ring of FAIR at GSI has been suggested by the PAX collaboration in 2004.[5] Since then, there has been much progress, both in understanding the physics potential of such a facility[1,2,6,7] and in studying the feasibility to effectively produce polarised antiprotons.[8] The modern understanding of the interaction of antiprotons with polarised atoms suggests that the spin-filtering technique allows us to reach a polarisation of the stored antiprotons at HESR-FAIR of $\sim$0.2–0.3.[8,9]

3. PAX accelerator set-up

Polarisation buildup in the HESR ring, operated at the lowest possible energy, as discussed in PAX LoI, does not allow one to achieve the optimum degree of polarisation for the antiproton beam. The goal of achieving the highest possible polarisation of antiprotons and optimisation of the figure of merit demands that one polarises antiprotons in a dedicated low-energy ring (APR, see Fig. 2).

[a]The estimates are based on the model calculations found in Refs. 3,4

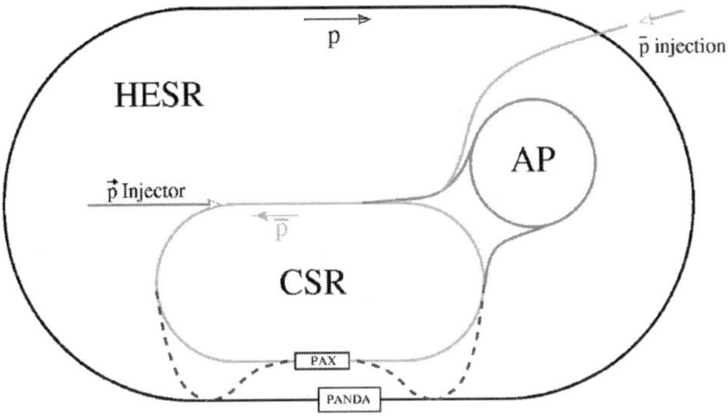

Fig. 2. The proposed accelerator setup at the HESR (black), with the equipment used by the PAX collaboration: CSR (light gray), APR, beam transfer lines and polarised proton injector (dark gray). By adding two transfer lines (dashed), an asymmetric collider is set up.

The transfer of polarised low-energy antiprotons into the HESR ring requires pre-acceleration to about 1.5 GeV/c in a dedicated booster ring (CSR). Simultaneously, the incorporation of this booster ring into the HESR complex opens up the possibility of building an asymmetric antiproton-proton collider, as shown in Fig. 2.

The kinematical regions covered by the PAX experiment, both in the fixed target and in the collider modes, are shown in Fig. 3.

With proper variation of the energy of the two colliding beams, the setup allows to measure transversity in the valence region of $x > 0.05$ with corresponding Q^2 =4–100 GeV^2. As long as the required luminosity in the collider mode might not be achievable, a high luminosity fixed target experiment could be conducted.

4. Expected precision of the experiment

To reveal rare reactions like the Drell-Yan process and the $\bar{p}p \to e^+e$ annihilation, the PAX detector has to be conceived as a large acceptance apparatus capable to unambiguously identify electron-positron pairs of large invariant mass and precisely measure their momenta. A clear particle identification is required to separate the electrons of the wanted processes from the large pion background. Reactions characterised by two-body hadronic final states like elastic scattering, present a higher cross-section and put less stringent constraints on the detector design. They can be identified by mea-

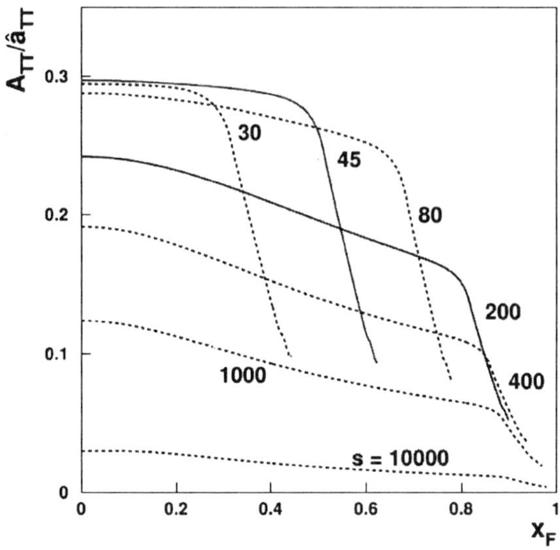

Fig. 3. The expected Drell-Yan asymmetry expected in PAX as a function of Feynman x_F for different values of s and $Q^2 = 16\,\text{GeV}^2$. The fixed target case represents antiprotons of 22 GeV/c colliding with a fixed polarised target ($s \sim 45\,\text{GeV}^2$).

suring scattering angles and momenta of the hadronic particles exploiting coplanarity and total momentum conservation.

The concept of the detector design has been simulated using GEANT. The achievable precision in h_1 expected from 1 year of data taking with an overall detection efficiency of 50% is displayed in Fig. 4. Assuming pessimistic values for the antiproton polarisation of 10% and for the luminosity of 5×10^{30} in the asymmetric collider mode results in the error bars depicted as an empty area in Fig. 4. In this case the invariant mass M of the lepton pair has been restricted to be in the region $M > M_{J/\psi}$ usually taken as "safe" region. It has been recently argued,[1] that the determination of the $h_1^q(x, Q^2)$ is not confined to the "safe" region, since the unknown quantities like unknown vector coupling $qq - J/\psi$ cancel in the ratios for A_{TT}, while the helicity structure remains. Cross sections increase dramatically when extending the measurement from $4\,\text{GeV/c}^2$ down to $2\,\text{GeV/c}^2$, thus enhancing sensitivity of PAX to A_{TT} (shaded area in Fig. 4).

In case higher antiproton polarisations are reached, the statistical errors are further reduced: the hatched area in Fig. 4 corresponds to the assumption $P = 30\%$.

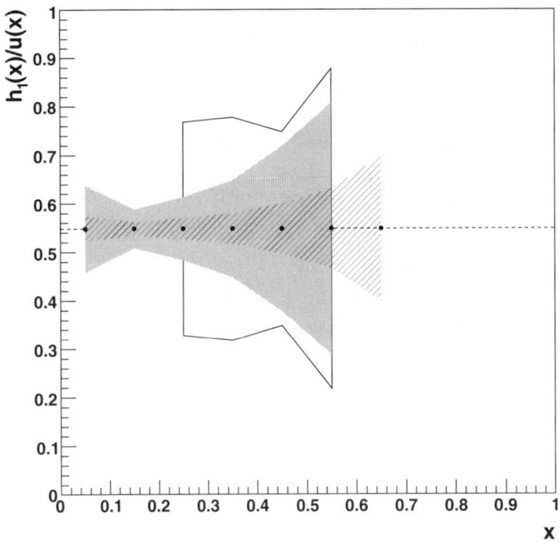

Fig. 4. Expected precision of the h_1 measurement after one year of data taking. The empty area represents the error bars in case when the invariant mass of the lepton pair $M > M_{J/\psi}$ and the antiproton polarisation equals to 10%, the shaded area — the case when $M > 2\,\mathrm{GeV/c^2}$, $P_{\bar{p}} = 10\%$, the hatched area — $M > 2\,\mathrm{GeV/c^2}$, $P_{\bar{p}} = 30\%$.

5. Outlook

The storage of polarised antiprotons at HESR will open unique possibilities to test QCD in hitherto unexplored domains. This will provide another cornerstone to the antiproton program at FAIR. Transversity is a fundamental but still elusive ingredient of the structure of the nucleon. SIDIS measurements at HERMES, COMPASS and Jlab are providing or will provide first results for h_1, however convoluted with Collins or interference fragmentation functions. These fragmentation functions can be directly studied at BELLE with high accuracy. Results from RHIC in *pp* interaction confirm that single spin asymmetries are large and can be interpreted as signals from the Sivers effect. Double polarised Drell-Yan in antiproton-proton interaction as proposed by PAX will provide in the future the best tool for a direct measurement of h_1.

References

1. M. Anselmino *et al.*, *Phys. Lett. B* **594**, 97 (2004).
2. A. Efremov, K. Goecke and P. Schweitzer, *Eur. Phys. J. C* **35**, 207 (2004).

3. T. Hippchen *et al.*, *Phys. Rev. C* **44**, 1323 (1991).
4. V. Mull and K. Holinde, *Phys. Rev. C* **51**, 2360 (1995).
5. PAX Letter-of-Intent, available from `http://www.fz-juelich.de/ikp/pax`.
6. S. Brodsky, *Testing Quantum Chromo Dynamics with antiprotons*, `http://lanl.arXiv.org/abs/hepph/0411046` (2004).
7. P. Zavada, *Proton transversity and intrinsic motion of the quarks*, `http://lanl.arXiv.org/abs/hepph/0412206` (2004).
8. F. Rathmann *et al.*, *Phys. Rev. Lett.* **014801**, 94 (2005).
9. F. Rathmann, Summary of the *Workshop on Spin Filtering in Storage Rings*, Heimbach, Germany, September 2005; available from `http://www.fz-juelich.de/ikp/pax`.

FUTURE DRELL-YAN MEASUREMENT @ COMPASS

M. Colantoni
(On behalf of the COMPASS Collaboration)
Sezione di Torino, Istituto Nazionale Fisica Nucleare,
Via P. Giuria 1, Torino, 10141, Italy

The COMPASS experiment is investigating possible physics aspects connect with the use of the polarized target (PT) together with the secondary (CERN–SPS) hadron beams. The main purpose of this project is to study the Drell–Yan reactions as a clear tool to measure the parton distribution functions, in particular the Boer-Mulders, the Sivers and the transversity functions.

Keywords: COMPASS; Drell–Yan; PDF; Transversity; Sivers function, Boer–Mulders function.

1. Introduction

The interest on the parton distribution functions (PDFs) has developed a large variety of researches both on theoretical and on experimental side. Nowadays, a large source of data has been provided with the deep inelastic scattering inclusive (DIS) and semi-inclusive (SIDIS). A complementary tool to access the PDFs is through the Drell-Yan (DY) mechanism. In this way it is possible to measure directly the PDFs without any convolution with the fragmentation functions, as it is for example in the case of the SIDIS. The DY reaction consists of an electromagnetic annihilation of quark and antiquark of colliding hadrons into a lepton pair production. In this context the PDF of the quarks of the interacting hadrons can be extracted. Thanks to the large lepton pair mass ($M^2 = Q^2$), this electromagnetic process can be factorized into terms describing the soft physics convoluted with the elementary cross-section:

$$d\sigma^{DY} \propto \sum_q f_q(x_1, k_{\perp 1}, Q^2) \otimes f_{\bar{q}}(x_2, k_{\perp 2}, Q^2) \cdot d\sigma^{q\bar{q} \to l^+ l^-}, \qquad (1)$$

where x_i is the Bjorken variable, $k_{\perp i}$ is the transverse momentum of the quark inside the hadron and Q^2 is the square of the four-momentum transferred, i.e. the square of invariant mass of the produced di-lepton pair.

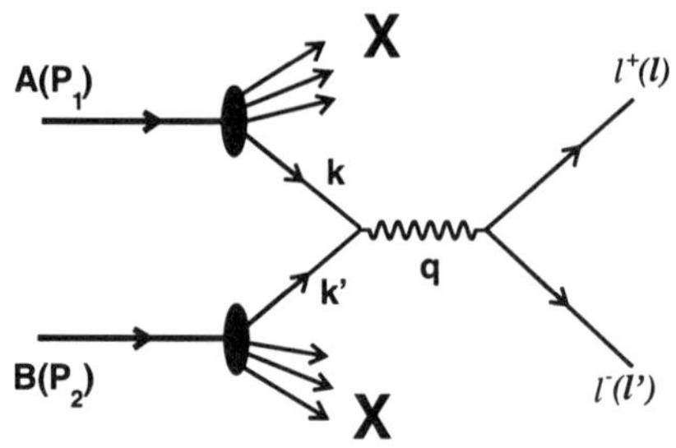

Fig. 1. Feynman diagram of the DY process

The introduction of the $k_\perp$ plays a fundamental role to explain the data on large p_T production of pions.[1] Without this inclusion the collinear model computation does not give results in agreement with the data. This fact open the way to many possible spin effect. For this purpose the COMPASS experiment[2] will perform a new generation of DY experiment with high intensity beam, large acceptance apparatus and a polarized target, not only to increase the statistics of the past experiment, limited by the low DY cross-section, but also to explore the effect due to the polarized target.

2. The PDFs

The complete knowledge of the nucleon spin structure at leading twist in case of perfect collinear quark is described by three distribution functions: the unpolarized PDF $f_1(x)$, that gives the probability to find an unpolarized quark with a fraction x of the longitudinal momentum of the unpolarized parent nucleon, the helicity PDF $g_1(x)$, that describes the difference between the number density of the quark with spin parallel and anti-parallel to the longitudinal polarized parent nucleon and the transversity PDF $h_1(x)$, that have the same meaning of $g_1(x)$, but in the case of transversely polarized nucleon. At the level of leading twist, if the partonic intrinsic motion $k_\perp$ inside the nucleon is considered, the nucleon is described by eight PDFs: $f_1(x, k_\perp^2)$, $g_{1L}(x, k_\perp^2)$, $h_1(x, k_\perp^2)$, $g_{1T}(x, k_\perp^2)$, $h_{1T}^\perp(x, k_\perp^2)$, $h_{1L}^\perp(x, k_\perp^2)$, $h_1^\perp(x, k_\perp^2)$, $f_{1T}^\perp(x, k_\perp^2)$, for a review see Ref.[3] The first three ones are integrated on $k_\perp$ become the PDFs before cited $f_1(x)$, $g_1(x)$, $h_1(x)$ respectively. A possible way to measure the transversity $h_1(x)$, the Boer–Mulders $h_1^\perp(x, k_\perp^2)$, Sivers function $f_{1T}^\perp(x, k_\perp^2)$ is presented.

3. Drell–Yan physics with transversely polarized target

The COMPASS future program with intense pion beams, involving the valence quark ($x_B > 0.1$) spin dependent PDFs in the energy scale of $Q^2 >> 1(GeV/c)^2$ is presented.

In COMPASS both unpolarized ($\pi + p \to l^+ + l^- + X$) and single transverse spin polarized ($\pi + p^{\rightarrow} \to l^+ + l^- + X$) DY processes are available. Depending on the pion beam charge, the PDFs for the u and for the d quark can be studied separately. They are not mixed up as in the case of the $\overline{p}p$, because the $\overline{u}u$ annihilation takes place mainly with the π^- beam, while the $\overline{d}d$ annihilation is accessible with the π^+ beam, negletting the non–valence quark contributions.

The Collins–Soper frame[4] (see Fig. 2) is commonly used to describe the DY processes where θ and ϕ are the polar and the azimuthal angle

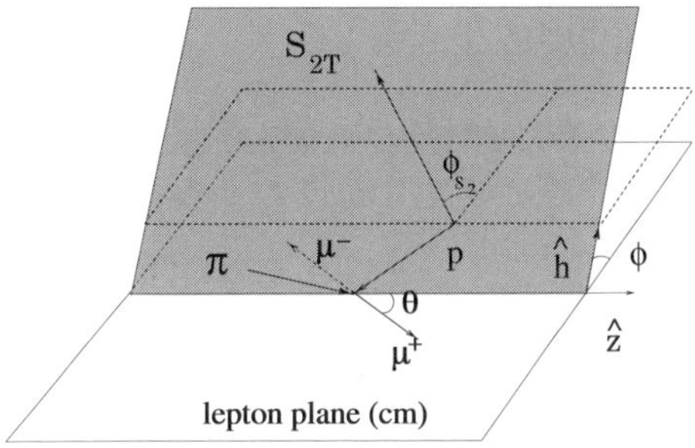

Fig. 2. The Collins-Soper frame.θ and Φ are the polar and the azimuthal

of the positive lepton produced in the reaction. The x-axis is chosen as the incoming hadron direction, the y-axis is perpendicular to the hadron plane, while the z-axis is chosen along the bisector between the μ^+ and the target hadron direction. The third angle ϕ_{S_2} is peculiar of the polarized DY reaction, in fact it represents the azimuthal spin vector of the polarized hadron target with respect to the di-lepton plane.

The effect of the transverse motion of the quark inside the nucleon can be studied through the Single Spin Asymmetry (SSA). In case of $\pi + p^{\rightarrow} \to l^+ + l^- + X$ the SSA can be expressed like:

$$A^{\sin(\Psi)} = \frac{\int d\Omega d\phi_{S_2} \int d^2 q_T (\frac{|q_T|}{M}) \sin\Psi [d\sigma(S_{2T}) - d\sigma(-S_{2T})]}{\int d\Omega d\phi_{S_2} \int d^2 q_T [d\sigma(S_{2T}) - d\sigma(-S_{2T})]} \tag{2}$$

where q_T are the transverse component of the virtual gamma exchange in the interaction in the Collins–Soper frame and $\Psi = (\phi \pm \phi_{S_2})$. The asymmetry, shown in Eq. 2, is obtained using the q_T weighting approach as suggested in Ref.[5–8]

Therefore using the Ψ weight variable it is possible to disentangle the two contributions present in this asymmetry:

$$A^{\sin(\phi+\phi_{S_2})} \propto \frac{\overline{h}_{1\overline{u}}^{\perp(1)}(x_\pi) \cdot h_{1u}(x_p)}{\overline{f}_{1\overline{u}}(x_\pi) \cdot f_{1u}(x_p)}; \; A^{\sin(\phi-\phi_{S_2})} \propto \frac{\overline{f}_{1\overline{u}}^{(1)}(x_\pi) \cdot f_{1T}^{\perp(1)}(x_p)}{\overline{f}_{1\overline{u}}(x_\pi) \cdot f_{1u}(x_p)} \quad (3)$$

The weighting approach can be also used in the unpolarized reaction $\pi + p \to l^+ + l^- + X$. In this case ratio R between the angular differential unpolarized cross-section and the unpolarized cross section can be constructed and its parameterization is:

$$\hat{R} = \frac{\int d^2 q_T [\frac{|q_T|^2}{\mathbf{M_\pi M_p}}][\frac{d\sigma^0}{d\Omega}]}{\int d^2 q_T \sigma^0} = \frac{3}{16\pi}(\gamma(1+\cos^2\theta) + \hat{k}\cos(2\phi)\sin^2\theta)) \quad (4)$$

in the coefficient $\hat{k}$ the PDFs appear as:

$$\hat{k} = 8\frac{\overline{h}_{1\overline{u}}^{\perp(1)}(x_\pi) \cdot h_{1u}^{\perp(1)}(x_p)}{\overline{f}_{1\overline{u}}^{\perp(1)}(x_\pi) \cdot f_{1u}^{\perp(1)}(x_p)} \quad (5)$$

where $\overline{h}_{1\overline{u}}^{\perp(1)}(x_\pi)$ and $h_{1u}^{\perp(1)}(x_p)$ are the Boer-Mulders functions for pion and proton respectively.

At this point there are three unknown functions to be extracted, but only two equations, nevertheless if one assume that:[9,10]

$$\frac{\overline{h}_{1u}^{\perp(1)}(x_{\pi^-})}{h_{1u}^{\perp(1)}(x_p)} = C_u \cdot \frac{\overline{f}_{1u}(x_{\pi^-})}{f_{1u}(x_p)}; \qquad C_u \approx 1 \quad (6)$$

the system of three equations can be solved.

Studying the dependence on $sin(\phi + \phi_{S_2})$ in polarized measurement of SSA, as reported in Eq. 3, together with the unpolarized measurement, the transversity and the first moment of the Boer-Mulders functions can be extracted. Studying the $sin(\phi - \phi_{S_2})$ dependence in the polarized reaction $\pi + p^{\rightarrow} \to l^+ + l^- + X$, the Sivers function can be deduced.

The measurement of this T-odd parton distribution function is very interesting because of its peculiar property:[11]

$$f_{1T}^{\perp}|_{DY} = -f_{1T}^{\perp}|_{DIS} \quad (7)$$

the Sivers function measured via the DY reaction and via the SIDIS should show a reversed sign, if measured in the same kinematical range. An experimental proof of the Eq. 7 would provide an exhaustive test of the Sivers mechanism within the QCD. In particular it would be a crucial test of the factorization approach for the description of this kind of processes, sensitive to the transverse parton momenta. One of the theoretical prediction for the Sivers function[12] for the COMPASS kinematical range is illustrated in Fig. 3.

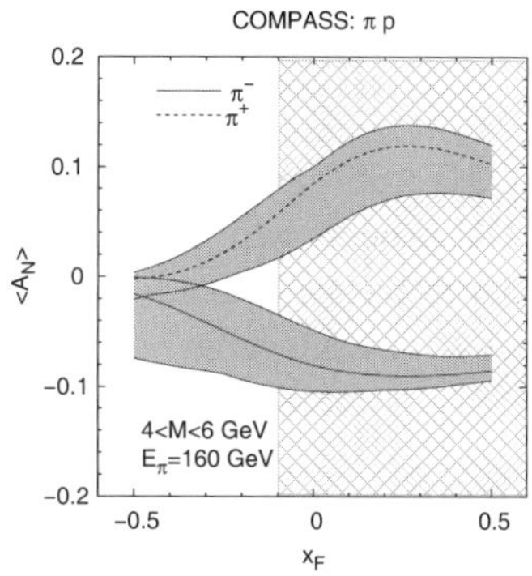

Fig. 3. Prediction[12] for the SSA contribution due to the Sivers function vs the x_F variable. In light-blue is highlighted the region covered by the COMPASS kinematical range

4. COMPASS apparatus and upgrade

To study the feasibility of the DY measurement with the COMPASS setup, a MonteCarlo (MC) event simulation, together with a beam test were performed in 2007. The MC simulations were done using the standard analysis COMPASS softwares.[13] To produce events dominated by the valence quark-antiquark annihilation in the reaction $\pi + p \to \mu^+ + \mu^- + X$ two event generators Pythia[14] and the DY-AB5[15] were used. Then the acceptance of the COMPASS setup for DY pair ($\mu^+\mu^-$) was studied. The events were generated in the invariant mass range of $4.0\ GeV/c^2 < M_{\mu^+\mu^-} < 9.0\ GeV/c^2$, to avoid the J/Ψ and Υ resonances, where the main production mechanism is not the quark-antiquark annihilation. Different pion beam momenta were tested in order to select events falling in the valence quark-antiquark region, but having cross-section large enough to provide a reasonable statistics in the dimuon mass region between $4.0\ GeV/c^2$ and $9.0\ GeV/c^2$. The target length and target material can to be optimized in order to maximize the rate of DY events together with the conditions to keep low the rate of background events and the effect multiple scattering in the target.

The very preliminary MC results together with the 2007 beam test shows the feasibility of this measurement. During this test a beam intensity of $2 \cdot 10^7 \pi$ per spill with the full COMPASS polarized target were used. The corrisponding luminosity reached was $L= 10^{31} cm^{-2} s^{-1}$.

A new trigger system, to select the di-muon pair, based on two scintillating hodoscopes dectectors, is under study for the COMPASS DY future program.

5. Conclusion

The DY program at COMPASS can be started after the 2010. For the first time the SSA in a DY process will be measured with the COMPASS experimental setup, which is basically ready to reconstruct the di–lepton muon pair. The statistical error achievable mainly depends on the choice of the target material.

References

1. D.L. Adams et al., Phys. Lett. B **264**, 462 (1991).
2. P. Abbon et al., Nucl. Instr. Meth. A **577**, 455 (2007).
3. V. Barone et al., Phys. Rep. **359**, 1 (2002).
4. J. Collins and D. Soper, Phys. Rev. D **16**, 2219 (1977).
5. A. M. Kotznian, P.J. Mulders, Phys. Lett. B **406**, 373 (1997).
6. D. Boer et al., Nucl. Phys. B **504**, 345 (1997).
7. D. Boer, P.J. Mulders, Phys. Rev. D **57**, 5780 (1998).
8. D. Boer et al., Phys. Lett. B **424**, 143 (1998).
9. D. Boer, Phys. Rev. D **60**, 014012 (1999).
10. A. N. Sissakian et al., Eur. Phys. Jour. C 147 (2006).
11. J. Collins, Phys. Lett. B **536**, 43 (2002).
12. S. Melis et al., Transversity 2008, 27-31 May 2008, Ferrara, Italy.
13. http://wwwcompass.cern.ch/compass/software/offline/welcome.html ♯software.
14. T. Sjöstrand et al., hep-ph/0308153.
15. A. Bianconi, Nucl. Instr. Meth. A,*(in press)*,0806.0946v1 [hep-ex].

MEASUREMENTS OF UNPOLARIZED AZIMUTHAL ASYMMETRIES AT COMPASS

W. Käfer
(On behalf of the COMPASS Collaboration)
Physikalisches Institut, Albert-Ludwigs-University Freiburg, 79104 Freiburg, Germany

Azimuthal Asymmetries in unpolarized SIDIS can be used to probe the transverse momentum of quarks inside the nucleon. Furthermore, they give access to the so-far unmeasured Boer-Mulders function. We report on the first measurement of azimuthal asymmetries of the SIDIS cross section from scattering of muons off a deuteron target.

Keywords: Deep Inelastic Scattering, DIS, Semi-inclusive Deep Inelastic Scattering, spin-independent azimuthal asymmetries, unpolarized Azimuthal Asymmetries, Cahn Effect, Boer-Mulders Effect, Boer-Mulders Function, COMPASS, CERN

1. Introduction and Theoretical Motivation

Semi Inclusive Deep Inelastic Scattering (SIDIS) reactions are an important tool to probe the structure of the nucleon. Of particular interest in unpolarized SIDIS is a possible dependence of the cross section on the azimuthal angle of the hadron as defined in Fig. 1.

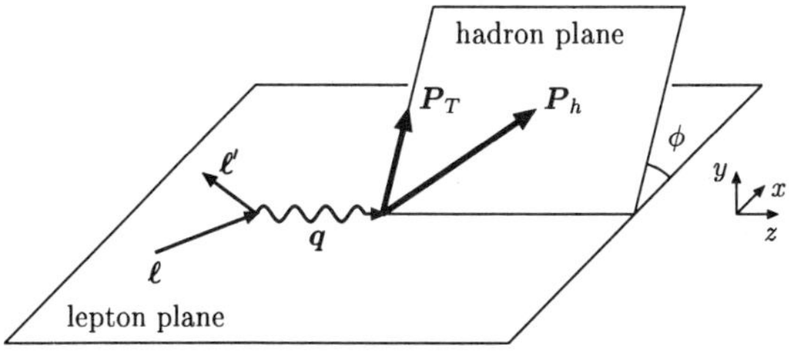

Fig. 1. Definition of the hadron azimuthal angle ϕ_h.

There are several effects which contribute to a $\cos\phi_h$ and $\cos 2\phi_h$ dependence of the cross section. The amplitudes of these modulations depend on the kinematic variables relevant for the SIDIS process, namely the Bjorken scaling variable x, the relative energy loss of the muon y, the negative 4-momentum of the virtual photon squared Q^2, the energy fraction of the hadron z and the transverse momentum of the hadron P_t^h. Furthermore, k_t is the transverse momentum of the quark with respect to the nucleon, and p_t the transverse momentum of the hadron with respect to the direction of the fragmenting quark.

The first azimuthal dependence discussed here is due to non-zero k_t: As Cahn pointed out in Ref. 1, an azimuthal modulation of the cross section is expected in one-photon exchange approximation, when the transverse momentum of the quark is taken into account. The Cahn effect contributes to a possible $\cos\phi_h$ and $\cos 2\phi_h$ asymmetry of the SIDIS cross section. It is kinematically suppressed by $\frac{k_t}{Q}$ for $\cos(\phi_q)$ and $\left(\frac{k_t}{Q}\right)^2$ for the $\cos 2\phi_q$ term. When going from the quark to hadron level, the unpolarized parton distribution functions (PDFs) $f_q(x, k_t)$ and fragmentation functions (FFs) $D_q^h(z, p_t)$ need to be taken into account. Assuming Gaussian distributions for the transverse momentum dependence of $f_q(x, k_t)$ and $D_q^h(z, p_t)$ and introducing $D_{\cos\phi_h}(y) = \frac{(2-y)\sqrt{1-y}}{1+(1+y)^2}$, the $\cos\phi_h$ term can be written as

$$\frac{\mathrm{d}^5\sigma}{\mathrm{d}x\,\mathrm{d}y\,\mathrm{d}z\,\mathrm{d}P_t^{h\,2}\,d\phi_h} \propto \sum_q f_q(x) D_q^h(z) \left[1 - 4D_{\cos\phi_h}(y)\frac{\langle k_t^2\rangle z P_t^h}{Q\langle (P_t^h)^2\rangle}\cos\phi_h\right]. \tag{1}$$

The $\cos\phi_h$ modulation allows the determination of $\langle k_t^2\rangle$. This has been done e.g. in Ref. 2 using results from previous experiments, arriving at $\langle k_t^2\rangle \approx 0.25\ \mathrm{GeV}^2/c^2$.

The second contribution to azimuthal asymmetries comes from the Boer-Mulders PDF[3] $h_1^\perp(x, k_t)$, convoluted with the Collins FF $H_1^\perp(x, p_t)$. A model calculation[4] shows that the Boer-Mulders contribution to the $\cos 2\phi_h$ modulation might be of similar magnitude than the contribution from the Cahn effect. In particular, a possible difference between positive and negative hadron asymmetries could be explained by the Boer-Mulders mechanism. It may also contribute to the $\cos\phi_h$ asymmetry,[3] but the size of this effect is so far unknown and no predictions exist.

Perturbative QCD (pQCD) introduces a third ϕ_h dependence at order α_s. However, QCD effects are only important at $P_t^h > 1$ GeV/c. Therefore they are expected to be small for COMPASS kinematics, where most of the statistics is at low transverse momentum. The perturbative contribution has been given in Refs. 5,6 at $\mathcal{O}(\alpha_s^1)$. Recently, higher order contributions were calculated in Ref. 7.

Since the muon beam used at the COMPASS experiment is naturally polarized, an additional modulation of the cross section is expected. In contrast to the Cahn, Boer - Mulders and pQCD contributions, the cross section for the beam asymmetry depends on $\sin\phi_h$.[8] This effect has recently been measured by the CLAS collaboration[9] on a proton target.

Radiative corrections and possible higher twist terms may contribute to the modulations, but these effects are considered to be small and therefore will not be discussed here.

The overall cross section for SIDIS on an unpolarized target is thus of the form:

$$\frac{\mathrm{d}\sigma}{\mathrm{d}\phi_h} \propto 1 + A_{\cos\phi_h}\cos\phi_h + A_{\cos 2\phi_h}\cos 2\phi_h + A_{\sin\phi_h}\sin\phi_h \,. \tag{2}$$

The amplitudes of these three modulations have been determined from data taken with the COMPASS experiment with a deuteron target.

For a proton target, three experiments have published results on azimuthal asymmetries in different kinematic regions: EMC,[10,11] the E665 collaboration[12] and the ZEUS experiment[13].

2. The COMPASS Experiment

The COMPASS experiment[14] is a fixed target experiment at CERN. It features a 160 GeV/c μ^+-beam, with a natural polarization of -80% and a polarized target, which consists of two cells. From 2002-2006, data was taken with a polarized ^{6}LiD target, while in 2007 a hydrogen target was used. The target can be polarized either in the longitudinal or the transverse direction with respect to the beam direction. These two configurations will be called CL and CT in the following.

The different magnetic fields needed to maintain the target polarization (a solenoid field for CL and a dipole field for CT), require changes of the magnetic configuration of the experiment. In particular, the beam line settings differ, since an additional kick is needed for the CT setup to compensate the additional dipole field. On the other hand, for CL, there is a strong interference between the target magnetic field and the magnetic field of the first spectrometer magnet, which is not present for CT.

3. Asymmetry Extraction

The data sample used for this analysis was taken in the year 2004. Data samples in both target configurations (CL and CT) have been used in order to get a better estimate of the systematic error generated by experimental conditions.

The final sample consists of about 5 million positive and 4 million negative hadrons in the kinematic range $Q^2 > 1$ GeV$^2/c^2$,$0.1 < y < 0.9$, $0.2 < z < 0.85$ and 0.1 GeV/c $< P_t^h <$ 1.5 GeV/c and contains data from both target configurations in roughly equal parts. The target polarization was canceled by event weighting, taking into account the fluxes and average polarizations:

$$N_{\text{unpol}} = P_2 N_1 + F P_1 N_2 \,. \tag{3}$$

where P_i indicates the absolute value of the polarizations for the two possible spin configurations ($\leftarrow, \rightarrow$ for CL and $\uparrow, \downarrow$ for CT), N_i the event number for the respective polarization and $F = \frac{F_1}{F_2}$ the corresponding ratio of the respective muon fluxes.

In order to correct for event losses caused by the non-uniform acceptance of the COMPASS spectrometer, full MC simulations have been performed in both CL and CT setup. In each case, the events were generated with LEPTO, transported through the COMPASS detector simulation program COMGEANT and the reconstruction software CORAL. From these MC samples, the acceptance of the COMPASS spectrometer $A(\phi_h)$ and the corrected count rates $N_{\text{corr}}(\phi_h)$ can be determined:

$$A(\phi_h) = \frac{N_{\text{rec}}(\phi_h)}{N_{\text{gen}}(\phi_h)} \qquad\qquad N_{\text{corr}}(\phi_h) = \frac{N_{\text{unpol}}(\phi_h)}{A(\phi_h)} \,. \tag{4}$$

The acceptance correction has been done in bins of x, z, and P_t^h separately, with the other two variables always integrated out. Fig. 2 shows a typical example of measured ϕ_h distribution and the corresponding acceptance. The corrected count rates are then fitted with a four-parameter fit containing the average count rate N_0 and the three amplitudes:

$$N_{\text{corr}} = N_0(1 + A_{\sin\phi_h} \sin\phi_h + A_{\cos\phi_h} \cos\phi_h + A_{\cos 2\phi_h} \cos 2\phi_h) \,. \tag{5}$$

4. Results

The amplitudes $A_{\cos\phi_h}, A_{\cos 2\phi_h}$ and $A_{\sin\phi_h}$ of the three modulations have been determined in dependence on x, z and P_t^h for positive and negative hadrons seperately. Fig. 3 shows the results for the three modulations for

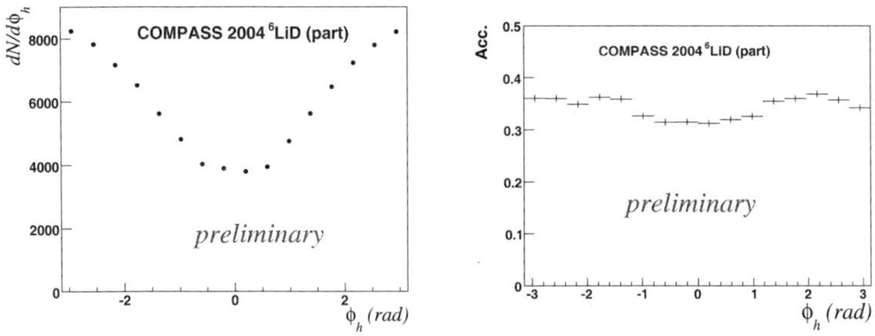

Fig. 2. Measured ϕ_h distribution before correction and acceptance in the CT case for $0.63 < z < 0.85$.

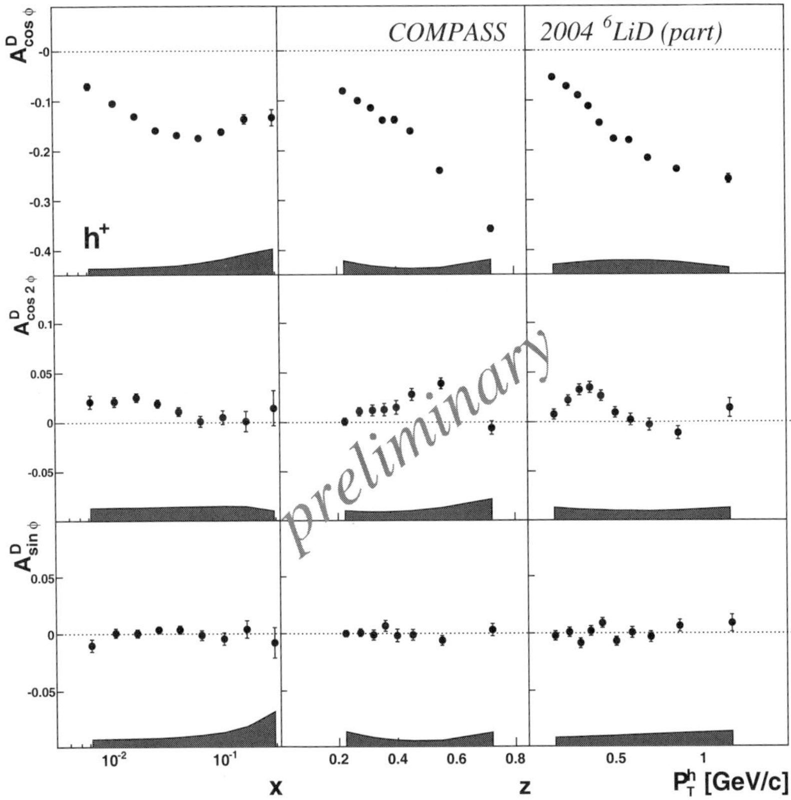

Fig. 3. Results for positive hadrons for $A_{\cos\phi_h}$, $A_{\cos 2\phi_h}$ and $A_{\sin\phi_h}$ in dependence of the kinematic variables x, z, P_t^h. The error bars correspond to the statistical errors, while the error bands at the bottom indicate systematic errors.

positive hadrons. A large $\cos\phi_h$ amplitude of up to 40% and a $\cos 2\phi_h$ amplitude of up to 5% is seen. The value for $A_{\sin\phi_h}$ are consistent with zero. Also for the negative hadrons, shown in Fig. 4, the $\sin\phi_h$ amplitude is consistent with zero, while the two cosine amplitudes show similar trends compared to the positive hadrons. However, the magnitude of the cosine modulations differs significantly for positive and negative hadrons.

Fig. 5 shows a recent prediction[15] for a deuteron target and the COMPASS kinematics, compared to the data. Only Cahn and perturbative QCD contributions have been considered. Calculation and data show similar behaviour in the region of large x, while there is disagreement in the region of small x. The strong disagreement in the low x region leads to a scale difference in z, although the slope, which is mainly due to the Cahn Effect, is

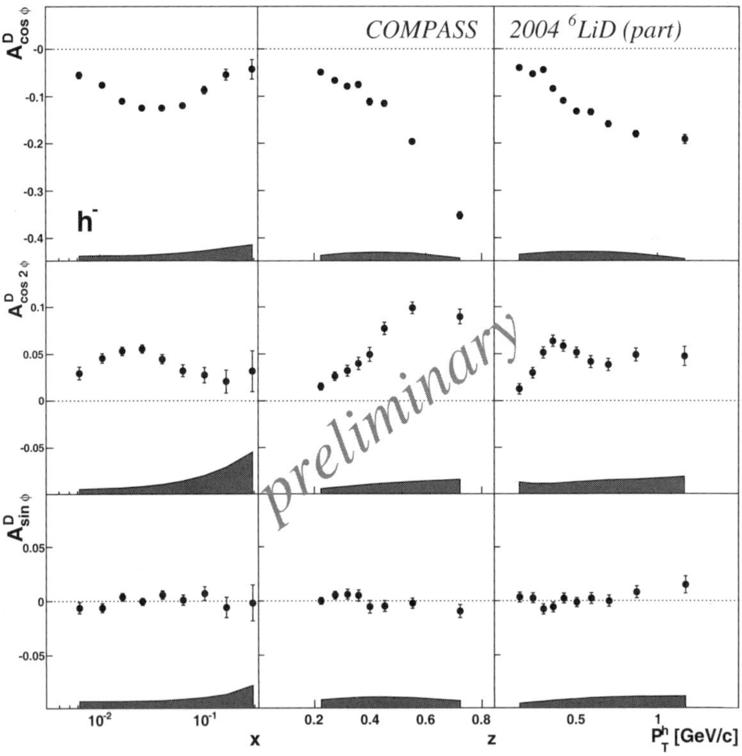

Fig. 4. Results for negative hadrons for $A_{\cos\phi_h}$, $A_{\cos 2\phi_h}$ and $A_{\sin\phi_h}$ in dependence of the kinematic variables x, z, P_t^h. The meaning of the error bars and bands is the same as in the previous figure.

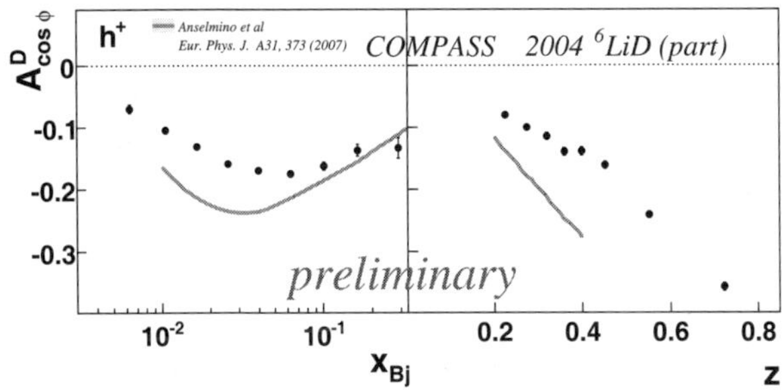

Fig. 5. Comparison of $A_{\cos\phi_h}$ for positive hadrons with model calculation from Ref. 15, which takes into account Cahn and perturbative QCD effects. Only statistical errors are shown.

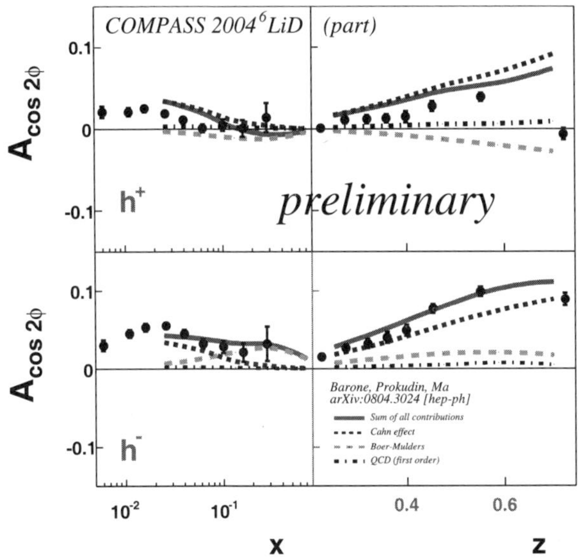

Fig. 6. Comparison of $A_{\cos 2\phi_h}$ with predictions for COMPASS kinematics from Ref. 4. The calculation takes into account the Cahn effect(dashed curve), perturbative QCD (dashed dotted) and Boer-Mulders (dotted curve). The continuous line describes the sum of these three. Again, only statistical errors are shown.

again similar. A model-calculation[4] for the $\cos 2\phi_h$ amplitude is compared to the COMPASS results in Fig. 6. Here also the Boer-Mulders mechanism has been included. Since $h_1^{\perp}(x, k_t)$ is presently not constrained by experimental data, it has been assumed to be proportional to the better known Sivers function. This assumption has e.g. been motivated in Ref. 16.

5. Systematic Tests

Several checks have been performed to determine the systematic error of the measurement. It turns out that the systematic error is due to two sources: the difference between the results in the setups CL and CT and the variation of the acceptance for different generated kinematic distributions. These differences are used as an estimate for the quality of the acceptance correction. To evaluate the second contribution, for each setup, CL and CT, two MC simulations with different LEPTO settings and thus different generated kinematic distributions have been performed. Several additional tests, such as splitting the data sample according to the event topology, target polarization and time of the measurement give no significant contribution to the systematic error.

6. Summary and Outlook

First results on unpolarized azimuthal asymmetries from COMPASS have been presented, extending the investigated kinematic region to low x. The data can be used to better determine the value of $\langle k_t^2 \rangle$. Also the differences between positive and negative hadrons allow to gain knowledge about $h_1^\perp(x, k_t)$, which can be further deepened with the data taken in 2007 with a NH_3 target.

References

1. R. N. Cahn, *Phys. Lett. B* **78**, 269 (1978).
2. M. Anselmino *et al.*, *Phys. Rev. D* **71**, 074006 (2005).
3. D. Boer and P. J. Mulders, *Phys. Rev. D* **57**, 5780 (1998).
4. V. Barone, A. Prokudin and B. Q. Ma, *Phys. Rev. D* (2008), accepted for publication, arXiv:0804.3024 [hep-ph].
5. H. Georgi and H. D. Politzer, *Phys. Rev. Lett.* **40**, 3 (1978).
6. A. Mendez, *Nucl. Phys. B* **145**, 199 (1978).
7. A. Daleo, D. de Florian and R. Sassot, *Phys. Rev. D* **71**, 023013 (2005).
8. A. Kotzinian, *Nuclear Physics B* **441**, 234 (1995).
9. CLAS Collaboration, *Phys. Rev. D* **69**, 112004 (2004).
10. European Muon Collaboration, *Phys. Lett. B* **130**, 118 (1983).
11. European Muon Collaboration, *Z. Phys. C* **34**, 277 (1987).
12. E665 Collaboration, *Phys. Rev. D* **48**, 5057 (1993).
13. ZEUS Collaboration, *Phys.Lett. B* **481**, 199 (2000).
14. COMPASS Collaboration, *Nucl. Instr. Meth.* **577**, 455 (2007).
15. M. Anselmino *et al.*, *Eur. Phys. J. A* **31**, 373 (2007).
16. M. Burkardt, *Phys. Rev. D* **72**, 094020 (2005).

MEASUREMENT OF AZIMUTHAL ASYMMETRIES OF THE UNPOLARIZED CROSS-SECTION AT HERMES

F. Giordano
(On behalf of the HERMES Collaboration)
INFN and Università degli studi di Ferrara,
Ferrara, 44100 Italy

In unpolarized Semi-Inclusive Deep Inelastic Scattering the $\cos\phi_h$ and $\cos 2\phi_h$ azimuthal asymmetries originate due to quark intrinsic transverse momentum κ_T and transverse spin s_T. The complicated issue of disentangling the physical azimuthal dependence of the cross-section from the acceptance and instrumental effects is discussed and preliminary Monte Carlo studies on a fully differential analysis corrected for instrumental and radiative effects are presented.

Keywords: Semi Inclusive DIS; Azimuthal modulation; Intrinsic transverse momentum; Intrinsic transverse spin.

1. Introduction

In lepton-nucleon Deep Inelastic Scattering (DIS), the structure of the nucleon is probed by the interaction of a high energy lepton and a target nucleon, via the exchange of one virtual boson. If at least one of the produced hadrons is detected in coincidence with the scattered lepton, the reaction is called Semi Inclusive Deep Inelastic Scattering (SIDIS):

$$l(\mathbf{k}) + N(\mathbf{P}) \to l'(\mathbf{k}') + h(\mathbf{P}_h) + X(\mathbf{P}_X) \tag{1}$$

where l (l') is the incident (scattered) beam, N is the nucleon target, h is the detected hadron, X is the target remnant and the quantities in parentheses the corresponding four-momenta.

Assuming one photon exchange, the leptoproduction cross-section can be expressed as the contraction of the leptonic ($L_{\mu\nu}$) and the hadronic ($W^{\mu\nu}$) tensors. The leptonic tensor can be exactly calculated within the Quantum ElectroDynamics (QED) framework, while the hadronic tensor can not be calculated from first principles. However, using symmetry arguments and conservation laws, $W^{\mu\nu}$ can be expressed, in a model-independent way, in terms of structure functions and, in case of unpolarized

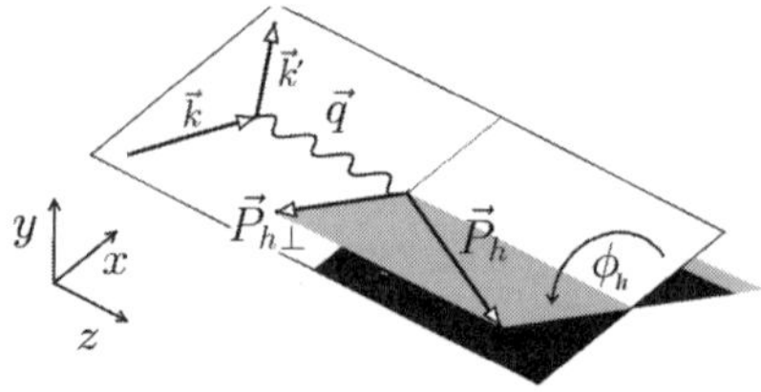

Fig. 1. Definition of the azimuthal angle ϕ_h between the scattering plane (white) and the hadron production plane (grey): $\phi_h = \frac{\vec{q}\times\vec{k}\cdot\vec{P_h}}{|\vec{q}\times\vec{k}\cdot\vec{P_h}|} cos^{-1}\left(\frac{\vec{q}\times\vec{k}\cdot\vec{q}\times\vec{P_h}}{|\vec{q}\times\vec{k}||\vec{q}\times\vec{P_h}|}\right)$.

scattering, the cross-section reads[1] :

$$\frac{d\sigma}{dx\,dy\,dz} = \frac{\alpha^2}{xyQ^2}\left(1+\frac{\gamma^2}{2x}\right)\{A(y)\,F_{UU,T} + B(y)\,F_{UU,L}\}, \tag{2}$$

where the subscripts UU stand for Unpolarized beam and target, while T (L) indicates the Transverse (Longitudinal) polarization of the virtual photon, α is the electromagnetic coupling constant, $\gamma = 2Mx/Q$ with M the target mass, $A(y) \sim (1-y+1/2y^2)$ and $B(y) \sim (1-y)$. Here Q^2 and y are respectively the negative squared four-momentum and the fractional energy of the virtual photon, x is the Bjorken scaling variable and z is the fractional energy of the produced hadron.

Equation (2) is the well known unpolarized cross-section integrated over the outgoing hadron momentum component transverse to the virtual photon direction ($P_{h\perp}$). In the unintegrated cross-section, an azimuthal dependence of the outgoing hadron direction becomes manifest and two new azimuthal dependent structure functions appear[1] :

$$\begin{aligned}\frac{d\sigma}{dx\,dy\,dz\,dP_{h\perp}^2\,d\phi_h} = \frac{\alpha^2}{xyQ^2}(1+\frac{\gamma^2}{2x})\{A(y)\,F_{UU,T} + B(y)\,F_{UU,L} +\\ C(y)\,\cos\phi_h F_{UU}^{\cos\phi_h} + B(y)\,\cos 2\phi_h F_{UU}^{\cos 2\phi_h}\},\end{aligned} \tag{3}$$

where ϕ_h is the azimuthal angle of the hadron plane around the virtual-photon direction (see Fig. 1) and $C(y) \sim (2-y)\sqrt{1-y}$.

Among the possible ones, two mechanisms are expected to give important contributions to the azimuthal dependence of the unpolarized cross-section in the hadron transverse momentum range accessed at HERMES:

the Cahn effect; a pure kinematical effect, generated by the non-zero intrinsic transverse motion of quarks, already pointed out by R. Cahn in 1978[2] ;

the Boer-Mulders effect, introduced more recently (1997) by D. Boer and P. J. Mulders[3]; an interesting mechanism generated when the quark transverse momentum couples to the quark transverse spin.

2. The HERMES experiment

The fixed target HERMES experiment ran for more than 10 years until 2007 at the electron-positron storage ring of HERA at DESY. The HERMES spectrometer[4] was a forward angle instrument consisting in two symmetric top-bottom halves above and below the horizontal plane defined by the lepton beam pipe. It was characterized by very high efficiency (around $98-99\%$) in electron-hadron separation, provided by a transition radiation detector, a preshower scintillation counter and an electromagnetic calorimeter. In addition, a dual-radiator Ring-Imaging CHerenkov (RICH) provided good hadron-type identification for momenta above 2 GeV/c.

3. Multidimensional analysis

To study the new structure functions $F_{UU}^{\cos\phi_h}$ and $F_{UU}^{\cos 2\phi_h}$ as defined in Eq. (3), a measure of the unpolarized cross-section azimuthal modulation is needed, which can be extracted via the so called $\langle\cos n\phi_h\rangle$-moments:

$$\langle\cos n\phi_h\rangle = \frac{\int \cos n\phi_h \, d^5\sigma}{\int d^5\sigma} \tag{4}$$

where $\int d^5\sigma$ is an abbreviation for $\int dx\, dy\, dz\, dP_{h\perp}^2\, d\phi_h \frac{d^5\sigma}{dx\, dy\, dz\, dP_{h\perp}^2\, d\phi_h}$, and with $n = 1, 2$.

Unfortunately, the extraction of the structure functions from data is challenging because in the moment they couple to a number of *experimental sources* of azimuthal cosine modulations, *e.g.* detector geometrical acceptance and higher order QED effects (*radiative effects*). Only for didactic purpose, the hypothesis that these contributions factorize is assumed in the following section. This hypothesis is not used in the final result extraction described in the next section. Within this hypothesis the event yields $N(\phi_h)$ can be expressed as:

$$N(\phi_h) = \int L\, \sigma_0\, A_{F_{UU}}\, \epsilon_{acc}\, \epsilon_{rad}\, dxdydzdP_{h\perp}^2 \tag{5}$$

where σ_0 includes all the ϕ_h-independent cross-section terms and $A_{F_{UU}} = (1 + 2\langle\cos\phi_h\rangle_{F_{UU}} \cos\phi_h + 2\langle\cos 2\phi_h\rangle_{F_{UU}} \cos 2\phi_h)$. Here L is the luminosity, ϵ_{acc} represents the spectrometer acceptance (geometry plus efficiency) and

ϵ_{rad} the radiative contribution. Each term in the latter equation depends on the 5 variables $x, y, z, P_{h\perp}, \phi_h$ and the integration is over the kinematical interval under study (typically one kinematical bin of the analysis).

In order to disentangle the azimuthal modulation related to the structure functions $F_{UU}^{\cos n\phi_h}$, the experimental results are compared to a Monte Carlo production (MC) including the complete simulation of the instrumental effects, but with $A_{F_{UU}} = 1$, so that the azimuthal distribution is generated only by acceptance and radiative contributions:

$$N^{MC}(\phi_h) = \int L^{MC}\,\sigma_0^{MC}\,\epsilon_{acc}^{MC}\,\epsilon_{rad}^{MC}\,dxdydzdP_{h\perp}^2, \tag{6}$$

One can argue that the ratio between the real (Eq. (5)) and the simulated (Eq. (6)) yields provides the $A_{F_{UU}}$ term, *i.e.* the ϕ_h modulation related to $F_{UU}^{\cos\phi_h}$ and $F_{UU}^{\cos 2\phi_h}$:

$$\frac{N}{N^{MC}}(\phi_h) \approx \frac{\int L\,\sigma_0\,A_{F_{UU}}\,\epsilon_{acc}\,\epsilon_{rad}\,dxdydzdP_{h\perp}^2}{\int L\,\sigma_0^{MC}\,\epsilon_{acc}^{MC}\,\epsilon_{rad}^{MC}\,dxdydzdP_{h\perp}^2}. \tag{7}$$

The integrals in the quotients of Eq. (7) do not allow the resolution of the correlation between the physical $A_{F_{UU}}$ and the instrumental ϵ_i contributions, unless the event sample is binned in all the relevant variables simultaneously. Only in this case, in each bin would equation (7) reduce to:

$$\frac{N}{N^{MC}}(\phi_h) \approx R_0\,A_{F_{UU}} \tag{8}$$

where in $R_0 = (L\,\sigma_0\,\epsilon_{acc}\,\epsilon_{rad})/(L^{MC}\,\sigma_0^{MC}\,\epsilon_{acc}^{MC}\,\epsilon_{rad}^{MC})$ the spurious ϕ_h dependencies cancel out. In the typical case the event sample is binned only in one of the kinematical variables (1-dimensional analysis), the quotients of equation (7) are integrated over the full range of all the other fundamental variables and $A_{F_{UU}}$ can not be correctly isolated.

The use of a Monte Carlo simulation to correct the detector acceptance must be justified. Indeed, differences between real and simulated ϕ_h-independent cross-sections (σ_0) can easily generate false azimuthal modulations when convoluted with detector acceptance. The possible false asymmetries can be estimated with two independent Monte Carlo event samples, generated with $A_{F_{UU}} = 1$, but with slightly different ϕ_h-independent cross-section models. The false cosine modulations were extracted fitting the ratio between the two MC yields. The comparison between the 1-dimensional and 5-dimensional (binned simultaneously in x, y, z, $P_{h\perp}$ and ϕ_h) analysis results are shown in Fig. (2). Here the moments are projected upon the different variables by weighting the moment in each bin with the corresponding 4π cross-section (*i.e.* Monte Carlo yields in 4π). For instance in

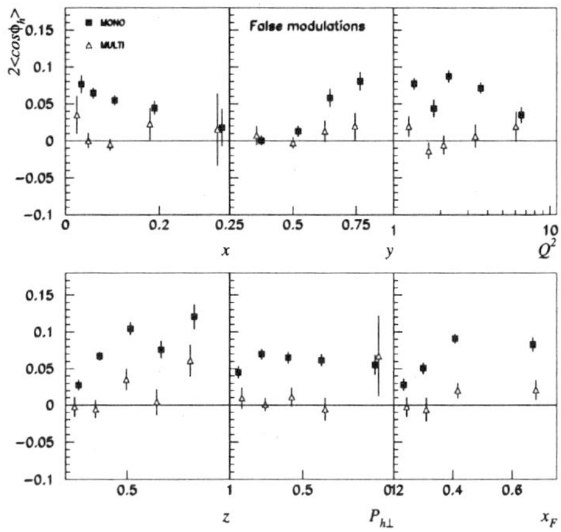

Fig. 2. False $\langle \cos\phi_h \rangle$ moments extracted using two Monte Carlo productions with different kinematical distributions: the solid squares are extracted in a 1-dimensional analysis, while the open triangles are extracted in a multidimensional analysis. Here x_F is the Feynmann variable.

case of the x variable one has:

$$\langle \cos\phi_h \rangle(x) = \frac{\sum_i \sigma^{4\pi}(x) \langle \cos\phi_h \rangle_i}{\sum_i \sigma^{4\pi}(x)}. \tag{9}$$

The false moments extracted in the multidimensional analysis (open triangles) are significantly smaller than those extracted in the 1-dimensional approach (closed squares), and are in general compatible with zero. As a result, the model-dependence of the extraction method is notably decreased in the multidimensional analysis.

3.1. *The unfolding extraction procedure*

For the extraction of the cosine moments an unfolding procedure was applied to the data yield distributions. The unfolding procedure is a standard algorithm that allows the extrapolation of experimental results, corrected for radiative and detector smearing, in a 4π geometry. This powerful tool allows relaxation of the factorization hypothesis used in last section.

A detailed Monte Carlo simulation of the experimental apparatus is used to define the *Smearing matrix* $S(i, j)$, providing the probability that an event originally in the *Born bin* j, corresponding to the original kinematics (free from the experimental distortions), is actually observed in the *measured bin* i.

The unknown *Born* yields distribution $B(j)$ is related to the *Measured* yield distribution $X(i)$ by:

$$X(i) = \sum_{j=1}^{n_b} S(i,j)B(j) + Bgk(i). \tag{10}$$

Assuming a non-singular $S(i,j)$ one then obtains:

$$B(j) = \sum_{i=1}^{n_b} S^{-1}(j,i)\big[X(i) - Bgk(i)\big], \tag{11}$$

where n_b is the total number of bins and $Bgk(i)$ is a vector that contains the events that are smeared in to the measured sample from outside the acceptance[5] . Since the Smearing matrix depends only on the description of the known detector and radiative effects, it provides a model-independent correction of the data. However, a possible source of systematic uncertainty in the unfolding procedure is the model of the background $Bgk(i)$.

Since the unfolding mixes different bins contents, the unfolded *Born* yields are statistically correlated, and the extraction of cosine moments from them requires a generalized least squares fit, which takes into account the non-diagonal covariance matrix. A linear regression can then be performed to linearize the problem and extract one ($\langle\cos\phi_h\rangle$, $\langle\cos 2\phi_h\rangle$) moment pair for each kinematical bin. These moment pairs represent results which are fully differential in the fundamental variables. The moment dependency on a single variable is obtained by projecting these results as in Eq. (9).

To estimate the goodness of the cosine moments unfolded in 5 dimensions, a Monte Carlo generator was used with the Cahn azimuthal dependence implemented in the cross-section. In particular, two Monte Carlo samples were generated: the first in a 4π geometry (4π-*level*), where the azimuthal modulation is due only to the Cahn model; the second within acceptance and with radiative effects turned on (*reconstructed-level*), where, as in real data, the Cahn model is convoluted with acceptance and radiative contributions. The *true* cosine moments are extracted (with a binned χ^2 fit) from the 4π yield distribution, and compared with the moments extracted after having unfolded the *reconstructed-level* yield distribution in Figs. (3) and (4). The unfolded moments are in a quite good agreement with the *true* moments, even if some systematic shifts can still be observed, expecially in the case of $\langle\cos 2\phi_h\rangle$.

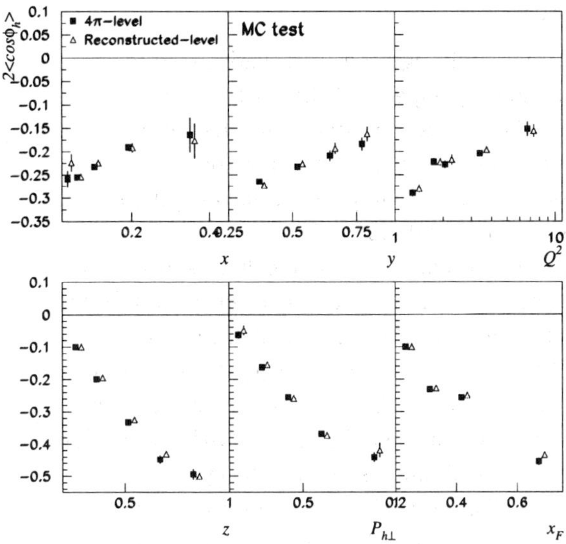

Fig. 3. $\langle \cos \phi_h \rangle$ moments extracted with a 5-dimensional analysis in the MC test described in the text: the unfolded moments (open triangles) are compared with the *true moments* (solid squares) generated in 4π.

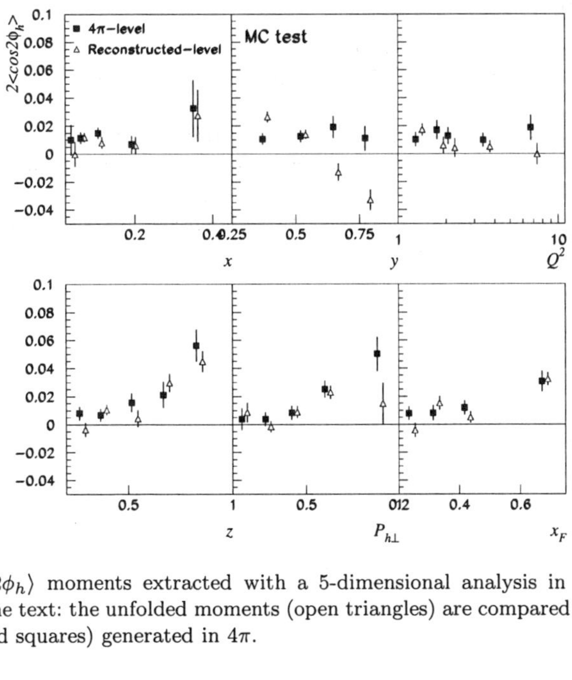

Fig. 4. $\langle \cos 2\phi_h \rangle$ moments extracted with a 5-dimensional analysis in the MC test described in the text: the unfolded moments (open triangles) are compared with the *true moments* (solid squares) generated in 4π.

4. Conclusion

The cross-section unintegrated over hadron transverse momentum gives access to new exciting aspects of the nucleon structure, which are currently under intense theoretical investigations. Because they are experimentally challenging, very few measurements have been performed to date, and the results available average out possible flavor dependence, except the very recent results from COMPASS.[6]

The Monte Carlo checks presented here show that a multidimensional analysis together with an unfolding correction can disentangle the physical azimuthal dependence of the cross-section from acceptance and instrumental effects. The extraction of the $\langle \cos n\phi_h \rangle$ azimuthal moments at HERMES can be performed using different target-type and can distinguish between positive and negative hadrons. Moreover, with RICH identification different hadron types can be distinguished, resulting in a flavor dependent measurement. Hence the discrimination power between different theoretical models in the HERMES kinematical regime can be enhanced, and flavor dependent information about quark intrinsic transverse momenta and spin-orbit correlation can be accessed.

References

1. A. Bacchetta, *JHEP.* **0702**, 093 (2007).
2. R. N. Cahn, *Phys. Lett. B* **78**, 269 (1978).
   R. N. Cahn, *Phys. Rev. D* **40**, 3107 (1989).
3. D. Boer, P. J. Mulders, *Phys. Rev. D* **57**, 5780 (1998).
4. K. Ackerstaff et al. (HERMES collaboration), *Nucl. Instr. Meth. A* **417**, 230 (1998).
5. G. Cowan, *Statistical Data Analysis* (Clarendon Press, Oxford, 1998).
6. W. Kafer, in this proceedings.

MEASUREMENT OF TRANSVERSITY VIA AN INTERFERENCE-FRAGMANTATION FUNCTION AT HERMES

R. Fabbri

(On behalf of the HERMES Collaboration)

DESY, Notkestrasse 85, 22607 Hamburg, Germany

A single-spin asymmetry has been measured in the azimuthal distribution of $\pi^+\pi^-$ pairs produced in semi-inclusive deep-inelastic scattering on a transversely polarized hydrogen target. For the first time, evidence is found for a correlation between the transverse target polarization and the azimuthal orientation of the plane containing the two pions. The corresponding single-spin asymmetry is expected to be related to the product of the poorly known quark transversity distribution function and an unknown naive T-odd chiral-odd dihadron fragmentation function.

Keywords: Nucleon spin; Transversity.

1. Introduction

Three fundamental parton distribution functions describe the structure of the nucleon at leading twist: the unpolarized distribution, the helicity distribution, and the transversity distribution. The last one describes the distribution of transversely polarized quarks in a transversely polarized nucleon.[1] Being a chiral-odd quantity, transversity can be investigated only in observables that involve an additional chiral-odd quantity, e.g., in single-spin cross section asymmetries (SSA) in semi-inclusive DIS electroproduction (SIDIS) of single hadrons or hadron pairs on transversely polarized hadrons.

SSA in the production of single charged-hadrons in semi-inclusive DIS on a transversely polarized hydrogen and deuterium target were recently measured[2] by the HERMES collaboration and by the COMPASS collaboration, respectively. The so-called Collins and Sivers mechanisms[3] have been identified as potential sources of the asymmetry. In the Collins mechanism the asymmetry arises by the correlation between the transverse polarization of the struck quark and the transverse momentum of the produced hadron. In the measured SSA transversity appears convoluted with the Collins function.

An independent experimental constraint on transversity is provided by analysing SSA in semi-inclusive electroproduction of $\pi^+\pi^-$ pairs on a transversely polarized target.[4] The underlying mechanism is the transfer of the transverse spin of the fragmenting quark to the relative orbital angular momentum of the produced pion pair. Consequently, this mechanism does not require transverse momentum of the pair. In this process transversity appears with the up-to-now unknown dihadron fragmentation function describing the interference between pion pairs in different partial wave states.

This paper reports a measurement of an azimuthal moment of a single-spin asymmetry in semi-inclusive $\pi^+\pi^-$ production on a transversely polarized hydrogen target.[5]

2. Transversity in Pion-Pair Production

In semi-inclusive deep-inelastic scattering the cross-section single-spin asymmetry for pion-pair production is given, at leading twist and in leading order in the strong-coupling α_S, by[6]

$$A_{UT} \sim \frac{1}{|S_\perp|} \frac{\sum_q e_q^2 \left[\sin(\phi_{R\perp} + \phi_S) \sin\theta\right] \cdot h_1^q(x) H_{1,q}^{\sphericalangle}(z, M_{\pi\pi}, \cos\theta)}{\sum_q e_q^2 \, f_{1,q}(x) \cdot D_{1,q}(z, M_{\pi\pi})} , \quad (1)$$

where f_1 is the unpolarized distribution function, D_1 is the unpolarized dihadron fragmentation function, and $H_1^{\sphericalangle}$ is the twist-two chiral-odd naive T-odd interference fragmentation (dihadron) function. Here, $x = Q^2/(P{\cdot}q)$, with P and q representing the four-momenta of the target nucleon and of the virtual photon, respectively. The fraction of energy ν of the virtual photon (in the target of mass frame) which is transferred to the pion pair is espressed by $z = (E_{\pi^+} + E_{\pi^-})/\nu = z_{\pi^+} + z_{\pi^-}$. The quantity $M_{\pi\pi}$ is the invariant mass of the pion pair. The azimuthal angles $\phi_{R\perp}$ and ϕ_S, and the polar angle θ are described in Fig. 1. The vector $\boldsymbol{S}_\perp$ is the component of the target spin $\boldsymbol{S}$ perpendicular to the virtual-photon direction. The summation runs over the quark and antiquark flavors q with charges e_q.

Other contributions (subleading twist) to the polarized and unpolarized cross sections appear with different angular dependences.

Both dihadron fragmentation functions D_1 and $H_1^{\sphericalangle}$ can be expanded in

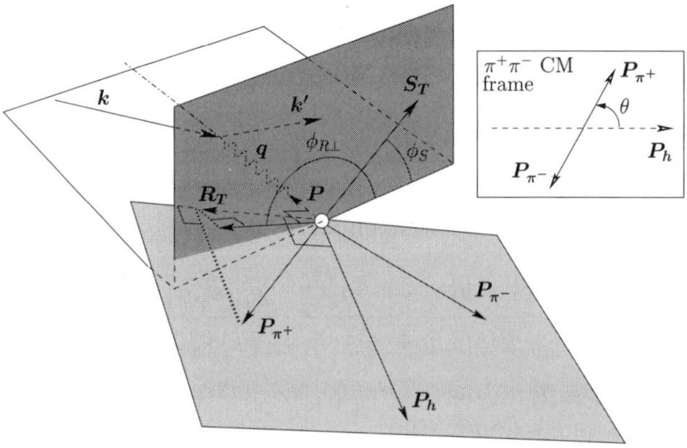

Fig. 1. Depiction of the azimuthal angles $\phi_{R\perp}$ and ϕ_S. Explicitly, $\phi_{R\perp} = \frac{(\boldsymbol{q}\times\boldsymbol{k})\cdot\boldsymbol{R_T}}{|(\boldsymbol{q}\times\boldsymbol{k})\cdot\boldsymbol{R_T}|}\arccos\frac{(\boldsymbol{q}\times\boldsymbol{k})\cdot(\boldsymbol{q}\times\boldsymbol{R_T})}{|\boldsymbol{q}\times\boldsymbol{k}||\boldsymbol{q}\times\boldsymbol{R_T}|}$ and $\phi_S = \frac{(\boldsymbol{q}\times\boldsymbol{k})\cdot\boldsymbol{S}_\perp}{|(\boldsymbol{q}\times\boldsymbol{k})\cdot\boldsymbol{S}_\perp|}\arccos\frac{(\boldsymbol{q}\times\boldsymbol{k})\cdot(\boldsymbol{q}\times\boldsymbol{S}_\perp)}{|\boldsymbol{q}\times\boldsymbol{k}||\boldsymbol{q}\times\boldsymbol{S}_\perp|}$. Here, $\boldsymbol{R_T} = \boldsymbol{R} - (\boldsymbol{R}\cdot\hat{\boldsymbol{P_h}})\hat{\boldsymbol{P_h}}$, with $\boldsymbol{R} \equiv (\boldsymbol{P}_{\pi^+} - \boldsymbol{P}_{\pi^-})/2$ and $\boldsymbol{P_h} \equiv \boldsymbol{P}_{\pi^+} + \boldsymbol{P}_{\pi^-}$. Both angles are evaluated in the virtual-photon-nucleon center-of-mass frame. The dotted lines represent projections of vectors. Also included is a description of the polar angle θ, which is evaluated in the center-of-mass frame of the pion pair.

terms of Legendre components in terms of $\cos\theta$. Hence,[6]

$$D_{1,q}(z, M_{\pi\pi}^2, \cos\theta) \simeq D_{1,q}(z, M_{\pi\pi}^2) + D_{1,q}^{sp}(z, M_{\pi\pi}^2)\cos\theta + D_{1,q}^{pp}(z, M_{\pi\pi}^2)\frac{1}{4}(3\cos^2\theta - 1) \qquad (2)$$

and

$$H_{1,q}^{\sphericalangle}(z, M_{\pi\pi}^2, \cos\theta) \simeq H_{1,q}^{\sphericalangle,sp}(z, M_{\pi\pi}^2) + H_{1,q}^{\sphericalangle,pp}(z, M_{\pi\pi}^2)\cos\theta, \qquad (3)$$

where the Legendre expansions are truncated to include only the s- and p-wave components, which is assumed to be a valid approximation in the range of the invariant mass $M_{\pi\pi} < 1.5$ GeV,[6] which is typical of the present experiment. The contributions to the fragmentation function $H_{1,q}^{\sphericalangle}$ are the interference between the s- and p-wave components of the $\pi^+\pi^-$ pair, $H_1^{\sphericalangle,sp}$, and the interference between two p-waves, $H_1^{\sphericalangle,pp}$.

In this paper the focus is on the sp-interference, since it has received the most theoretical attention. In particular, in Ref. 7 this function has been predicted to change sign at a very specific value of the invariant mass of the $\pi^+\pi^-$ pair, close to the mass of the ρ^0 meson. However, other models[8] predict a completely different behavior.

The single-spin asymmetry $A_{UT} \equiv \frac{1}{|\mathbf{S}_\perp|}\sigma_{UT}/\sigma_{UU}$ contains components of a simultaneus Fourier and Legendre expansion. The amplitude

$A_{UT}^{\sin(\phi_{R\perp}+\phi_S)\sin\theta}$ of the modulation of interest here, which is related to the product of transversity and the fragmentation function $H_1^{\sphericalangle,sp}$, is defined as:

$$A_{UT}^{\sin(\phi_R+\phi_S)\sin\theta} \equiv \frac{2}{|\boldsymbol{S}_\perp|}\frac{\int \mathrm{d}\cos\theta\,\mathrm{d}\phi_R\,\mathrm{d}\phi_S\,\sin(\phi_R+\phi_S)\,\mathrm{d}\sigma_{UT}^7/\sin\theta}{\int \mathrm{d}\cos\theta\,\mathrm{d}\phi_R\,\mathrm{d}\phi_S\,\mathrm{d}\sigma_{UU}^7}. \quad (4)$$

Using eqs. (1)- (3), it can be written as

$$A_{UT}^{\sin(\phi_R+\phi_S)\sin\theta} = \frac{-(1-y)\sqrt{1-4\frac{M_\pi^2}{M_{\pi\pi}^2}}}{2(1-y+\frac{y^2}{2})}\cdot\frac{\sum_q e_q^2\, h_1^q(x)\, H_{1,q}^{\sphericalangle,sp}(z,M_{\pi\pi})}{\sum_q e_q^2\, f_1^q(x)\, D_{1,q}(z,M_{\pi\pi})}, \quad (5)$$

with M_π being the pion mass. Due to the factor e_q^2, the amplitude is expected to be *up*-quark dominated.

The measurement reported here is the first evidence of a non-null naive T-odd chiral-odd dihadron fragmentation function $H_1^{\sphericalangle,sp}$ that can provide access to transversity.

3. The HERMES Experiment

The data presented here were recorded during the 2002-2004 running period of the HERMES experiment,[9] using the 27.6 GeV positron beam and a transversely polarized hydrogen gas target in the HERA storage ring at DESY. The open-ended target cell was fed by an atomic-beam source based on Stern-Gerlach separation and RF transitions of hydrogen hyperfine states. The nuclear polarization of the atoms was flipped at 1–3 min. time intervals, while both this polarization and the atomic fraction inside the target cell were continuously measured. The average value of the proton polarization $|\boldsymbol{S}|$ was 0.75 ± 0.05.

Scattered positrons and coincident hadrons are detected by the HERMES spectrometer. Its acceptance spans the scattering-angle range $40 < |\theta_y| < 140$ mrad and $|\theta_x| < 170$ mrad, corresponding to an almost full coverage in ϕ_S from 0 to 2π with only small gaps at $1.40 < \phi_S < 1.74$ rad and $4.54 < \phi_S < 4.88$ rad. Positrons are identified with an efficiency exceeding 98% and a hadron contamination of less than 1% using an electromagnetic calorimeter, a transition-radiation detector, a preshower scintillation counter, and a dual-radiator ring-imaging Čerenkov detector, mainly used here to identify charged pions with momentum $|\boldsymbol{P_\pi}| > 1$ GeV.

4. Extraction of the Asymmetries

Events were selected with the kinematic requirements $W^2 > 4\ \mathrm{GeV}^2$, $0.1 < y < 0.85$, and $Q^2 > 1\ \mathrm{GeV}^2$, where W is the invariant mass of the initial photon-nucleon system and $y = (P \cdot q)/(P \cdot k)$, with k representing the four-momentum of the incident lepton. A constraint was placed on the missing energy: $\Delta E \equiv (M_X^2 - M_p^2)/2M_p > 2$ GeV, with M_X the missing mass and M_p the proton mass. This avoids contributions from exclusive two-pion production, where factorization in distribution functions and fragmentation functions cannot be applied. All possible combinations of detected $\pi^+\pi^-$ pairs were included for each event, in contrast to keeping only the combination with the largest energy fraction z, a choice for which fragmentation functions are not defined.

The results reported here are extracted from the experimentally measured single-spin asymmetry

$$A_{U\perp} \equiv \frac{1}{|\boldsymbol{S}_\perp|}\frac{N^\uparrow - N^\downarrow}{N^\uparrow + N^\downarrow}, \tag{6}$$

where $N^{\uparrow(\downarrow)}$ is the luminosity-normalized number of semi-inclusive $\pi^+\pi^-$ pairs detected while in the target $\uparrow(\downarrow)$ spin state (with respect to the beam direction). To eliminate the effects of the natural polarization of the HERA positron beam, a data set with a negligible net beam polarization was selected. The asymmetry is evaluated as a function of the Bjorken scaling variable x, z, the invariant mass $M_{\pi\pi}$ of the $\pi^+\pi^-$ pair, and the angles $\phi_{R\perp}$, ϕ_S, and θ.

A two-dimensional binned χ^2 fit was performed in $(\phi_{R\perp} + \phi_S)$ versus $\theta' \equiv ||\theta - \pi/2| - \pi/2|$ with a function of the form:

$$A_{U\perp} = \sin(\phi_{R\perp} + \phi_S)\frac{a \sin\theta'}{1 + b\frac{1}{4}(3\cos^2\theta' - 1)}, \tag{7}$$

where $a \equiv A_{U\perp}^{\sin(\phi_{R\perp}+\phi_S)\sin\theta}$ is a free parameter of the fit, while b is varied to study the influence of the unknown contribution $D_{1,q}^{pp}$ to the polarization-averaged 2-hadron cross section. The fit is evaluated as a function of θ', which corresponds to an antisymmetrization of the fit around $\theta = \pi/2$. This has the advantage that the contributions to $A_{U\perp}$ containing $D_{1,q}^{sp}$ and $H_{1,q}^{\sphericalangle,pp}$ drop out (see, e.g., Eqs. (2) and (3)), reducing the statistical uncertainty on $a \equiv A_{U\perp}^{\sin(\phi_{R\perp}+\phi_S)\sin\theta}$, the modulation amplitude of interest that approximates $A_{UT}^{\sin(\phi_{R\perp}+\phi_S)\sin\theta}$ defined in Eq. 5. The value of the fit parameter a depends on the value of b. Therefore, a systematic uncertainty was assigned to the extracted value of a by studying its responce to variation of b within its positivity limits.[5]

The amplitude value extracted from events summed over the experimental acceptance is $A_{U\perp}^{\sin(\phi_{R\perp}+\phi_S)\sin\theta} = 0.018 \pm 0.005_{\mathrm{stat}} \pm 0.002_{\mathrm{b-scan}}$ with an

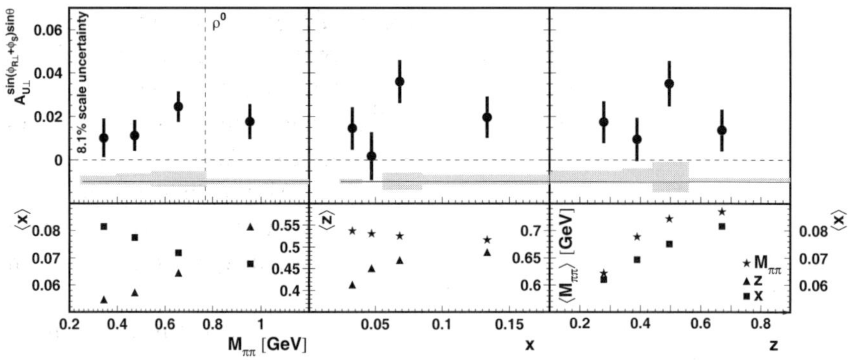

Fig. 2. The top panels show $A_{U\perp}^{\sin(\phi_{R\perp}+\phi_S)\sin\theta}$ versus the invariant mass $M_{\pi\pi}$ of the $\pi^+\pi^-$ pair, x, and z. The bottom panels show the average values of the variables that were integrated over. For the dependence on x and z, $M_{\pi\pi}$ was constrained to the range 0.50 GeV $< M_{\pi\pi} <$ 1.0 GeV, where the signal is expected to be largest. The error bars show the statistical uncertainty. A scale uncertainty of 8.1% arises from the uncertainty in the target polarization. Other contributions to the systematic uncertainty are summed in quadrature and represented by the asymmetric error band.

additional 8.1% relative scale uncertainty coming from the uncertainty in the determination of the target polarization. The ranges selected in x and $M_{\pi\pi}$ are $0.023 < x < 0.400$ and $0.5 < M_{\pi\pi} < 1.0$ GeV. The mean values of the kinematic variables are $\langle x \rangle = 0.07$, $\langle y \rangle = 0.64$, $\langle Q^2 \rangle = 2.35$ GeV^2, $\langle z \rangle = 0.43$, and $\langle |\boldsymbol{P}_{h\perp}| \rangle = 0.42$ GeV, where $\boldsymbol{P}_{h\perp}$ is the component of the momentum $\boldsymbol{P}_h$ of the pion pair transverse to the virtual-photon direction.

The values of the amplitudes $A_{U\perp}^{\sin(\phi_{R\perp}+\phi_S)\sin\theta}$ extracted are shown in Fig. 2 as functions of $M_{\pi\pi}$, x, and z. They are positive over the entire range of all three variables. The reduced-χ^2 values for these fits are in the range 0.70–1.38. The $M_{\pi\pi}$ dependence of the measured amplitude shows no sign change at the ρ^0 mass, contrary to the prediction in Ref. 7. The dependence on $M_{\pi\pi}$ and z is consistent with the model calculation of Ref. 8, but the size is substantially lower, indicating that the model probably overestimates interference effects.

These amplitudes are not influenced by the addition in the fit of terms of the form $\sin\phi_S$ (which appears at subleading twist in the polarized cross section σ_{UT}), or of the term $\cos\phi_{R\perp}\sin\theta$ (which appears at subleading twist in the unpolarized cross section σ_{UU}). These angular combinations exhaust the possibilities up to subleading twist. There was also no influence from the addition to the fit of a constant term, the latter being consistent with zero.

Tracking corrections that were applied for the deflections of the scattered particles caused by the vertical 0.3 T target holding field had also only a negligible effect on the extracted asymmetries.

The fully differential asymmetry depends on 9 kinematic variables: x, y, z, $\phi_{R\perp}$, ϕ_S, and θ, $M_{\pi\pi}$ and $\boldsymbol{P_{h\perp}}$ ($\mathrm{d}\boldsymbol{P_{h\perp}} = |\boldsymbol{P_{h\perp}}|\,\mathrm{d}|\boldsymbol{P_{h\perp}}|\,\mathrm{d}\phi_h$). Due to the limited statistics, it is not possible to measure the asymmetry $A_{U\perp}$ fully differential in all relevant variables. Combined with the fact that the HERMES spectrometer does not have a full 4π acceptance, this implies that the measured number of events is always convolved with the experimental acceptance. A systematic uncertainty was estimated based on a Monte Carlo study.[5]

The largest effect was seen when comparing the amplitudes in 4π and in HERMES acceptance. Using a particular choice for transversity and for each of the dihadron fragmentation functions, the extracted amplitudes in the HERMES acceptance were up to 43% underestimated for the z dependence when compared to amplitudes extracted in a 4π coverage. The effect was negligible for the x-dependence. The incomplete integration over $\boldsymbol{P_{h\perp}}$ was found to have only a small influence on the extracted amplitudes. In view of the large uncertainties above it can be neglected.

The interpretation of the measured amplitudes can, in principle, be complicated by the experimental condition that the target nucleons are polarized transverse to the beam axis instead of a direction transverse to the virtual-photon direction. These *beam-axis* asymmetries can receive contributions not only from the transverse component of the nucleon spin with respect to the virtual-photon direction but also from a small longitudinal component proportional to $\sin\theta_{\gamma^*}$, where θ_{γ^*} is the angle between the directions of the virtual photon and the incoming lepton beam. Such a contribution to the amplitude presented here can occur only when a $\sin\phi_{R\perp}$ amplitude exists in the corresponding asymmetry A_{UL}, i.e., the *photon-axis* asymmetry in dihadron lepto-production with an unpolarized beam on a longitudinally polarized target.[10] Such an amplitude exists at subleading twist,[11] but was found to be small for pairs of unidentified hadrons.[12] In addition, $\langle\sin\theta_{\gamma^*}\rangle$ was typically less than 0.09,[12] leading to an insignificant difference in the presented amplitude for lepton-axis and photon-axis asymmetries.

Besides this contribution, no other twist-3 effects are present in the measured amplitude. Besides this twist-3 contribution, modifications due to even higher twist and NLO effects are unknown for dihadron production in DIS. However, the dominant NLO contribution to the "longitudinal" cross section σ_L are known to be up to 30% for the unpolarized *inclusive* DIS cross section in these kinematic conditions.[13]

5. Summary

A measurement of $A_{U\perp}^{\sin(\phi_{R\perp}+\phi_S)\sin\theta}$ of the transverse-target-spin asymmetry in the leptoproduction of $\pi^+\pi^-$ pairs has provided the first evidence that the naive T-odd chiral-odd dihadron fragmentation function $H_1^{\sphericalangle}$ and in particular $H_1^{\sphericalangle,sp}$ is nonzero. The amplitude is positive in the whole range in the invariant mass of the $\pi^+\pi^-$ pairs in contrast to a previous expectation[7] of a sign change around the mass of the ρ^0 meson. The fragmentation function for which this measurement has provided the first strong evidence also determines fragmentation in *pp* collisions at RHIC which will be exploited to measure transversity. The BELLE collaboration plans to extract dihadron fragmentation functions from their data. Such results could then be combined with DIS and *pp* data to extract transversity in the proton.

References

1. V. Barone et al., Phys. Rept. **359**, 1 (2002).
2. HERMES Collaboration, A. Airapetian et al., Phys. Rev. Lett. **94**, 012002 (2005); M. Diefenthaler (for the HERMES Collaboration), Proc. of the 15th International Workshop on Deep Inelastic Scattering, and QCD, Munich, Germany, April 2007; COMPASS Collaboration, V. Y. Alexakhin et al., Phys. Rev. Lett. **94**, 202002 (2005); COMPASS Collaboration, E. S. Ageev et al., Nucl. Phys. B **765** 31 (2007); COMPASS Collaboration, M. Alekseev et al., hep-ex/0802.2160.
3. J. C. Collins, Nucl. Phys. B **396**, 161 (1993); D. W. Sivers, Phys. Rev. D **41**, 83 (1990).
4. A. V. Efremov et al., Phys. Lett. B **284**, 394 (1992); J. C. Collins et al., Nucl. Phys. B **420**, 565 (1994).
5. HERMES Collaboration, A. Airapetian et al., JHEP 06, 017 (2008).
6. A. Bacchetta and M. Radici, Phys. Rev. D **67**, 094002 (2003).
7. R. L. Jaffe et al., Phys. Rev. Lett. **80**, 1166 (1998).
8. A. Bacchetta and M. Radici, Phys. Rev. D **74**, 114007 (2006); J. She et al., Phys. Rev. D **77**, 014035 (2008).
9. HERMES Collaboration, K. Ackerstaff et al., Nucl. Instrum. Meth. A **417**, 230 (1998).
10. M. Diehl and S. Sapeta, Eur. Phys. J. C **41**, 515 (2005).
11. A. Bacchetta and M. Radici, Phys. Rev. D **69**, 074026 (2004).
12. P. B. van der Nat and K. Griffioen (for the HERMES Collaboration), Proc. of 6th international spin physics symposium, Trieste, Italy, October 2004.
13. L. W. Whitlow et al., Phys. Lett. B **250**, 193 (1990).

RELATING LEADING AND HIGHER TWIST CONTRIBUTIONS TO NUCLEON SPIN STRUCTURE

P.G. Ratcliffe

Dipartimento di Fisica e Matematica, Università degli Studi dell'Insubria, via Valleggio 11, 22100 Como, Italy

O.V. Teryaev

Bogoliubov Laboratory of Theoretical Physics, Joint Institute for Nuclear Research, Dubna 141980, Russia

Various methods of relating (naively) twist-2 Sivers function and twist-3 quark–gluon correlators are reviewed. The applications of Sivers function at large transverse momenta provides a justification (up to colour factors) of practical methods of calculations used earlier by Torino group. These factors correspond to colour correlations between partons inside hadrons and the ones participating in hard processes. The particular case of such correlations is represented by sign change between SIDIS and Drell–Yan processes.

Keywords: twist, effective, colour.

1. Introduction

The twist expansion is one of the cornerstones of the applications of QCD to hadronic physics. Usually, one performs the analysis at the leading-twist (LT) level neglecting the existence of higher twists (HT). However, the very separation into leading and higher twist is rather more mathematical than physical and there are some examples of the relations between LT and HT.

We address here the particular, but very important, case of single-spin asymmetries (SSA), which represent one of the most subtle and intriguing effects in QCD. In the simplest inclusive processes parity conservation requires a transversely polarised beam or target. Inasmuch as the transverse polarisation component is not enhanced by Lorentz boosts, one immediately encounters the necessity of describing twist-3 effects. This can be achieved via use of either local[1] or non-local[2–5] operators.

The latter approach also permits the description of the imaginary phases required to produce T-odd effects, such as SSA. These phases mimic true T(CP) violation (see *e.g.* Ref. 6) and allow T-odd effects in a T-conserving theory, such as QCD. The phases emerging from gluon loops describing initial- and final-state interactions (ISI and FSI) in hard subprocesses are suppressed by powers of light-quark masses and the QCD coupling constant.[7] However, deeper analysis[8] shows that quark masses should be substituted by hadronic mass scales. Moreover, ISI and FSI between the hard and soft regions of QCD factorisation, which is just the physical picture corresponding to twist three, lead to SSA free of both suppression factors.[9] The imaginary phase is generated by gluon correlations with soft quarks; the situation when the gluon is instead soft was also considered later.[10]

An alternative description of SSA effects is provided by a T-odd transverse-momentum dependent (TMD) distribution function, first introduced by Sivers.[11] It happens to be related to twist-3 correlators and currently there are several ways to establish such a relation.

2. Ways from Sivers function to twist three

i) *Effective function*
As soon as there is no kinematical variable whose cut produces an imaginary phase in the hadron–parton transition amplitude, this may simply become an effective function,[6] so that the phase also emerges owing to the ISI and FSI involving hard subprocess. The first case of the appearance of an effective T-odd distribution was found[12] for soft-gluon SSA in the Drell–Yan (DY) process integrated over transverse momenta.[13]

ii) *Relations at the matrix-element level between Sivers-function moments and twist 3*
This "empirically" found effective function was later identified[14] with the *first moment* of the Sivers function, which plays a special role in what follows.

iii) *Unsuppressed twist 3 at low transverse momentum and the BHS model*
The role of FSI between the hard and soft regions of semi-inclusive deeply inelastic scattering (SIDIS) was clearly revealed in the model of Brodsky, Hwang and Schmidt,[15] where it was interpreted[16] as a manifestation of the Sivers function. The crucial role of hard processes in defining this function was made manifest by the discovery of a sign difference between SIDIS and DY. This is all qualitatively similar to earlier findings[17] in the twist-3 case. However, the apparent difference between the FSI arising in twist-3 interactions is the absence of true power suppression. The situation is, though,

even more peculiar. In the hard Abelian process of semi-inclusive production of a real photon by a deeply virtual photon (SIDVCS, the semi-inclusive counterpart of the well-known DVCS[18] process) an overall suppression as Mp_T/Q^2 was reported at Transversity 2005 in Como[19] to be compensated by a gluonic pole in the quark–gluon correlator, which is approached at low $p_T \ll Q$ as the gluon momentum fraction is defined by kinematics $x_g \sim p_T^2/Q^2$, indicating the possibility to obtain unsuppressed (in Q) twist-3 effects. However, the very application of twist 3 at low p_T is, strictly speaking, problematic and a rigorous proof is lacking.

iv) *Matching of Sivers function and twist 3*
Fortunately, analogous conclusions that the Sivers function and gluonic poles describe similar physics for different p_T have been reached within the framework of a general proof.[20] These analyses imply a picture in which the Sivers function is limited to the low-p_T region, where a special type of factorisation[21] is assumed valid and either the continuation of the twist-3 result to lower p_T[19] or matching[20] of high- and low-p_T results is adopted.

v) *Sivers function at large transverse momentum*
One may ask if it is possible to extend the application of the Sivers function to the region originally allocated for twist 3. To do so, we suggested a different, complementary approach, to apply the Sivers function at *high* p_T. This is of special importance for hadronic processes where p_T is the only hard scale. We present general quantitative relations between the Sivers function and gluonic poles, using master formulæ[22] for the latter, and find that besides the sign there are important process-dependent colour factors (*cf.* Refs. 23,24, where such colour factors were calculated by considering gauge links) modifying the Sivers function and underlining its effective nature.

3. From the Sivers function to gluonic poles

To prove the relation between twist three and the Sivers function, we shall not attempt to obtain the latter as some special limit[19,20] of a twist-3 contribution, but instead transform an approximation of it to the form[22] appearing in the twist-3 calculation. That is, we shall provide *a posteriori* proof of the following factorised formula involving the Sivers function

$$d\Delta\sigma \sim \int d^2k_T\, dx\, f_S(x,k_T)\, \mathrm{Tr}\big[\gamma_\rho H(xP,k_T)\big]\, \epsilon^{\rho s P k_T}, \qquad (1)$$

where other (unpolarised and collinear) distribution or fragmentation functions should normally also be present. This leads to the dependence of H on the respective transverse momenta. This, and the presence of only one TMD distribution function, makes a notable difference with respect to standard TMD factorisation,[21,25] where all non-perturbative ingredients are transverse-momentum dependent, so that our approach would mean the justification of the Generalised Parton Model.[26]

We shall now prove that at large transverse momenta it is related to the contribution of gluonic poles in twist-3 factorisation. This will allow us to justify the applicability of the Sivers function in an extended kinematical region and to estimate the accuracy of such an application, depending on the degree to which the gluonic-pole contribution may be considered dominant. To achieve this goal, we expand the subprocess coefficient function H in powers of k_T, retaining only the first non-vanishing term:

$$d\Delta\sigma \sim \int d^2k_T \, dx \, f_S(x,k_T) \, \mathrm{Tr}\left[\gamma_\rho \frac{\partial H(xP,k_T)}{\partial k_T^\alpha}\right]_{k_T=0} k_T^\alpha \, \epsilon^{\rho s P k_T}. \tag{2}$$

This is a natural step in selecting twist-3 terms,[3,4] justified by the expected rapid decrease of the Sivers function with respect to any hard scale determining the k_T dependence of the hard kernel, and is a crucial element of our proof.

The k_T integration now only includes soft parts and may thus be expressed via moments of the Sivers function, which may otherwise appear due to particular definitions of some measurable asymmetries.[24] Let us first average over the directions of k_T using the standard relation

$$\left\langle k_T^\mu k_T^\nu \right\rangle = -\tfrac{1}{2} g_T^{\mu\nu} \left\langle k_T^2 \right\rangle, \tag{3}$$

where $g_T^{\mu\nu} \equiv g^{\mu\nu} - P^\mu n^\nu - n^\mu P^\nu$ is defined with respect to the same light-cone vectors P and n ($P{\cdot}n = 1$) that define the direction of k_T, *i.e.* $P{\cdot}k_T = 0 = n{\cdot}k_T$. On substitution into (2), one obtains

$$d\Delta\sigma \sim -M \int dx \, f_S^{(1)}(x) \, \mathrm{Tr}\left[\gamma_\rho \frac{\partial H(xP,k_T)}{\partial k_T^\alpha}\right]_{k_T=0} \left(\epsilon^{\rho s P \alpha} - P^\alpha \epsilon^{\rho s P n}\right), \tag{4}$$

where

$$f_S^{(1)}(x) = \int d^2k_T \, f_S(x,k_T) \, \frac{k_T^2}{2M^2}. \tag{5}$$

The final step now exploits the following kinematic identity:[3]

$$\epsilon^{\rho s P \alpha} = P^\alpha \epsilon^{\rho s P n} - P^\rho \epsilon^{\alpha s P n}, \tag{6}$$

which results from the vanishing of any totally antisymmetric fifth-rank tensor in four-dimensional space. This allows the two terms in (4) to be recombined:

$$d\Delta\sigma \sim M \int dx\, f_{\mathrm{S}}^{(1)}(x)\ \mathrm{Tr}\left[\not{P}\frac{\partial H(xP, k_T)}{\partial k_T^\alpha}\right]_{k_T=0} \epsilon^{\alpha s P n}. \tag{7}$$

The above-mentioned dependence of H on the momenta of other particles, described by collinear distribution and fragmentation functions, is now crucial as only these momenta may carry the index α while the derivative is calculated. If other detected particles are absent, as in the case of DIS, the asymmetry is, of course, identically zero. The key observation is that this expression exactly coincides with the recently obtained master formula[22] for the contribution of twist-3 gluonic poles in high-p_T processes. The Sivers distribution can then be identified with the gluonic pole strength $T(x,x)$ multiplied by a process-dependent colour factor. In turn, the sign of the Sivers function is fixed according to which of ISI or FSI is relevant:

$$f_{\mathrm{S}}^{(1)}(x) = \sum_i C_i \frac{1}{2M} T(x,x), \tag{8}$$

where C_i is a relative colour factor, defined with respect to an Abelian subprocess (say SIDVCS discussed above, where it is just C_F), which is naturally absorbed into the definition of the quark–gluon correlator.[3] As discussed below, this is also the factor appearing in low-p_T SIDIS and DY at the Born level.

The relation established is a principal result of this paper. It completes the *a posteriori* proof of (1) and relates the twist-3 factorisation to a *modified* (by colour factors) factorisation in terms of Sivers function. The second moment of the Sivers function enters the original expression (1) with a factor M instead of $1/M$, indicating its twist-3 nature. This may be seen immediately by defining the Sivers function in coordinate (impact-parameter) space, in a manner similarly to earlier discussions[27] of the Collins fragmentation function:

$$\langle P, s|\bar\psi(0)\gamma_\rho\psi(z)|P, s\rangle \sim M\epsilon^{\rho s P z} \int dx\ e^{izx} f_{\mathrm{S}}^{(1)}(x). \tag{9}$$

Note too that higher Sivers-function moments enter with higher derivatives of the coefficient function and therefore correspond to higher twist (5, 7, 9, ...). The entire k_T-dependent Sivers function thus corresponds to a resummed infinite tower of higher twists. This property has also been studied in coordinate space,[27] where k_T-dependent functions represent a complete similarity with non-local quark condensates. The latter manifest a similar resummation of an infinite tower of higher twists (see *e.g.* Ref. 28 and refs. therein), but for vacuum rather than hadronic matrix elements.

4. Colour factors and the transition from large to small transverse momenta

Let us consider some particular applications of this relation, starting with high-p_T SIDIS. In this case there are only final-state interactions, while the colour factors differ for mesons produced in fragmentation of quarks $(-1/2N_c)$. This shows that there is a specific enhancement in the latter, which is of special importance for K^- mesons.

Matching of the large- and small-p_T descriptions is an intriguing question as there is no sharp border. Clarification may be obtained by comparison with the perturbative Sivers-function calculation:[20] there is a hard (in addition to soft) gluonic-pole contribution (when the gluon carries a finite momentum fraction) restoring the correct "Abelian" colour factor C_F. Moreover, the entire surviving contribution is just due to hard poles.

It is interesting to consider the possibility of describing these effects within the framework of a model, such as that of.[15] To do so, one may modify it by considering the emission of an extra hard gluon, in order to allow for high p_T. Moreover, to distinguish the colour factors, one should attribute some colour-charge exchange to both the FSI and to the emission of the extra hard gluon. Another possibility may be to consider the *electric* instead of colour charge, following the recent idea of Collins and Qiu.[29]

Generalising their approach, one may consider the emission of a hard charged gluon, balancing the large transverse momentum. In this case, the hard gluonic pole leading to the Abelian colour factor (or to the electric-charge factor corresponding to the charge of the quark from the polarised target in the approach of Ref. 29) in SIDIS corresponds to FSI that occur *before* emission of this extra gluon. At the same time, the description of the twist-3 mechanism at high p_T in terms of an effective Sivers function related to soft poles emerges when the FSI happen after emission of the extra gluon. Even in SIDIS factorisation, in the sense of Ref. 29, is therefore broken by the emission of charged gluons; that is, in the sense of the present paper, it is modified by the (colour) charge factor.

We note the similarity with the SIDVCS calculation,[19] where the entire contribution associated with the Sivers function is also due to hard poles, since for finite p_T there are no soft poles at all. With decreasing p_T, the hard pole becomes soft and produces the Sivers function. We therefore expect that, with decreasing p_T, the increasing contributions of softening hard poles should smoothly alter the colour factor of quark fragmentation contributions to C_F while the gluonic contributions disappear, both effects occurring at typical hadronic scales. It would be challenging to verify this theoretical picture by experimental observation. We might also note that the Abelian colour factor is recovered if the fragmentation probabilities from quark and gluon jets coincide. This is natural, since for small p_T there is no way to distinguish or resolve quarks and collinearly emitted gluons.

To experimentally verify such a picture, it would be of major importance to distinguish between mesons originating from either quark or gluon fragmentation at large p_T. While a complete separation is impossible, there are methods that can help. Firstly, one may use jet shape, which differs for quark and gluon jets owing to the different spins of the fragmenting objects. This difference in spin can also be seen in the tensor polarisation of vector mesons.[30] However, most promising would seem to be exploration of the different z-dependences in quark and gluon fragmentation functions. The faster decrease of the latter should result in dramatic variations of SSA, so that at low z gluon fragmentation would be dominant with a colour factor $N_c/2$, while at large z one would expect a sign change and transition to quark fragmentation with a factor $-1/2N_c$.

Note that at high p_T there is thus a simple and direct connection between the Sivers functions for SIDIS and DY processes. Indeed, the relation is then

$$f_{\mathrm{S}}^{\mathrm{SIDIS},i} = -f_{\mathrm{S}}^{\mathrm{DY},i} \quad (i = q, g), \tag{10}$$

which holds separately for quark and gluon contributions. In DY this entails dilepton production by quarks or gluons from an unpolarised hadron, while in SIDIS it corresponds to meson production via quark or gluon fragmentation. The observable asymmetries are therefore proportional to distribution and fragmentation functions respectively. This relation is also very interesting to test experimentally, especially since the (unpolarised) gluon distribution function is much better known than that for fragmentation.

Let us now turn to hadronic processes, starting with the simplest: direct-photon production. There are only initial-state interactions with gluons, resulting in the very simple relation

$$F_{\mathrm{S}}^{hh\to\gamma X} = \frac{N_c}{2}\, f_{\mathrm{S}}^{\mathrm{DY}}. \tag{11}$$

Exploration of this process in various kinematical regions can provide information on the gluon Sivers function.[31] There is little doubt that this is also an effective function, related to the three-gluon correlators considered earlier in relation to pion SSA[32] and the DIS structure function g_2.[33,34] The generalisation of our approach to the case of three-gluon correlators is thus an important task. Quark–gluon processes are more complicated: a list of colour factors relevant for twist-3 subprocesses may be found in

Refs. 23,24. Let us only mention that FSI for pions produced in quark fragmentation may be re-expressed in a manifestly gauge-invariant manner via the summation formula

$$t^a S t^a = -\frac{1}{2N_c} S + \frac{1}{2} \cdot \mathrm{Tr}\, S. \tag{12}$$

The first term corresponds to the usual Sivers function[26] with colour factor $-1/2N_c$ and the second to the Abelian Compton subprocess, with s- and u-channel diagrams contributing with the same factors while the t-channel is absent. Both terms are separately gauge invariant.

Such a dramatic modifications of hadronic processes may be considered as another way of describing what, in the model approach[29] discussed above, was termed the violation of factorisation in hadronic processes. There the calculation was performed in an Abelian model similar to that of Ref. 15.

The *essential* point elucidated in Ref. 29 is the proportionality of the contribution of the Sivers function to the electric charge of the quark from the other (unpolarised) hadron. This observation has a direct counterpart in our approach. The colour factor is defined by the colour charge of the parton participating in the ISI and FSI. This charge is, generally speaking, independent of the properties of the polarised hadron emitting the gluon that participates in the ISI and FSI and, in this sense, breaks factorisation. SIDIS and DY processes at low p_T are exceptional: the colour charge of the quark participating in the FSI in SIDIS is the same as that of the quark originally emitted by polarised hadron. By the same token, the colour charge of the antiquark participating in the ISI in DY processes at low p_T is just the opposite, which explains the Collins sign rule. At the same time, the emission of a hard gluon changes these colour charges in high-p_T SIDIS, DY processes and, needless to say, other hadronic processes. This modification of colour charge alters the effective Sivers function.

5. Discussion and conclusions

We have suggested and proved here a method of applying the Sivers distribution at large transverse momenta. We have shown that the Sivers function is, in effect, none other than an expression of the contribution of gluonic poles. It is therefore process dependent and this dependence includes, besides the sign related to the ISI and FSI responsible for the imaginary phase, a colour factor. The *validity* of factorisation in terms of twist-3 correlators thus leads to a *violation* or *modification* in terms of the Sivers function.

The colour factor is defined by the colour charge of the initial and final partons participating in the hard scattering. Its simplest manifestation is provided by the sign difference between SIDIS and DY processes at low p_T. At the same time, at high p_T and in hadronic reaction these factors are much more complicated.

Such a picture is complementary to that considered previously, in which matching between the Sivers function and twist-3 matrix elements occurred in the region where, strictly speaking, factorisation formulæ were not valid. The matching between various p_T regions now takes the form of a p_T-dependent colour factor. We have studied this dependence in SIDIS, where it is a sort of colour separation, similar to the phenomenon of colour transparency. That is, at low p_T one cannot distinguish between mesons originating from quark and gluon fragmentation. As soon as one is able to do so at larger p_T, they enter the Sivers asymmetry with dramatically different colour factors, in both sign and scale.

This complementary method of establishing a relation between the Sivers function and twist-3 matrix elements lends support to the possibility of global fits of Sivers functions,[19] including lepton–hadron and hadron–hadron processes, as well as DIS, where twist three also contributes. We have shown that transverse moments of the Sivers function correspond to increasing twists, starting at three. The entire function corresponds to a resummed tower of twists, which is just the object to be considered at low p_T. Our result is especially relevant for those hadronic processes in which p_T is the only hard scale. We have proved that direct-photon production is described by a colour factor $N_c/2$ and have suggested a method to rearrange the colour factors in a manifestly gauge-invariant way.

Acknowledgments

O.V.T. is grateful to the Cariplo Science Foundation for support during his stay at the University of Insubria in Como and to the Department of Physics and Mathematics of the University for kind hospitality. This work was partially supported by the Deutsche Forschungsgemeinschaft, grant 436 RUS 113/881/0, the Russian Foundation for Basic Research (Grant 03-02-16816) and the Russian Federation Ministry of Education and Science (Grant MIREA 2.2.2.2.6546).

References

1. E.V. Shuryak and A.I. Vainshtein, Nucl. Phys. **B201**, 141 (1982).
2. A.P. Bukhvostov, É.A. Kuraev and L.N. Lipatov, JETP Lett. **37**, 482 (1983).
3. A.V. Efremov and O.V. Teryaev, Sov. J. Nucl. Phys. **39**, 962 (1984).
4. P.G. Ratcliffe, Nucl. Phys. **B264**, 493 (1986).

5. I.I. Balitsky and V.M. Braun, Nucl. Phys. **B311**, 541 (1989).
6. O.V. Teryaev, in proc. of the *Circum-Pan-Pacific RIKEN Symp. on High Energy Spin Physics* (Wako, Nov. 1999), eds. S. Kumano, T.-A. Shibata, Y. Watanabe and K. Yazaki; RIKEN Rev. **28**, 101 (2000).
7. G.L. Kane, J. Pumplin and W. Repko, Phys. Rev. Lett. **41**, 1689 (1978).
8. A.V. Efremov and O.V. Teryaev, Sov. J. Nucl. Phys. **36**, 140 (1982).
9. A.V. Efremov and O.V. Teryaev, Phys. Lett. **B150**, 383 (1985).
10. J. Qiu and G. Sterman, Phys. Rev. Lett. **67**, 2264 (1991).
11. D. Sivers, Phys. Rev. **D41**, 83 (1990).
12. D. Boer, P.J. Mulders and O.V. Teryaev, Phys. Rev. **D57**, 3057 (1997).
13. N. Hammon, O. Teryaev and A. Schäfer, Phys. Lett. **B390**, 409 (1997).
14. D. Boer, P.J. Mulders and F. Pijlman, Nucl. Phys. **B667**, 201 (2003).
15. S. J. Brodsky, D. S. Hwang and I. Schmidt, Phys. Lett. **B530**, 99 (2002).
16. J. C. Collins, Phys. Lett. **B536**, 43 (2002); A.V. Belitsky, X. Ji and F. Yuan, Nucl. Phys. **B656**, 165 (2003).
17. O.V. Teryaev, in proc. of the *10th. Int. Workshop on Deep Inelastic Scattering—DIS 2002* (Kraków, Apr.–May 2002), eds. A. Eskreys, J. Kwieciński and J. Szwed; Acta Phys. Polon. **B33**, 3749 (2002).
18. X. Ji, Phys. Rev. **D55**, 7114 (1997).
19. O.V. Teryaev, in proc. of the *Int. Workshop on Transverse Polarisation Phenomena in Hard Processes—Transversity 2005* (Como, Sept. 2005), eds. V. Barone and P.G. Ratcliffe (World Sci., 2006), p. 276.
20. X. Ji, J.-W. Qiu, W. Vogelsang and F. Yuan, Phys. Rev. Lett. **97**, 082002 (2006).
21. J.C. Collins and A. Metz, Phys. Rev. Lett. **93**, 252001 (2004).
22. Y. Koike and K. Tanaka, Phys. Lett. **B646**, 232 (2007); hep-ph/0703169.
23. C.J. Bomhof and P.J. Mulders, in proc. of the *XVII Int. Symp. on High-Energy Spin Physics—SPIN 2006* (Kyoto, Oct. 2006), eds. K. Imai, T. Murakami, N. Saito and K. Tanida; AIP Conf. proc. **915**, 563 (2007).
24. C.J. Bomhof, P.J. Mulders, W. Vogelsang and F. Yuan, Phys. Rev. **D75**, 074019 (2007); A. Bacchetta *et al.*, hep-ph/0703153.
25. X.-d. Ji, J.-p. Ma and F. Yuan, Phys. Rev. **D71**, 034005 (2005).
26. M. Anselmino *et al.*, Phys. Rev. **D71**, 014002 (2005).
27. O.V. Teryaev, in proc. of the *X Adv. Research Workshop on High Energy Spin Physics—SPIN-03* (Dubna, Sept. 2003), eds. A.V. Efremov and O.V. Teryaev; Phys. Part. Nucl. **35**, S24 (2004).
28. A.P. Bakulev and S.V. Mikhailov, Phys. Rev. **D65**, 114511 (2002).
29. J. Collins and J.-W. Qiu, Phys. Rev. **D75**, 114014 (2007).
30. A. Schäfer, L. Szymanowski and O.V. Teryaev, Phys. Lett. **B464**, 94 (1999).
31. I. Schmidt, J. Soffer and J.-J. Yang, Phys. Lett. **B612**, 258 (2005).
32. X. Ji, Phys. Lett. **B289**, 137 (1992).
33. A.V. Belitsky, A.V. Efremov and O.V. Teryaev, Phys. Atom. Nucl. **58**, 1253 (1995).
34. A.V. Belitsky, X. Ji, W. Lu and J. Osborne, Phys. Rev. **D63**, 094012 (2001).

DIS AT LOW AND HIGH TRANSVERSE MOMENTUM: MATCHES AND MISMATCHES

A. Bacchetta[1], D. Boer[2], M. Diehl[3], and P.J. Mulders[2]

[1] *Jefferson Lab, 12000 Jefferson Ave., Newport News, VA 23606, USA*
[2] *Vrije Universiteit Amsterdam, 1081 HV Amsterdam, The Netherlands*
[3] *Deutsches Elektronen-Synchroton DESY, 22603 Hamburg, Germany*

This talk reports on recent work where we studied the connection between the description of semi-inclusive DIS at high transverse momentum (based on collinear factorization) and low transverse momentum (based on transverse-momentum-dependent factorization). We used power counting to determine the leading behavior of the structure functions at intermediate transverse momentum in the two descriptions. When the power behaviors are different, two distinct mechanisms are present and there can be no matching between them. When the power behavior is the same, the two descriptions must match. An explicit calculation however shows that for some observables this is not the case, suggesting that the transverse-momentum-dependent-factorization description beyond leading twist is incomplete.

1. Introduction

The cross section for polarized semi-inclusive DIS can be written in terms of 18 structure functions.[1] Each of them depends, among other variables, on the square of the transverse momentum of the outgoing hadron, $P_{h\perp}^2$, with respect to the virtual photon direction. For theoretical considerations, it is often preferable to consider the transverse momentum $q_T^2 \approx P_{h\perp}^2/z^2$. The problem involves three scales, namely the scale of nonperturbative QCD dynamics, which we represent by the nucleon mass M, the transverse momentum q_T, and the photon virtuality Q, which we require to be large compared with M.

At high q_T ($q_T \gg M$) the structure functions can be described using collinear factorization, i.e., in terms of collinear distribution and fragmentation functions together with perturbative radiation. At low-q_T ($q_T \ll Q$) the structure functions can be described using Collins–Soper TMD factorization,[2,3] i.e., in terms transverse-momentum-dependent (TMD) parton distribution and fragmentation functions. The low- and high-q_T domains overlap for $M \ll q_T \ll Q$ (intermediate q_T), where both descriptions can hence be applied.

Studying the relation between the high-q_T and low-q_T regimes is important both from the theoretical and phenomenological point of view. We observe that in some cases the calculations in the two regimes have to give the same result at intermediate q_T, i.e., they have to match. If this does not occur, we can make the important conclusion that there is some flaw in the formalism. We observe in other cases that the two calculations describe different mechanisms and therefore do not have to match. Both of them have to be taken into consideration independently in the overlap region.

2. Matches and mismatches: general discussion

To assess whether the high-q_T and low-q_T calculations have to match or not in the intermediate-q_T region it is sufficient to study the power behavior of the structure functions in the two regimes.

It is important to realize that the power expansions are done in two different ways in the two descriptions. At low q_T, first we expand in $(q_T/Q)^{n-2}$ and neglect terms with n bigger than a certain value (so far, analyses have been carried out only up to $n = 3$, i.e., twist-3). To study the behavior at intermediate q_T we further expand in $(M/q_T)^k$. Conversely, at high q_T we first expand in $(M/q_T)^n$ (also in this case, analyses are available up to $n = 3$, i.e., twist-3). To study the intermediate-q_T region, we further expand in $(q_T/Q)^{k-2}$.

We can encounter two different situations. For simplicity, we will refer to them as type-I and type-II observables.

2.1. *Type-I observables*

Consider, e.g., a structure functions described by two contributions

$$F = A\,\frac{M^2}{M^2 + q_T^2} + B\,\frac{q_T^2}{Q^2}\,\frac{M^2}{M^2 + q_T^2}. \tag{1}$$

At low transverse momentum, term B is neglected from the very beginning because it is of order $(q_T/Q)^2$ (twist-4). The remaining term is

$$F_{\text{low}}^{\text{twist-2}} = A\,\frac{M^2}{M^2 + q_T^2} = A\,\frac{M^2}{q_T^2} + \mathcal{O}\left(\frac{M^4}{q_T^4}\right), \tag{2}$$

where the second step identifies the leading term at $q_T \gg M$.

At high transverse momentum, both terms A and B are twist-2 and are taken into consideration. However, the second term is neglected if a further expansion in q_T/Q is performed, to study the regime of intermediate transverse momentum, i.e.,

$$F_{\text{high}}^{\text{twist-2}} = A\,\frac{M^2}{q_T^2} + B\,\frac{M^2}{Q^2} = A\,\frac{M^2}{q_T^2} + \mathcal{O}\left(\frac{q_T^2}{Q^2}\right). \tag{3}$$

Therefore, the *leading* terms in the two expansions are the same. In this case, the calculations at high and low transverse momentum must yield exactly the same result at intermediate transverse momentum.[4,5] If a mismatch occurs, it means that one of the calculations is incorrect or incomplete.

2.2. *Type-II observables: expected mismatches*

Consider the example of a structure functions composed by two terms

$$F = A\,\frac{M^4}{M^4 + q_T^4} + B\,\frac{q_T^2}{Q^2}\,\frac{M^2}{M^2 + q_T^2}. \tag{4}$$

At low transverse momentum, term B is neglected from the very beginning because it is of order $(q_T/Q)^2$ (twist-4). What is left is

$$F_{\text{low}}^{\text{twist-2}} = A\,\frac{M^4}{M^4 + q_T^4} = A\,\frac{M^4}{q_T^4} + \mathcal{O}\left(\frac{M^8}{q_T^8}\right), \tag{5}$$

where in the second step we expanded in M/q_T.

At high transverse momentum, the term A is now twist-4 and it is usually neglected. Only the second term is kept and gives

$$F_{\text{high}}^{\text{twist-2}} = B\,\frac{q_T^2}{Q^2}\,\frac{M^2}{q_T^2} \tag{6}$$

In this case, if the calculations at high and low transverse momentum are performed at their respective leading twist, they correspond to two different contributions to the cross section and will not lead to the same result at intermediate transverse momentum. In order to "match", the calculations would have to be carried out in both regimes up to the sub-subleading order. We could call this situation an "expected mismatch", since it is simply due to the difference between the two expansions.

3. Matches and mismatches: semi-inclusive DIS case

In Tab. 1 we list the power behavior of the structure functions at intermediate transverse momentum, as obtained from the limits of the low-q_T and high-q_T calculation. For details of the calculation, we refer to Ref. 6. In the names of the structure functions, the first and second subscript respectively

Table 1. Behavior of SIDIS structure functions in the region $M \ll q_T \ll Q$. Empty fields indicate that no calculation is available (in this case, twist 4 indicates observables that are zero when calculated up to twist-three accuracy). Yes/no in parentheses: expected answers based on analogy, rather than actual calculation.

observable	low-q_T calculation twist	low-q_T calculation power	high-q_T calculation twist	high-q_T calculation power	power match	exact match
$F_{UU,T}$	2	$1/q_T^2$	2	$1/q_T^2$	yes	yes
$F_{UU,L}$	4		2	$1/Q^2$		
$F_{UU}^{\cos\phi_h}$	3	$1/(Q\,q_T)$	2	$1/(Q\,q_T)$	yes	no
$F_{UU}^{\cos 2\phi_h}$	2	$1/q_T^4$	2	$1/Q^2$	no	
$F_{LU}^{\sin\phi_h}$	3	$1/(Q\,q_T)$	2	$1/(Q\,q_T)$	yes	(no)
$F_{UL}^{\sin\phi_h}$	3	$1/(Q\,q_T)$			(yes)	(no)
$F_{UL}^{\sin 2\phi_h}$	2	$1/q_T^4$			(no)	
F_{LL}	2	$1/q_T^2$	2	$1/q_T^2$	yes	yes
$F_{LL}^{\cos\phi_h}$	3	$1/(Q\,q_T)$	2	$1/(Q\,q_T)$	yes	no
$F_{UT,T}^{\sin(\phi_h-\phi_S)}$	2	$1/q_T^3$	3	$1/q_T^3$	yes	yes
$F_{UT,L}^{\sin(\phi_h-\phi_S)}$	4		3	$1/(Q^2\,q_T)$		
$F_{UT}^{\sin(\phi_h+\phi_S)}$	2	$1/q_T^3$	3	$1/q_T^3$	yes	(yes)
$F_{UT}^{\sin(3\phi_h-\phi_S)}$	2	$1/q_T^3$	3	$1/(Q^2\,q_T)$	no	
$F_{UT}^{\sin\phi_S}$	3	$1/(Q\,q_T^2)$	3	$1/(Q\,q_T^2)$	yes	(no)
$F_{UT}^{\sin(2\phi_h-\phi_S)}$	3	$1/(Q\,q_T^2)$	3	$1/(Q\,q_T^2)$	yes	(no)
$F_{LT}^{\cos(\phi_h-\phi_S)}$	2	$1/q_T^3$			(yes)	(yes)
$F_{LT}^{\cos\phi_S}$	3	$1/(Q\,q_T^2)$			(yes)	(no)
$F_{LT}^{\cos(2\phi_h-\phi_S)}$	3	$1/(Q\,q_T^2)$			(yes)	(no)

specifies the polarization of the beam and the target. When present, the third subscript refers to the polarization of the photon.

In summary, the calculation at high q_T is done using standard collinear factorization, as done in, e.g., Ref. 7,8 and in Ref. 9 for the subleading-twist sector. To obtain the power behavior at intermediate q_T, we need to perform an expansion in q_T/Q. The calculation has no fundamental difficulties and allows us to fill in the third column of Tab. 1. The blank entries correspond to the structure functions that have not yet been computed in the high-transverse-momentum regime.

The calculation at low q_T is done using TMD factorization.[2,3] The behavior of the TMD functions at intermediate transverse momentum can be calculated perturbatively by considering diagrams as the ones depicted in Fig. 1. We calculated the power behavior of all twist-2 and twist-3 TMD functions, which allowed us to fill in the second column of Tab. 1. Two struc-

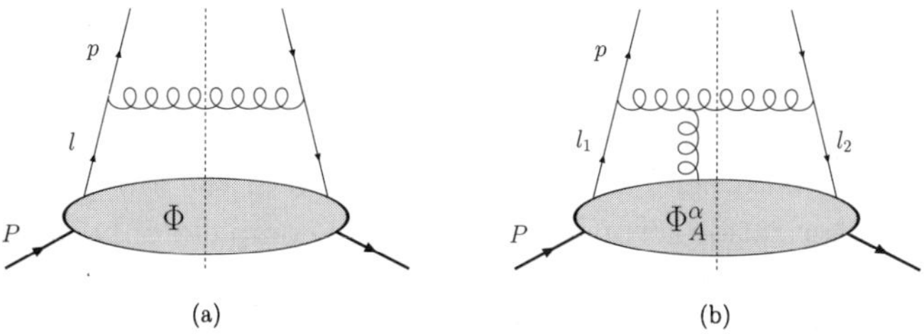

Fig. 1. Example diagrams for the calculation of the high-p_T behavior TMD parton distribution functions. Φ_A^α represent the quark-gluon-quark correlator. The dashed lines represent the final-state cut.

ture functions cannot be calculated as they require twist-4 contributions, which are beyond the current limits of the TMD factorization framework.

The fourth column of Tab. 1 is obtained by comparing the second and third column. The structure functions with a "yes" are type-I observables, those with a "no" are type-II. The values in parentheses are expectations based on analogy with similar structure functions, since the high-q_T calculations are not available.

Beside studying the power behavior, we also calculated the explicit form of some of the TMD functions at intermediate transverse momentum, namely the ones requiring only the evaluation of diagrams analogous to that of Fig. 1(a). A calculation of the Sivers function, requiring the evaluation of diagrams like that of Fig. 1(b), was already performed in Refs. 5,10.

The explicit calculations allows us to check if for type-I observables the explicit expressions obtained from high and low transverse momentum exactly match or not. The results are listed in the fifth column of Tab. 1. The entries in parentheses are conjectures based on analogy rather than actual calculation.

3.1. *Type-I structure functions*

For type-I structure functions ("yes" in the column "power match"), we know from power counting that the two calculations describe the same physics and should therefore exactly match. In these cases, the high-q_T calculation corresponds to the perturbative tail of the low-q_T effect. The two mechanisms need not be distinguished. Using resummation it should be possible to construct expressions for these observables that are valid at any q_T, as was done for the Drell-Yan analog of $F_{UU,T}$ in Ref. 4.

Only five of these structure functions have been calculated explicitly: $F_{UU,T}$, F_{LL} and $F_{UT,T}^{\sin(\phi_h-\phi_S)}$ (Sivers structure function) present an exact matching,[5,10] while in our work we showed that $F_{UU}^{\cos\phi_h}$ and $F_{LL}^{\cos\phi_h}$ do not match. In analogy to these results, we expect that also $F_{UT}^{\sin(\phi_h+\phi_S)}$ (Collins structure function) and $F_{LT}^{\cos(\phi_h-\phi_S)}$ will match exactly, while problems will occur with all the others, since they are twist-3 in the low-q_T regime, and the TMD factorization formalism is probably complete only at twist 2.

The structure function $F_{UT,T}^{\sin(\phi_h-\phi_S)}$, related to the Sivers function, is an example of a match between high- and low-q_T. Some of the consequences of the calculation are:

- the leading (twist-3, in this case) contribution of the high-q_T calculation corresponds to the tail of the Sivers function at intermediate q_T, it is not a competing effect and should not be summed to the Sivers function;
- it is conceivable to construct an expression that extends the high-q_T calculation to $q_T \approx M$, through a smooth merging into the Sivers function;
- since the structure function falls as $1/q_T^3$, it is safe to use q_T-weighted asymmetries to extract the Sivers function.

As an example of a mismatch we consider the structure function $F_{UU}^{\cos\phi_h}$, related to the Cahn effect. We show in this case the main steps of the calculation to explain the nature of the problem. In the low-q_T formalism, the expression for this observable is[1]

$$F_{UU}^{\cos\phi_h} = \frac{2M}{Q}\,\mathcal{C}\biggl[-\frac{\hat{\boldsymbol{h}}\cdot\boldsymbol{k}_T}{M_h}\Bigl(xh\,H_1^{\perp} + \frac{M_h}{M}\,f_1\frac{\tilde{D}^{\perp}}{z}\Bigr) - \frac{\hat{\boldsymbol{h}}\cdot\boldsymbol{p}_T}{M}\Bigl(xf^{\perp}D_1 + \frac{M_h}{M}\,h_1^{\perp}\frac{\tilde{H}}{z}\Bigr)\biggr], \tag{7}$$

where the convolution means

$$\mathcal{C}\bigl[wfD\bigr] = \sum_a x e_a^2 \int d^2\boldsymbol{p}_T\, d^2\boldsymbol{k}_T\, d^2\boldsymbol{l}_T\, \delta^{(2)}\bigl(\boldsymbol{p}_T - \boldsymbol{k}_T + \boldsymbol{l}_T + \boldsymbol{q}_T\bigr) \times w(\boldsymbol{p}_T,\boldsymbol{k}_T)\, f^a(x,p_T^2)\, D^a(z,k_T^2)\, U(l_T^2)\,. \tag{8}$$

The term U denotes the so-called soft factor. It is obtained in the factorization proof for twist-two observables. Here we assume we can use it also for twist-three observables. The terms with $h_1^{\perp}$ and $H_1^{\perp}$ fall off as $1/p_T^3$ or $1/k_T^3$ and are power suppressed compared to the terms with $f^{\perp}$ and $\tilde{D}^{\perp}$ when $q_T \gg M$. For intermediate q_T we therefore have

$$F_{UU}^{\cos\phi_h} = -\frac{2q_T}{Q}\sum_a x e_a^2 \left[x f^{\perp a}(x,q_T^2)\,\frac{D_1^a(z)}{z^2} - f_1^a(x)\,\frac{\tilde{D}^{\perp a}(z,q_T^2)}{z}\right] \tag{9}$$

at leading power. In this case there is no leading contribution from the soft factor taken at large transverse momentum. The tail of the functions at $q_T \gg M$ can be calculated perturbatively and yields

$$x f^{\perp q}(x, p_T^2) = \frac{\alpha_s}{2\pi^2} \frac{1}{2\boldsymbol{p}_T^2} \left[\frac{L(\eta^{-1})}{2} f_1^q(x) + \left(P'_{qq} \otimes f_1^q + P'_{qg} \otimes f_1^g\right)(x) \right],$$

$$\frac{\tilde{D}^{\perp q}(z, k_T^2)}{z} = -\frac{\alpha_s}{2\pi^2} \frac{1}{2z^2 \boldsymbol{k}_T^2} \left[\frac{L(\eta_h^{-1})}{2} D_1^q(z) - 2C_F D_1^q(z) \right.$$
$$\left. + \left(D_1^q \otimes P'_{qq} + D_1^g \otimes P'_{gq}\right)(z) \right], \qquad (10)$$

where $L(y) = 2C_F \ln y - 3C_F$, and P'_{qq}, P'_{gq}, P'_{qg} are kernels specific to the functions under consideration. The parameters η and η_h are related to the choice of a nonlightlike gauge (Wilson line) in the calculation of the functions and fulfill the relation $\sqrt{\eta\eta_h} = q_T^2/Q^2$. Putting these ingredients together we arrive at

$$F_{UU}^{\cos\phi_h} = -\frac{1}{Q\, q_T} \frac{\alpha_s}{2\pi^2 z^2} \sum_a x e_a^2 \left[f_1^a(x)\, D_1^a(z)\, L\!\left(\frac{Q^2}{q_T^2}\right) \right.$$
$$+ f_1^a(x) \left(D_1^a \otimes P'_{qq} + D_1^a \otimes P'_{gq}\right)(z)$$
$$\left. + \left(P'_{qq} \otimes f_1^a + P'_{qg} \otimes f_1^g\right)(x)\, D_1^a(z) - 2C_F\, f_1^a(x)\, D_1^a(z) \right]. \quad (11)$$

This expression differs from the one obtained at high q_T by the last term $2C_F f_1^a(x)\, D_1^a(z)$. At this point we are forced to conclude that the description of twist-3 structure functions is incomplete in the TMD-factorization formalism. However, it is interesting to note that adding a term $f_1^a(x)\, D_1^a(z) z^{-2} U(q_T^2)/2$ within brackets in Eq. (9) would be sufficient to cure this problem. It is however not clear how such an expression would be obtained from a factorized formula.

3.2. *Type-II structure functions*

For type-II structure functions ("no" in the column "power match") the low-q_T and high-q_T calculations at leading order pick up two different components of the full structure function. They therefore describe two different mechanisms and do not match.

An example of a type-II observable is the structure function $F_{UU}^{\cos 2\phi_h}$, related at low q_T to the Boer–Mulders function.[11] Some studies of this structure functions have recently appeared.[12,13] However, some considerations have to be kept in mind:

- the leading contribution from the high-q_T calculation (often referred to as a pQCD or radiative correction) is a competing effect that has to be taken into account;
- it is at present not possible to construct an expression that extends the high-q_T calculation to $q_T \approx M$, since this requires a smooth merging into unknown twist-4 contributions in TMD factorization;
- Using q_T-weighted asymmetries to extract the Boer–Mulders function is *not* a good idea, since the high-q_T mechanism dominates the observable;
- a solution to the above problems could be to consider observables that are least sensitive to the effect of radiative corrections, for instance by considering specific combinations of structure functions.

We stress that the above considerations apply not only to semi-inclusive DIS, but also to Drell–Yan and e^+e^- annihilation.[14] Drell–Yan data have been already used to extract[15] the Boer–Mulders function, without taking into account radiative corrections, while the extraction[16] of the Collins function from e^+e^- relies on the cancellation of radiative effects through the construction of suitable experimental observables.[14,17]

References

1. A. Bacchetta, M. Diehl, K. Goeke, A. Metz, P. J. Mulders and M. Schlegel, *JHEP* **02**, 093 (2007).
2. J. C. Collins and D. E. Soper, *Nucl. Phys.* **B193**, 381 (1981).
3. X. Ji, J.-P. Ma and F. Yuan, *Phys. Rev.* **D71**, 034005 (2005).
4. J. C. Collins, D. E. Soper and G. Sterman, *Nucl. Phys.* **B250**, 199 (1985).
5. X. Ji, J. Qiu, W. Vogelsang and F. Yuan, *Phys. Rev. Lett.* **97**, 082002 (2006).
6. A. Bacchetta, D. Boer, M. Diehl and P. J. Mulders, *JHEP* **08**, 023 (2008).
7. A. Mendez, *Nucl. Phys.* **B145**, 199 (1978).
8. Y. Koike, J. Nagashima and W. Vogelsang, *Nucl. Phys.* **B744**, 59 (2006).
9. H. Eguchi, Y. Koike and K. Tanaka, *Nucl. Phys.* **B763**, 198 (2007).
10. Y. Koike, W. Vogelsang and F. Yuan, *Phys. Lett.* **B659**, 878 (2008).
11. D. Boer and P. J. Mulders, *Phys. Rev.* **D57**, 5780 (1998).
12. V. Barone, A. Prokudin and B.-Q. Ma, arXiv:0804.3024 [hep-ph] (2008).
13. B. Zhang, Z. Lu, B.-Q. Ma and I. Schmidt, arXiv:0807.0503 [hep-ph] (2008).
14. D. Boer, arXiv:0804.2408 [hep-ph] (2008).
15. B. Zhang, Z. Lu, B.-Q. Ma and I. Schmidt, *Phys. Rev.* **D77**, 054011 (2008).
16. M. Anselmino *et al.*, *Phys. Rev.* **D75**, 054032 (2007).
17. K. Abe *et al.*, *Phys. Rev. Lett.* **96**, 232002 (2006).

PRETZELOSITY DISTRIBUTION FUNCTION $h_{1T}^{\perp}$

H. Avakian,[1] A. V. Efremov,[2] P. Schweitzer,[3,4] F. Yuan[5,6]

[1] *Thomas Jefferson National Accelerator Facility, Newport News, VA 23606, U.S.A.*
[2] *Joint Institute for Nuclear Research, Dubna, 141980 Russia*
[3] *Inst. f. Theoret. Physik II, Ruhr-Universität Bochum, D-44780 Bochum, Germany*
[4] *Departement of Physics, University of Connecticut, Storrs, CT 06269, U.S.A.*
[5] *RIKEN BNL Research Center, Building 510A, BNL, Upton, NY 11973, U.S.A.*
[6] *Nuclear Science Division, Berkeley National Laboratory, Berkeley, CA 94720, U.S.A.*

The 'pretzelosity' distribution $h_{1T}^{\perp}$ is discussed. Theoretical properties, model results, and perspectives to access experimental information on this leading twist, transverse momentum dependent parton distribution function are reviewed. Its relation to helicity and transversity distributions is highlighted.

Keywords: semi-inclusive deep inelastic scattering (SIDIS), single spin asymmetry (SSA), transverse momentum dependent distribution function (TMD)

1. Introduction

SIDIS allows to access information on TMDs that are defined in terms of light-front correlators[1–6] (with process-dependent paths[7–10])

$$\phi(x, \vec{p}_T)_{ij} = \int \frac{\mathrm{d}z^- \mathrm{d}^2\vec{z}_T}{(2\pi)^3}\, e^{ipz} \langle N|\bar{\psi}_j(0)\, \mathcal{W}(0, z, \text{path})\, \psi_i(z)|N\rangle \bigg|_{\substack{z^+=0 \\ p^+=xP^+}} \tag{1}$$

where light-cone components $p^{\pm} = (p^0 \pm p^3)/\sqrt{2}$ are along the virtual photon momentum, and transverse vectors like $\vec{p}_T$ are perpendicular to it. Different TMDs are given by traces of the correlator (1) with specific γ-matrices. There are 8 leading-twist TMDs, i.e. they give rise to effects that are not power suppressed in the hard scale, and each of them contains independent information about the nucleon structure. All leading-twist TMDs can be accessed unambiguously in SIDIS with polarized leptons and nucleons by observing the azimuthal distributions of produced hadrons.

In this note we will focus on the leading-twist TMD pretzelosity $h_{1T}^{\perp a}$, on which interesting results have been obtained from model calculations. This TMD is responsible for a SSA $\propto \sin(3\phi - \phi_S)$ in SIDIS that could be accessed in experiments — most promisingly at Jefferson Lab.

2. Properties of $h_{1T}^{\perp a}$

Let us list briefly, what we know about the pretzelosity distribution.

I. It can be projected out from the correlator (1) by tracing it with $i\sigma^{j+}\gamma_5$ where it appears as the coefficient of the structure $S_T^k(p_T^j p_T^k - \frac{1}{2}\vec{p}_T^{\,2}\delta^{jk})$, and it has a probabilistic interpretation.[6]

II. It requires nucleon wave-function components with two units orbital momentum difference,[11] and 'measures' the deviation of the 'nucleon shape' from a sphere.[12] (That is why it is called 'pretzelosity'!)

III. It is expected to be suppressed at small and large[11,13,14] x with respect to parton distributions like $f_1^a(x)$, $g_1^a(x)$, $h_1^a(x)$.

IV. It satisfies the positivity condition[15] $2|h_{1T}^{\perp(1)a}(x)| \leq f_1^a(x) - g_1^a(x)$. The above and Soffer bound[16] imply: $|h_{1T}^{\perp(1)a}(x)| + |h_1^a(x)| \leq f_1^a(x)$.

V. In the limit of a large number of colors N_c in QCD it was shown that[17] $(h_1^{\perp u} + h_1^{\perp d})/(h_1^{\perp u} - h_1^{\perp d}) \sim \mathcal{O}(1/N_c)$ for $xN_c \sim \mathcal{O}(1)$ and $p_T \sim \mathcal{O}(1)$. The same pattern is predicted also for antiquarks.[17]

VI. It was observed in the bag model[18] that $h_{1T}^{\perp(1)q}(x) = g_1^q(x) - h_1^q(x)$ which is confirmed in many[18–22] (but not all[22,23]) models.

VII. In simple (spectator-type) models, it has been related to chirally odd generalized parton distributions.[22]

VIII. In SIDIS with unpolarized electrons (U) and transversely polarized nucleons (T) it gives rise (in combination with Collins fragmentation function[24,25]) to an azimuthal modulation of the produced hadrons proportional to $\sin(3\phi - \phi_S)$. Here ϕ (ϕ_S) is the azimuthal angle of the produced hadron (nucleon polarization vector) with respect to the virtual photon. The corresponding SSA is given by

$$A_{UT}^{\sin(3\phi-\phi_S)} = \frac{C_{\text{Gauss}} \sum_a e_a^2 x\, h_{1T}^{\perp(1)a}(x)\, H_1^{\perp(1/2)a}(z)}{\sum_a e^a\, x f_1^a(x)\, D_1^a(z)} \tag{2}$$

with $h_{1T}^{\perp(1)}(x) = \int d^2\vec{p}_T \frac{\vec{p}_T^{\,2}}{2M_N^2} h_{1T}^{\perp}(x, \vec{p}_T^{\,2})$ the 'transverse moment' of pretzelosity. Unless the DIS counts are weighted with adequate powers of transverse hadron momenta[5] it is necessary to assume a model for the distribution of transverse parton momenta. In (2) the Gauss model is assumed. The factor C_{Gauss} contains the dependence on Gauss model parameters and is a slowly varying function of these parameters that can be well approximated for practical purposes[18] by its maximum $C_{\text{Gauss}} \leq C_{\max} = 3/(2\sqrt{2})$. Extractions of the (1/2)-moment of the Collins function $H_1^{\perp(1/2)}(z) = \int d^2\vec{K}_T \frac{|\vec{K}_T|}{2zm_h} H_1^{\perp}(z, \vec{K}_T^{\,2})$ from data[26–30] were reported elsewhere.[31–33]

3. Pretzelosity in the bag model

In this Section we review the pretzelosity calculation[18] in the MIT bag model, in which the nucleon consists of 3 non-interacting quarks confined in a spherical cavity.[34–37] The momentum space wave function is given by

$$\varphi_m(\vec{k}) = i\sqrt{4\pi}NR_0^3 \begin{pmatrix} t_0(|\vec{k}|)\chi_m \\ \vec{\sigma}\cdot\hat{k}\, t_1(|\vec{k}|)\chi_m \end{pmatrix}, \tag{3}$$

with $N = \omega^{3/2}(2R_0^3(\omega-1)\sin^2\omega)^{-1/2}$, $\hat{k} = \vec{k}/|\vec{k}|$. We fix the bag radius R_0 in terms of the proton mass M_N as $R_0 M_N = 4\omega$ with $\omega \approx 2.04$ the lowest root of the bag eigen-equation. Finally $t_i(\kappa) = \int_0^1 u^2 du j_i(uR_0\kappa) j_i(u\omega)$ where j_i are spherical Bessel functions. The bag model wave function (3) contains both S (represented by t_0) and P (represented by t_1) waves.

With the above wave functions, one obtains the following results[18]

$$f_1(x,k_\perp) = A\left[t_0^2 + 2t_0t_1\frac{k_z}{k} + t_1^2\right] \tag{4}$$

$$g_1(x,k_\perp) = A\left[t_0^2 + 2t_0t_1\frac{k_z}{k} + t_1^2(\frac{2k_z^2}{k^2} - 1)\right] \tag{5}$$

$$h_1(x,k_\perp) = A\left[t_0^2 + 2t_0t_1\frac{k_z}{k} + t_1^2\frac{k_z^2}{k^2}\right] \tag{6}$$

$$h_{1T}^{\perp}(x,k_\perp) = A\left[-2\,\frac{M_N^2}{k^2}\,t_1^2\right] \tag{7}$$

with $A = 16\omega^4/[\pi^2(\omega-1)j_0^2(\omega)\,M_N^2]$ and the flavor dependence given by (we assume an $SU(6)$ spin-flavor symmetric proton wave function):

$$\begin{aligned} f_1^q(x,k_\perp) &= N_q f_1(x,k_\perp)\,, \quad N_u = 2\,, \quad N_d = 1 \\ g_1^q(x,k_\perp) &= P_q\, g_1(x,k_\perp)\,, \quad P_u = \frac{4}{3}\,, \quad P_d = -\frac{1}{3}\,. \end{aligned} \tag{8}$$

The flavor dependence of h_1^q and $h_{1T}^{\perp q}$ is analog to g_1^q. The momenta k_z and k are defined as $k_z = xM_N - \omega/R_0$, and $k = \sqrt{k_z^2 + k_\perp^2}$.

Since $h_{1T}^{\perp} \propto t_1^2$ it is proportional to the square of the P-wave[11] and thus sensitive to the quark orbital angular momentum in the proton. The results (4-8) satisfy[18] the positivity conditions in point III, and are consistent[18] with the predictions from the large-N_c limit discussed in point IV of Sec. 2.

From the results in Eqs. (4-7) we find that out of the 4 TMDs $f_1(x,k_\perp)$, $g_1(x,k_\perp)$, $h_1(x,k_\perp)$, $h_{1T}^{\perp(1)}(x,k_\perp) \equiv k_\perp^2/(2M_N^2)\,h_{1T}^{\perp}(x,k_\perp)$ only 2 are linearly independent. In QCD the different TMDs are, of course, all independent of each other. But in the bag model all TMDs are expressed in terms of t_0 and t_1 in Eq. (3), which naturally gives rise to *bag model relations*

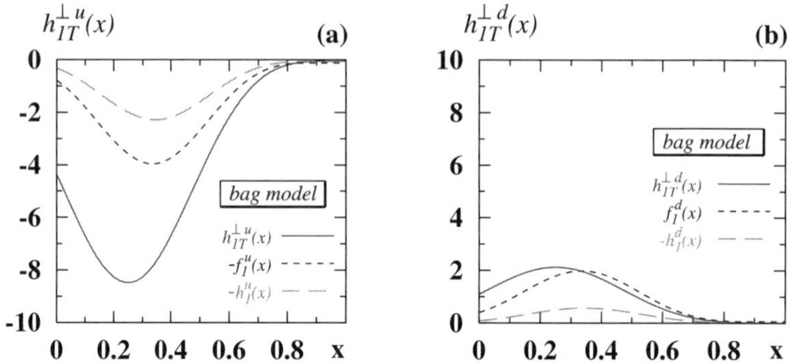

Fig. 1. The parton distribution function $h_{1T}^{\perp q}(x)$ vs. x from the bag model[18] in comparison to $f_1^q(x)$ and $h_1^q(x)$ from this model, The results refer to a low scale.[36] Notice that $h_{1T}^{\perp q}(x)$ is not constrained by positivity.

among different TMDs. The most interesting (and possibly more general, see Sec. 4) relation is[18]

$$h_{1T}^{\perp(1)q}(x, k_\perp) = g_1^q(x, k_\perp) - h_1^q(x, k_\perp)\,. \tag{9}$$

Figs. 1a and 1b show results for $h_{1T}^{\perp q}(x) = \int \mathrm{d}^2\vec{k}_\perp\, h_{1T}^{\perp q}(x, k_\perp)$ at the low bag model scale. The pretzelosity distributions $h_{1T}^{\perp q}(x)$ have opposite signs compared to transversity and are larger than $h_1^q(x)$ in the bag model, even larger than $f_1^q(x)$. However, $h_{1T}^{\perp q}(x)$ is not constrained by positivity bounds.

The bag[18] (and spectator[19]) model results satisfy the positivity bounds for $h_{1T}^{\perp(1)q}(x)$ in point III of Sec. 2. In Fig. 2 the results for $h_{1T}^{\perp(1)q}(x)$ from both models are compared. We observe good qualitative agreement.

The moduli of the transverse moments $h_{1T}^{\perp(1)q}(x)$ of the pretzelosity distribution functions are about 3 or more times smaller than those of the transversity distribution functions $h_1^q(x)$ with the exception of very small-x where both models are strictly speaking not applicable.[18]

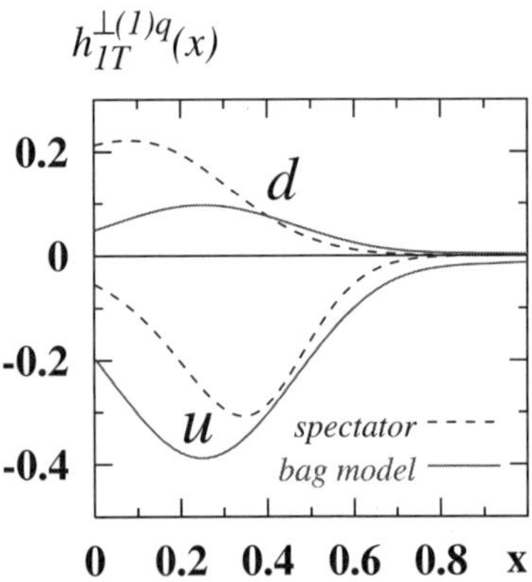

Fig. 2. The $h_{1T}^{\perp(1)q}(x)$ vs. x at low scale from bag model[18] and spectator model.[19]

4. How general is the relation in Eq. (9)?

Eq. (9) is remarkable from the point of view of the observation that[38] 'the difference of $g_1^q(x)$ and $h_1^q(x)$ is a measure for relativistic effects in nucleon'. This difference is just the transverse moment of pretzelosity!

It is clear that this relation cannot be strictly valid in QCD, where all TMDs are independent. However, could it nevertheless allow to gain a reasonable estimate for $h_{1T}^{\perp(1)}(x)$ in terms of transversity and helicity? Until clarified by experiment, we can address this question only in models.

Interestingly, the relation (9) does not hold only in the bag model,[18] but is found[18] to be satisfied also in the spectator model.[19] In fact, it was conjectured[18] that (9) could hold in a large class of *relativistic quark models*, which was subsequently confirmed in the constituent quark model[20] and the relativistic model of the proton.[21] But (9) is not satisfied in a different than[19] version of the spectator model.[23]

The limitations of (9) are nicely illustrated in the 'quark target model' where in addition to quarks there are also gluons, and (9) is not satisfied.[22] Thus, the explicit inclusion of gluon degrees of freedom spoils this relation. Of course, as stressed above, we do not expect (9) to be valid in QCD.

It would be interesting to know the necessary and sufficient conditions for the relation (9) to hold in a model.

5. Preliminary COMPASS data & prospects at JLab

At COMPASS the $\sin(3\phi-\phi_S)$ and other SSAs were measured on a deuteron target.[39] By saturating the positivity bound for $h_{1T}^{\perp(1)}(x)$ (point III in Sec. 2) we estimated[18] the maximum effect for $A_{UT}^{\sin(3\phi-\phi_S)}$. For that information

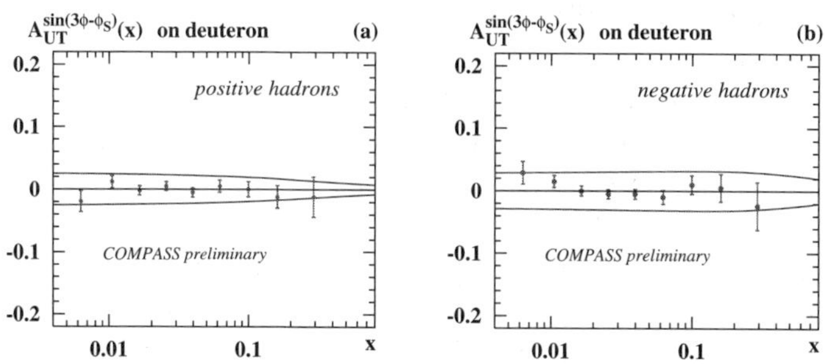

Fig. 3. The transverse target SSA $A_{UT,\pi}^{\sin(3\phi-\phi_S)}$ for deuteron estimated on the basis of the positivity bound vs. preliminary COMPASS data.[39]

on Collins effect[32,33] and parameterizations[40,41] were used. The results are shown in Fig. 3 and compared to the preliminary data.[39]

At small x the preliminary data favor that pretzelosity does not reach the bound. Whether due to the expected suppression at small x or opposite signs of u and d-flavors, see Sec. 2, cannot be concluded.

The important observation is that preliminary COMPASS data[39] do not exclude a sizeable effect in the region $x > 0.1$, see Fig. 3, where JLab can measure with precision. This is demonstrated in Fig. 4 showing the π^+ SSA from a proton target in the kinematics of CLAS with 12 GeV beam upgrade (with error projections for 2000 hours run time[42]).

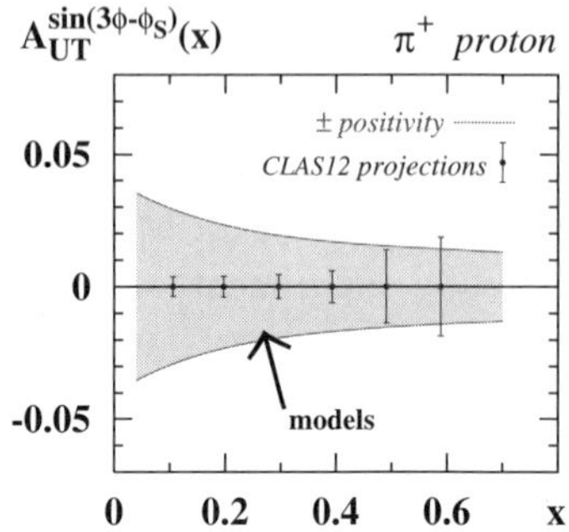

Fig. 4. SSA $A_{UT}^{\sin(3\phi-\phi_S)}$ in π^+ production from proton in the kinematics of CLAS 12 as function of x (error projections from[42]). The shaded areas indicate the range allowed by positivity. The bag[18] and spectator[19] models predict a negative SSA.

It could be even more promising to look at SSAs due to Collins effect, like $A_{UT}^{\sin(3\phi-\phi_S)}$, in kaon production. The statistics for kaon production is lower than for pion production, but the SSA might be larger as it is suggested by a model[43] of the Collins function. With a RICH detector at CLAS the kaon SSAs could be measured in the valence-x region.[44]

6. Conclusions

We reviewed the bag model calculation of pretzelosity.[18] An interesting observation is the relation (9) which connects helicity and transversity distributions to $h_{1T}^{\perp(1)}(x)$, and is valid in many[18–22] (though not all[22,23]) models, but not in QCD where all TMDs are independent.[45] Nevertheless (9) could turn out a useful approximation. In view of the numerous novel TMDs, well-motivated approximations are welcome.[46]

In the bag model $h_{1T}^{\perp u}$ is proportional to *minus* the square of the p-wave component of the nucleon wave function, and therefore manifestly negative ($h_{1T}^{\perp d}$ has opposite sign dictated by SU(6) symmetry, and large N_c[17]). As a consequence of (9) we have $|h_1^q(x)| > |g_1^q(x)|$. This is found in models.[47,48]

Forthcoming analyzes and experiments at COMPASS, HERMES and JLab[42,49,50] will provide valuable information on the pretzelosity distribution function, and deepen our understanding of the nucleon structure.

Acknowledgements

We thank the organizers of the *Transversity 2008* for the opportunity to present our work supported by: BMBF, German-Russian collaboration (DFG-RFFI, 436 RUS 113/881/0), the EIIIHP project RII3-CT-2004-506078, the Grants RFBR 06-02-16215 and 07-02-91557, RF MSE RNP.2.2.2.2.6546 (MIREA), the Heisenberg-Landau Program of JINR, the contract DE-AC05-06OR23177, RIKEN, Brookhaven National Laboratory, and the U.S. Department of Energy DE-AC02-98CH10886.

References

1. J. C. Collins and D. E. Soper, Nucl. Phys. B **193**, 381 (1981) [Erratum-ibid. B **213**, 545 (1983)].
2. J. C. Collins, D. E. Soper and G. Sterman, Nucl. Phys. B **250**, 199 (1985).
3. X. D. Ji, J. P. Ma and F. Yuan, Phys. Rev. D **71**, 034005 (2005). X. D. M. Ji, J. P. M. Ma and F. Yuan, Phys. Lett. B **597**, 299 (2004).
4. J. C. Collins and A. Metz, Phys. Rev. Lett. **93**, 252001 (2004).
5. D. Boer and P. J. Mulders, Phys. Rev. D **57**, 5780 (1998).
6. P. J. Mulders and R. D. Tangerman, Nucl. Phys. B **461** (1996) 197 [Erratum-ibid. B **484** (1997) 538].
7. S. J. Brodsky, D. S. Hwang and I. Schmidt, Phys. Lett. B **530**, 99 (2002); Nucl. Phys. B **642**, 344 (2002).
8. J. C. Collins, Phys. Lett. B **536**, 43 (2002) [arXiv:hep-ph/0204004].
9. A. V. Belitsky, X. Ji and F. Yuan, Nucl. Phys. B **656**, 165 (2003). X. D. Ji and F. Yuan, Phys. Lett. B **543**, 66 (2002). D. Boer, P. J. Mulders and F. Pijlman, Nucl. Phys. B **667**, 201 (2003)
10. I. O. Cherednikov and N. G. Stefanis, Phys. Rev. D **77**, 094001 (2008); I. O. Cherednikov and N. G. Stefanis, Nucl. Phys. B **802**, 146 (2008).
11. M. Burkardt, arXiv:0709.2966 [hep-ph].
12. G. A. Miller, Phys. Rev. C **76**, 065209 (2007).
13. H. Avakian, S. J. Brodsky, A. Deur and F. Yuan, Phys. Rev. Lett. **99**, 082001 (2007).
14. S. J. Brodsky and F. Yuan, Phys. Rev. D **74** (2006) 094018.
15. A. Bacchetta, M. Boglione, A. Henneman and P. J. Mulders, Phys. Rev. Lett. **85**, 712 (2000).
16. J. Soffer, Phys. Rev. Lett. **74** (1995) 1292.
17. P. V. Pobylitsa, arXiv:hep-ph/0301236.
18. H. Avakian, A. V. Efremov, P. Schweitzer and F. Yuan, arXiv:0805.3355.
19. R. Jakob, P. J. Mulders and J. Rodrigues, Nucl. Phys. A **626**, 937 (1997).
20. B. Pasquini, S. Cazzaniga and S. Boffi, arXiv:0806.2298 [hep-ph]. B. Pasquini, arXiv:0807.2825 [hep-ph].
21. A. V. Efremov, P. Schweitzer, O. V. Teryaev and P. Zavada, in preparation (P. Zavada, talk at *SPIN Praha 2008*, Prague, 20-26 July 2008).
22. S. Meissner, A. Metz and K. Goeke, Phys. Rev. D **76**, 034002 (2007).
23. A. Bacchetta, F. Conti and M. Radici, arXiv:0807.0323 [hep-ph].

24. J. C. Collins, Nucl. Phys. B **396**, 161 (1993).
25. A. V. Efremov, L. Mankiewicz and N. A. Tornqvist, Phys. Lett. B **284** (1992) 394. J. C. Collins, S. F. Heppelmann and G. A. Ladinsky, Nucl. Phys. B **420** (1994) 565.
26. A. Airapetian *et al.* [HERMES Coll.], Phys. Rev. Lett. **94**, 012002 (2005).
27. V. Y. Alexakhin *et al.* [COMPASS Coll.], Phys. Rev. Lett. **94**, 202002 (2005).
28. M. Diefenthaler [HERMES Coll.], AIP Conf. Proc. **792** (2005) 933, arXiv:0706.2242 and arXiv:hep-ex/0612010.
29. E. S. Ageev *et al.* [COMPASS Coll.], Nucl. Phys. B **765** (2007) 31
30. M. Alekseev *et al.* [COMPASS Coll.], arXiv:0802.2160 [hep-ex].
31. W. Vogelsang and F. Yuan, Phys. Rev. D **72** (2005) 054028.
32. A. V. Efremov, K. Goeke and P. Schweitzer, Phys. Rev. D **73**, 094025 (2006).
33. M. Anselmino, M. Boglione, U. D'Alesio, A. Kotzinian, F. Murgia, A. Prokudin and C. Turk, Phys. Rev. D **75**, 054032 (2007).
34. A. Chodos, R. L. Jaffe, K. Johnson, C. B. Thorn and V. F. Weisskopf, Phys. Rev. D **9**, 3471 (1974). R. L. Jaffe, Phys. Rev. D **11**, 1953 (1975).
35. L. S. Celenza and C. M. Shakin, Phys. Rev. C **27**, 1561 (1983) [Erratum-ibid. C **39**, 2477 (1989)].
36. M. Stratmann, Z. Phys. C **60**, 763 (1993).
37. F. Yuan, Phys. Lett. B **575**, 45 (2003).
38. R. L. Jaffe and X. D. Ji, Nucl. Phys. B **375** (1992) 527, and Phys. Rev. Lett. **67**, 552 (1991).
39. A. Kotzinian [on behalf of the COMPASS Coll.], arXiv:0705.2402 [hep-ex].
40. M. Glück, E. Reya and A. Vogt, Eur. Phys. J. C **5**, 461 (1998). M. Glück, E. Reya, M. Stratmann and W. Vogelsang, Phys. Rev. D **63**, 094005 (2001).
41. S. Kretzer, Phys. Rev. D **62** (2000) 054001.
42. H. Avakian, *et al.*, JLab LOI 12-06-108 (2008).
43. A. Bacchetta, L. P. Gamberg, G. R. Goldstein and A. Mukherjee, Phys. Lett. B **659**, 234 (2008) [arXiv:0707.3372 [hep-ph]].
44. *CLAS 12 Rich Detector Workshop*, 28-29 January 2008, Jefferson Lab, Newport News, VA, U.S.A. `http://conferences.jlab.org/CLAS12`
45. K. Goeke, A. Metz and M. Schlegel, Phys. Lett. B **618**, 90 (2005).
46. H. Avakian, A. V. Efremov, K. Goeke, A. Metz, P. Schweitzer and T. Teckentrup, Phys. Rev. D **77**, 014023 (2008).
47. P. Schweitzer *et al.*,Phys. Rev. D **64** (2001) 034013.
K. Goeke *et al.*, Acta Phys. Polon. B **32**, 1201 (2001).
M. Wakamatsu and T. Kubota, Phys. Rev. D **60**, 034020 (1999).
48. A. V. Efremov, O. V. Teryaev and P. Zavada, Phys. Rev. D **70** (2004) 054018. P. Zavada, Eur. Phys. J. C **52**, 121 (2007), Phys. Rev. D **55**, 4290 (1997), Phys. Rev. D **67**, 014019 (2003).
49. H. Avakian *et al.*, JLab E05-113, "Semi-Inclusive Pion Production with a Longitudinally Polarized Target at 6 GeV".
50. H. Avakian *et al.*, JLab PR12-07-107, "Studies of Spin-Orbit Correlations with Longitudinally Polarized Target".

TRANSVERSE MOMENTUM-SPIN CORRELATIONS

L.P. Gamberg, G.R. Goldstein and M. Schlegel

Department of Physics, Penn State University-Berks, Reading, PA 19610, USA

Department of Physics and Astronomy, Tufts University, Medford, MA 02155 USA

Theory Division, Jefferson Lab, Newport News, VA 23608 USA

The flavor dependence of the naive time reversal odd ("T-odd") Boer-Mulders $h_1^\perp$ and Sivers $f_{1T}^\perp$ parton distribution functions as well as the "T-even" but chiral odd function $h_{1L}^\perp$ are explored in the spectator framework. These transverse momentum dependent parton distribution functions are of significance for the analysis of azimuthal asymmetries in semi-inclusive deep inelastic scattering. In this context we also consider the Collins fragmentation function $H_1^\perp$. With this input, we calculate the leading twist unpolarized $\cos(2\phi)$ asymmetry, and $\sin(2\phi)$ single spin asymmetry for a longitudinally polarized target in semi-inclusive deep inelastic scattering.

Keywords: TMDs, T-odd, TSSAs

1. Introduction

Transverse momentum dependent (TMD) parton distributions (PDFs) have gained considerable attention in recent years. Theoretically they can account for non-trivial transverse spin and momentum correlations such as single spin asymmetries (SSAs) in hard scattering processes when transverse momentum scales are on the order of quarks in hadrons, namely $P_T \sim k_\perp \ll \sqrt{Q^2}$. For example the Sivers function $f_{1T}^\perp$[1] which accounts for the observed SSA in semi-inclusive deep inelastic scattering (SIDIS) for a transversely polarized proton target by the HERMES collaboration[2] describes correlations of the intrinsic quark transverse momentum and the transverse nucleon spin. The corresponding SSA on a deuteron target measured by COMPASS[3] vanishes, indicating a flavor dependence of the Sivers function. Another leading twist "T-odd" PDF, the chiral-odd Boer-Mulders function $h_1^\perp$[4] describes correlations between the transverse spin of a quark and its transverse momentum within the nucleon. Theoretically, twist two "T-odd" PDFs are of particular interest as they emerge from the color gauge invariant definition of the quark-gluon-quark correlation function.[5–7] The gauge link not only ensures a color gauge invariant definition of correlation functions, but it also describes initial/final (ISI/FSI) state interactions[8–10] which can generate SSAs.[8,11] Assuming factorization of leading twist SIDIS spin observables in terms of the "T-even" and "T-odd" TMD PDFs and fragmentation functions (FFs),[12] Refs.[4] shows how spin observables in SIDIS can be expressed in terms of convolutions of these functions.

Here we report on the flavor dependence of the twist-2 "T-odd" Boer-Mulders function $h_1^\perp$ and Sivers $f_{1T}^\perp$ TMD PDFs,[13] as well as work with and A. Mukherjee[14] where we calculate Collins fragmentation function $H_1^\perp$.[15] The Boer-Mulders and Collins functions are particularly important for the analysis of the azimuthal $\cos(2\phi)$ asymmetry in unpolarized SIDIS. We employ the factorized approach used in Refs.[11,16] In the partonic picture the Boer-Mulders function is convoluted with the "T-odd" (and chiral-odd) Collins fragmentation function $H_1^\perp$. Although these azimuthal asymmetries were measured in SIDIS by the ZEUS collaboration[17] and in Drell-Yan (DY),[18] little is known about the Boer-Mulders function. Of particular interest is the sign for different flavors u and d since this significantly affects predictions for these asymmetries. We also consider the flavor dependence of the "T-even" function $h_{1L}^\perp$, which is of interest in the transverse momentum and quark spin correlations in a longitudinally polarized target.

2. TMD Correlators in the Spectator Framework

Transverse momentum quark distribution and fragmentation functions contain essential non-perturbative information about the partonic structure of hadrons. In principle their moments are calculable from first principles in lattice QCD. A great deal of understanding has also been gained from model calculations using the spectator framework.

The starting point for the analysis is (cf.[16])is the definition of the un-integrated, color gauge invariant, quark-quark correlator

$$\Phi_{ij}(x,\vec{p}_T^{\,2})=\int\frac{d\xi^- d^2\xi_T}{(2\pi)^3}\mathrm{e}^{ip\cdot\xi}\langle P,S|\,\bar{\psi}_j(0)\,\mathcal{U}^n_{(0,\infty)}|X\rangle\langle X|\,\mathcal{U}^n_{(\infty,\xi)}\,\psi_i(\xi)\,|P,S\rangle|_{\xi^+=0} \tag{1}$$

where the sum over states $|X\rangle$ is implicit and where the gauge link is $\mathcal{U}^C_{(\zeta,\xi)}=\mathcal{P}\exp\left\{-ig\int_\zeta^\xi d\eta\cdot A(\eta)\right\}$ and C designates the process dependent (see e.g.[19]) integration path with endpoints ζ and ξ. In an arbitrary gauge there is an un-suppressed contribution at light cone infinity from the trans-

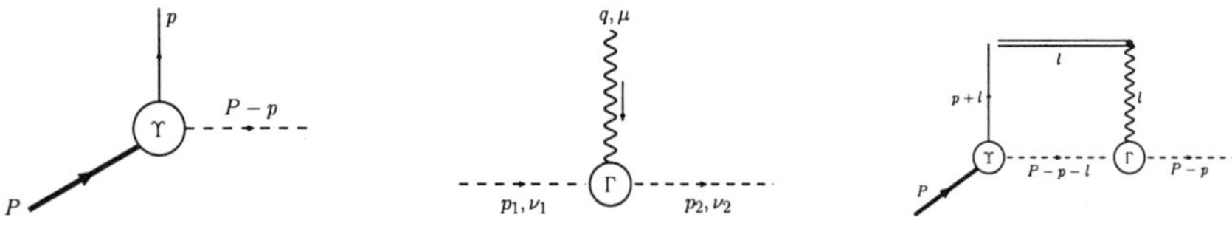

Fig. 1. Vertices for the proton-quark-diquark, gluon-diquark, and contribution of the gauge link in the one-gluon approximation.

verse component of the color field.[6,7] We work in Feynman gauge where the transverse Wilson line vanishes.[6] In the spectator model the sum over a complete set of intermediate on-shell states $|X\rangle$ is represented by a single one-particle diquark state $|\, dq;\, p_{dq}, \lambda\rangle$, where p_{dq} is the diquark momentum and λ its polarization. Since the diquark is "built" from two valence quark it can be a spin 0 (scalar diquark) or a spin 1 (axial-vector diquark) particle. Applying a translation in Eq. (1) we can integrate out ξ and perform the momentum integration over the diquark momentum p_{dq} to obtain the form of the correlator,

$$\Phi_{ij}(p,P,S) = \sum_{\lambda} \frac{\delta((P_s)^2 - m_s^2)\Theta(P_s^0)}{(2\pi)^3} \langle P,S|\,\bar{\psi}_j(0)\mathcal{U}^n_{(0,\infty)}|\, dq;\, P_s,\lambda\rangle \times \langle\, dq;\, P_s,\lambda|\,\mathcal{U}^n_{(\infty,\xi)}\,\psi_i(0)\,|P,S\rangle. \tag{2}$$

The essence of the diquark spectator model for TMD PDFs is to calculate the matrix elements in Eq. (2) by the introduction of effective nucleon-diquark-quark vertices $\Upsilon_s(N)$ (scalar) and $\Upsilon^{\mu}_{ax}(N)$ (axial-vector), which are represented in the Left Panel of Fig. 1. For example, the matrix element for the axial vector diquark is

$$\langle P_s;\lambda|\,\psi_i(0)\,|P,S\rangle = i\frac{g_{ax}(p^2)\varepsilon^*_\mu(P_s;\lambda)}{\sqrt{3}}\,\frac{\left[(\not{p}+m_q)\gamma_5\left[\gamma^\mu - R_g\frac{P^\mu}{M}\right]u(P,S)\right]_i}{p^2-m_q^2+i0}, \tag{3}$$

where the polarization vector of the axial-vector diquark is ε_μ and $u(P,S)$ denotes the nucleon spinor where M and m_q are nucleon and quark masses respectively. We consider the diquark as an on-shell particle with mass m_s, momentum $P_s = P - p$ where the polarization sum for the axial-vector diquark is

$$\sum_{\lambda} \varepsilon^*_\mu(P_s;\lambda)\varepsilon_\nu(P_s;\lambda) = -g_{\mu\nu} + \frac{(P_s)_\mu(P_s)_\nu}{m_s^2}.$$

The unpolarized TMD f_1 is obtained by inserting these expressions into

Eq. (2) and projecting from the quark-quark correlator

$$f_1(x,\vec{p}_T^2) = \frac{1}{4}\int dp^- \left(\mathrm{Tr}\left[\gamma^+\Phi\left(p,P,S\right)\right] + \mathrm{Tr}\left[\gamma^+\Phi\left(p,P,-S\right)\right]\right) \quad (4)$$

where the "+" sign of the γ-matrix denotes the usual light cone component ($a^\pm = 1/\sqrt{2}(a^0 \pm a^3)$) and $p^+ = xP^+$. Another "T-even" but chiral-odd function of interest is the distribution of transversely polarized quarks in a longitudinally polarized target,

$$\frac{4\lambda_P p_T^i h_{1L}^\perp(x,\vec{p}_T^2)}{M} = \int dp^- \left\{\mathrm{Tr}\left[\gamma^+\gamma^i\gamma_5\Phi(p;P,S_L)\right] - \mathrm{Tr}\left[\gamma^+\gamma^i\gamma_5\Phi(p;P,-S_L)\right]\right\}$$

where λ_P is the target helicity, and S_L is the spin 4-vector in longitudinal direction, i.e. $S_L = (-\frac{\lambda_P}{M}P^-, \frac{\lambda_P}{M}P^+, \vec{0}_T)$. The analytical expressions for f_1 and h_{1L} in the scalar and axial vector diquark sectors are given in Ref.[13] By contrast, "T-odd" PDFs cannot be generated by simply considering the tree-level diagram (Left Panel of Fig. 1). In the spectator framework the "T-odd" PDFs[8] are generated by the gauge link in Eq. (1).[9–11] The so-called leading contribution can be obtained by expanding the exponential of the gauge link up to first order in the quark gluon coupling. This contribution results in a box diagram as shown in the Right Panel of Fig. 1 which contains an imaginary part/phase which is necessary for the existence of the "T-odds". We restrict ourselves to the case where one gluon models the FSIs. The contribution from the gauge link in the box diagram Fig. 1 is given by the double line and the eikonal vertex $\frac{i}{[l\cdot v+i0]} \times (-ie_q v^\lambda)$, where l is the loop momentum, e_q the charge of the quark and $n = v$ is a light cone vector representing the direction of the Wilson line. In the box diagram we specify the gluon-diquark interaction for a scalar and axial-vector diquark with a general axial-vector-vector coupling that models the composite nature of the diquark through an anomalous magnetic moment κ. In the notation of Fig. 1 (Center Panel) the gluon-diquark vertices are $\Gamma_s^\mu = -ie_{dq}(p_1+p_2)^\mu$ and $\Gamma_{ax}^{\mu\nu_1\nu_2} = -ie_{dq}\left[g^{\nu_1\nu_2}(p_1+p_2)^\mu + (1+\kappa)\left(g^{\mu\nu_2}(p_2+q)^{\nu_1} + g^{\mu\nu_1}(p_1-q)^{\nu_2}\right)\right]$. For $\kappa = -2$ the vertex Γ_{ax} reduces to the standard γWW-vertex. The matrix elements in the *one gluon approximation* are

$$\langle P_{s;sqd}|\,\mathcal{U}^{n_-}_{(\infty,0)}\psi_i(0)\,|P,S\rangle = -ie_q e_{dq}\int \frac{d^4l}{(2\pi)^4}\, g_{sc}((l+p)^2)$$
$$\times\mathcal{D}^{sc}(P_s-l)\frac{\left[(\not{k}-\not{l}+m_q)u(P,S)\right]_i v\cdot(2P_s-l)}{[l\cdot v+i0]\,[l^2+i0]\,\left[(l+p)^2-m_q^2+i0\right]},$$

$$\langle P_{s;adq},\lambda|\,\mathcal{U}^{n_-}_{(\infty,0)}\psi_i(0)\,|P,S\rangle = -ie_q e_{dq}\int \frac{d^4l}{(2\pi)^4}\,\frac{g_{ax}\left((p+l)^2\right)}{\sqrt{3}}\varepsilon^*_\sigma(P_s,\lambda)$$

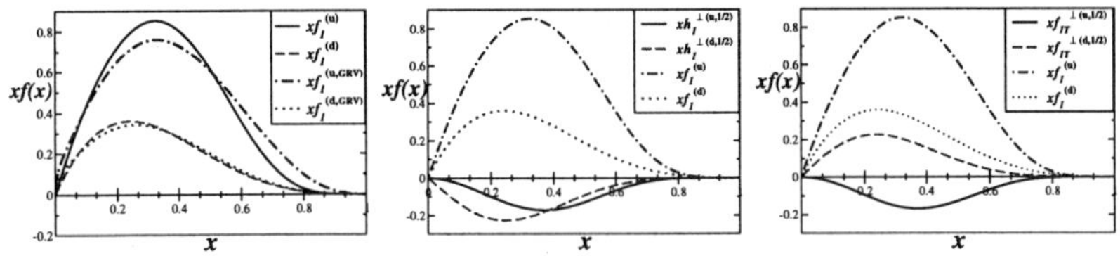

Fig. 2. Left Panel: The unpolarized u- and d-quark distributions functions versus x compared to the low scale parameterization of the unpolarized u- and d- quark distributions.[21] Center Panel: The half-moment of the Boer Mulders function versus x compared to the unpolarized u- and d- quark distribution functions. Right Panel: The half moment of the Sivers functions compared to the unpolarized u- and d- quark distributions.

$$\mathcal{D}^{ax}_{\rho\eta}(P_s-l)\frac{\left[g^{\sigma\rho}\, v\cdot(2P_s-l)+(1+\kappa)\left(v^{\sigma}\left(P_s+l\right)^{\rho}+v^{\rho}\left(P_s-2l\right)^{\sigma}\right)\right]}{\left[l\cdot v+i0\right]\left[l^2+i0\right]\left[(l+p)^2-m_q^2+i0\right]}$$
$$\times\left[\not{p}\not{A}\not{l}+m_q)\gamma_5\left(\gamma^{\eta}-R_g\frac{P^{\eta}}{M}\right)u(P,S)\right]_i, \qquad (5)$$

In these expressions $\mathcal{D}(P)$ denotes the propagator of the scalar and axial-vector diquark. We obtain the Boer-Mulders function $h_1^{\perp}$ by inserting Eqs. (5) (and the tree-level scalar and axial vector matrix elements (3)) into Eq. (2) and projecting from quark-quark correlator

$$\frac{4\epsilon_T^{ij}p_T^j h_1^{\perp}(x,\vec{p}_T^{\,2})}{M}=\int dp^-\left(\mathrm{Tr}\left[i\sigma^{i+}\gamma_5\Phi(p,P,S)\right]+\mathrm{Tr}\left[i\sigma^{i+}\gamma_5\Phi(p,P,-S)\right]\right) \quad (6)$$

where $\epsilon_T^{ij}\equiv\epsilon^{-+ij}$ and $\epsilon^{0123}=+1$. It is observed that the resulting numerator in Eq. (6) contains (at most) the fourth power of the loop integral Thus, the integration over the light cone component l^+ results in an integral that is potentially ill-defined. This happens when $g(p^2)$ is a holomorphic function in p^2 and at least one of the Minkowski indices is light-like in the minus direction, e.g. $\alpha_1=-,\alpha_2,...,\alpha_i\in\{+,\perp\}$ resulting in an integral of the form $\int dl^+\,\delta(l^+)\Theta(-l^+)$, implying that $l^+=0$ and $l^-=\infty$. This signals a light cone divergence (see Ref.[20]). One can handle such divergences by introducing phenomenological form factors with additional poles[13]

$$g_{ax}(p^2)=\frac{(p^2-m_q^2)f(p^2)}{\left[p^2-\Lambda^2+i0\right]^n}.$$

For $n\geq 3$ there are enough powers of l^+ to eliminate this divergence. $f(p^2)$ is a covariant Gaussian[13] which cuts off the p_T integrations and Λ is an arbitrary mass scale fixed by fitting f_1 to data (see below and Ref.[13] for details). The resulting analytic expressions for the Boer-Mulders function for the scalar and axial-vector diquark are given in.[13]

In a similar manner, the Sivers-function is projected from the trace of the quark-quark correlator (2),

$$\frac{M S_T^i \epsilon_T^{ij} p_T^j f_{1T}^\perp(x, \vec{p}_T^2)}{4} = \int dp^- \left(\mathrm{Tr}\left[\gamma^+ \Phi(p; P, S_T)\right] - \mathrm{Tr}\left[\gamma^+ \Phi(p; P, -S_T)\right]\right) \tag{7}$$

While in the scalar diquark approximation, $h_1^\perp$ and $f_{1T}^\perp$ exactly coincide,[10] the different Dirac structure for the chiral even $f_{1T}^\perp$ and chiral odd $h_1^\perp$ in the axial-vector diquark sector (cf. Eqs. (6) and (7)) result in different analytical expressions.[13] We fix the model parameters such as masses and normalization constants by comparing the model result for the unpolarized f_1 for u and d quarks to the low-scale ($\mu^2 = 0.26$GeV) data parameterization of GRV.[21] Note that PDFs for u and d quarks are given by linear combinations of PDFs for an axial vector and scalar diquark, $u = \frac{3}{2} f^{sc} + \frac{1}{2} f^{ax}$ and $d = f^{ax}$.[16] For "T-odd" PDFs we fix the sign and the strength of the final state interactions, the product of the charges of the diquark and quark, by comparing $f_{1T}^\perp$ for u and d quarks in the diquark model with the existing data parameterizations (e.g. Refs.[22]). We display the "one-half" moment $f_{1T}^{\perp(q,1/2)}(x) = \int d^2 p_T \frac{|\vec{p}_T|}{M} f_{1T}^{\perp(q)}(x, \vec{p}_T^2)$, for u- and d- quark Sivers functions $f_{1T}^{\perp(q)}$ and Boer-Mulders functions $h_1^{\perp(q)}$ (where $q = u, d$) along with the unpolarized u- and d- quark pdfs in Figs. 2. The u- and d- quark Sivers functions are negative and positive respectively while the u- and d- quark Boer Mulders functions are both negative over the full range in Bjorken-x. These results are in agreement impact parameter distortion picture of Burkardt.[23] Choosing $\kappa = 1$ we find reasonable agreement with extractions reported in.[22] It is worth noting (see Fig. 2) that the resulting u-quark Sivers function and Boer-Mulders function are nearly equal, even with the inclusion of the axial vector spectator diquark.[10]

Sometime ago it was observed that both kinematic and dynamical[24] sub-leading twist effects could give rise to an $\cos 2\phi$ azimuthal asymmetry going like p_T^2/Q^2 (where Q is a hard scale) when transverse momentum scales are on the order of the intrinsic momentum scales of partons, $P_T \sim p_T$. By contrast, taking into account the existence of "T-odd" TMDs and fragmentation functions it was pointed out by Boer and Mulders[4] that at leading twist a convolution of the Boer-Mulders and the Collins functions would give rise to non-trivial azimuthal asymmetries in unpolarized SIDIS. Thus, having studied the flavor dependence of the $h_1^\perp$ we consider the spin independent "T-odd" $\cos 2\phi$ contribution for π^+ and π^- production to the unpolarized cross section

$$\frac{d\sigma}{dxdydzd\phi_h dP_{h\perp}^2}=\frac{2\pi\alpha^2}{xyQ^2}\left[\left(1-y+\frac{1}{2}y^2\right)F_{UU,T}+(1-y)\cos(2\phi_h)\,F_{UU}^{\cos 2\phi_h}\right] \tag{8}$$

where the structure function $F_{UU}^{\cos 2\phi_h}$ involves a convolution of the Boer-Mulders and Collins fragmentation function

$$F_{UU}^{\cos 2\phi_h}=\mathcal{C}\left[-\frac{2\hat{\boldsymbol{h}}\cdot\boldsymbol{k}_T\hat{\boldsymbol{h}}\cdot\boldsymbol{p}_T-\boldsymbol{k}_T\cdot\boldsymbol{p}_T}{MM_h}h_1^\perp H_1^\perp\right], \tag{9}$$

where $\mathcal{C}$ represents the momentum convolution integral.[4] The Collins mechanism[15] describes the spin transfer of an initial state transverse polarization to the final state fragmenting quark. The Collins function is a measure of the spin asymmetry in the azimuthal distribution of the out going hadron about the jet axis of the fragmenting quark. We calculated the Collins functions in the spectator framework[14] where the fragmentation functions are calculated from the color gauge invariant quark-quark correlation function

$$\Delta(z,k_T)=\frac{1}{2z}\int\frac{d\xi^+d^2\boldsymbol{\xi}_T}{(2\pi)^3}\mathrm{e}^{ik\cdot\xi}\langle 0|\mathcal{U}^n_{(+\infty,\xi)}\psi(\xi)|h,X\rangle\langle h,X|\bar{\psi}(0)\mathcal{U}^n_{(0,+\infty)}|0\rangle\Big|_{\xi^-=0} \tag{10}$$

with $k^- = P_h^-/z$ (the sum over states $|X\rangle$ is implicit). The Collins fragmentation functions are projected from Eq. 10 as

$$2\frac{\epsilon_T^{ij}k_{Tj}}{M_h}H_1^\perp(z,k_T^2)=\mathrm{Tr}[\Delta(z,k_T)i\sigma^{i-}\gamma_5]. \tag{11}$$

In Refs.[25] it was shown that the fragmentation correlators are the same in both semi-inclusive DIS and e^+e^- annihilation, as was also observed earlier in the context of a specific model calculation[25] similar to the one under consideration here. Again, we work in Feynman gauge where the transverse gauge link vanishes.[6] Employing a pseudo-scalar pion-quark coupling and Gaussian form factors at the pion-quark vertex a nonzero Collins function is generated by means of the dynamics of ISI interactions from gluons.

3. Azimuthal Asymmetries

Utilizing these results we calculate the double "T-odd" contribution to the azimuthal asymmetry (cf. Eq. (9)),

$$A_{UU}^{\cos 2\phi}\equiv\frac{\int\cos 2\phi\; d\sigma}{\int\; d\sigma}. \tag{12}$$

In Fig. 3 we display the $A_{UU}^{\cos 2\phi}(P_T)$ in the range of HERMES[2] and future JLAB kinematics[26] as well as x and z dependence in the range $0.5 <$

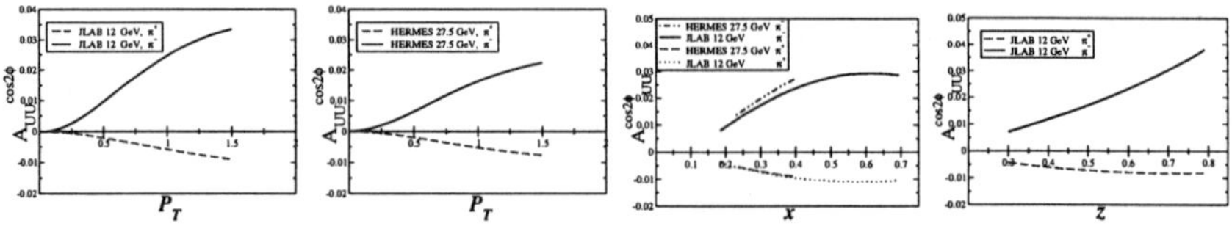

Fig. 3. Left Panel: The $\cos 2\phi$ asymmetry for π^+ and π^- as a function of P_T at JLAB 12GeV kinematics. Center Left Panel: The $\cos 2\phi$ asymmetry for π^+ and π^- as a function P_T for HERMES kinematics. Center Right Panel: The $\cos 2\phi$ asymmetry for π^+ and π^- as a function of x at JLAB 12GeV and HERMES kinematics. Right Panel: The $\cos 2\phi$ asymmetry for π^+ and π^- as a function z for JLAB kinematics.

$P_T < 1.5$ GeV/c. It should be noted that this asymmetry was measured at HERA by ZEUS, but at very low x and very high Q^2 [17] where other QCD effects dominate. Having calculated the chiral-odd but "T-even" parton distribution $h_{1L}^{\perp}$ we use this result together with the result of Ref.[14] for the Collins function to give a prediction for the $\sin(2\phi)$ single spin asymmetry A_{UL} for a longitudinally polarized target. A decomposition into structure functions of the cross section of semi-inclusive DIS for a longitudinally polarized target reads

$$\frac{d\sigma_{UL}}{dxdydzd\phi_h dP_{h\perp}^2} = \frac{2\pi\alpha^2(1-y)}{xyQ^2} S_{\parallel} \left[\sin(2\phi_h)\, F_{UL}^{\sin(2\phi)} + \frac{(2-y)}{\sqrt{1-y}} \sin(\phi_h)\, F_{UL}^{\sin\phi} \right], \tag{13}$$

$S_{\parallel}$ is the projection of the spin vector on the direction of the virtual photon. In a partonic picture the structure function $F_{UL}^{\sin(2\phi)}$ is leading twist (while $F_{UL}^{\sin\phi}$ is sub-leading) and given by a convolution of the TMD $h_{1L}^{\perp}$ and the Collins function

$$F_{UL}^{\sin(2\phi)} = \mathcal{C}\left[-\frac{2\hat{\boldsymbol{h}}\cdot\boldsymbol{k}_T\,\hat{\boldsymbol{h}}\cdot\boldsymbol{p}_T - \boldsymbol{k}_T\cdot\boldsymbol{p}_T}{MM_h}\, h_{1L}^{\perp} H_1^{\perp} \right]. \tag{14}$$

We display the results for the single spin asymmetry $A_{UL}^{\sin(2\phi)}$ in Fig. 4 using the kinematics of the upcoming JLab 12 GeV upgrade. We note that the π^- asymmetry is large and positive due to the model assumption $H_1^{\perp(dis-fav)} \approx -H_1^{\perp(fav)}$ This asymmetry has been measured at HERMES for longitudinally polarized protons and deuterons.[27] The data show that for the proton target at HERMES 27.5 GeV kinematics both π^+ and π^- asymmetries are consistent with 0 down to a sensitivity of about 0.01. These asymmetries could be non-zero, but with magnitudes less than 0.01 or 0.02. These results are considerably smaller than our predictions for the JLab upgrade. For the deuteron target the results are consistent with 0 for π^+ and

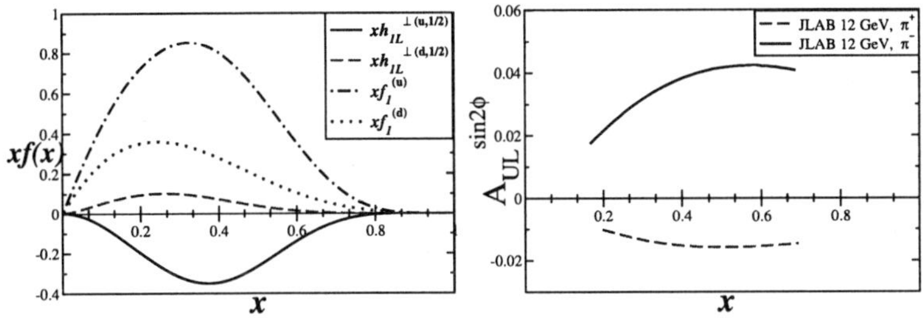

Fig. 4. Left Panel: The half-moment of $xh_{1L}^{\perp(1/2)}$ versus x compared to the unpolarized u- and d- quark distribution functions. Right Panel: The $\sin 2\phi$ asymmetry for π^+ and π^- as a function of x at JLAB 12GeV kinematics.

π^-. This SIDIS data for polarized deuterons could reflect the near cancellation of u- and d-quark $h_{1L}^{\perp}$ functions and/or the large unfavored Collins function contributions. There is also CLAS preliminary data[28] at 5.7 GeV that shows slightly negative asymmetries for π^+ and π^- and leads to the extraction of a negative $xh_{1L}^{\perp(u)}$. This suggests that the unfavored Collins function (for $d \to \pi^+$) is not contributing much. Data from the upgrade should help resolve these phenomenological questions.

In summary we have explored the flavor dependence leading twist "T-odd" TMD parton distribution functions, $h_1^{\perp}$ (Boer-Mulders) and $f_{1T}^{\perp}$ (Sivers) as well as the chiral odd but "T-even" function $h_{1L}^{\perp}$ in the diquark spectator model taking into account both axial-vector and scalar contributions. Our results are in agreement with other approaches that predict negative $h_1^{\perp(u)}$ and $h_1^{\perp(d)}$. We also note that the resulting u-quark Sivers function and Boer-Mulders function are nearly equal, even with the inclusion of the axial vector spectator diquark. This near equality $h_1^{\perp} \sim f_{1T}^{\perp}$ was obtained from models without axial di-quarks,[10] hinting at some more general mechanism that preserves the relation. We used our results for $h_1^{\perp}$ as one ingredient in the factorized formula for the azimuthal asymmetry $A_{UU}^{(\cos(2\phi))}$ in unpolarized lepto-production of positively and negatively charged pions, as well as $h_{1L}^{\perp}$ as input for the single-spin asymmetry $A_{UL}^{(\sin(2\phi))}$ for a longitudinally polarized target in SIDIS.[13] A key additional ingredient for determining such asymmetries is the Collins fragmentation function $H_1^{\perp}$. Using the most current expressions that were obtained in a similar spectator model,[14] we provide estimates for $A_{UU}^{(\cos(2\phi))}$ and $A_{UL}^{(\sin(2\phi))}$. The latter has already been measured at HERMES .[27] and preliminarily by CLAS.[28] There are important differences in the kinematic regions explored, but there remain discrepancies that may be resolved in the future at Jefferson Lab.

Acknowlegements

We thank A. Bacchetta and A. Mukherjee for use of results from.[14] L.G. acknowledges support from the U.S. Deparment of Energy under contract No. DE-FG02-07ER41460.

References

1. D. W. Sivers, *Phys. Rev.* **D41**, p. 83 (1990).
2. A. Airapetian *et al.*, *Phys. Rev. Lett.* **94**, p. 012002 (2005).
3. V. Y. Alexakhin *et al.*, *Phys. Rev. Lett.* **94**, p. 202002 (2005).
4. D. Boer and P. J. Mulders, *Phys. Rev.* **D57**, 5780 (1998).
5. J. C. Collins, *Phys. Lett.* **B536**, 43 (2002).
6. A. V. Belitsky, X. Ji and F. Yuan, *Nucl. Phys.* **B656**, 165 (2003).
7. D. Boer, P. J. Mulders and F. Pijlman, *Nucl. Phys.* **B667**, 201 (2003).
8. S. J. Brodsky, D. S. Hwang and I. Schmidt, *Phys. Lett.* **B530**, 99 (2002).
9. X.-d. Ji and F. Yuan, *Phys. Lett.* **B543**, 66 (2002).
10. G. R. Goldstein and L. Gamberg (2002), Proceedings of ICHEP 2002: North Holland, Amsterdam (2003) p. 452.
11. L. P. Gamberg, G. R. Goldstein and K. A. Oganessyan, *Phys. Rev.* **D67**, p. 071504 (2003).
12. P. J. Mulders and R. D. Tangerman, *Nucl. Phys.* **B461**, 197 (1996).
13. L. P. Gamberg, G. R. Goldstein and M. Schlegel, *Phys. Rev.* **D77**, p. 094016 (2008).
14. A. Bacchetta, L. P. Gamberg, G. R. Goldstein and A. Mukherjee, *Phys. Lett.* **B659**, 234 (2008).
15. J. C. Collins, *Nucl. Phys.* **B396**, 161 (1993).
16. R. Jakob, P. J. Mulders and J. Rodrigues, *Nucl. Phys.* **A626**, 937 (1997).
17. S. Chekanov *et al.*, *Eur. Phys. J.* **C51**, 289 (2007).
18. J. S. Conway *et al.*, *Phys. Rev.* **D39**, 92 (1989).
19. C. J. Bomhof and P. J. Mulders, *Nucl. Phys.* **B795**, 409 (2008).
20. L. P. Gamberg, D. S. Hwang, A. Metz and M. Schlegel, *Phys. Lett.* **B639**, 508 (2006).
21. M. Gluck, E. Reya and A. Vogt, *Eur. Phys. J.* **C5**, 461 (1998).
22. M. Anselmino *et al.*, *Phys. Rev.* **D72**, p. 094007 (2005).
23. M. Burkardt, *Phys. Rev.* **D72**, p. 094020 (2005).
24. R. N. Cahn, *Phys. Lett.* **B78**, p. 269 (1978).
25. A. Metz, *Phys. Lett.* **B549**, 139 (2002).
26. H. Avakian *et al.*, *approved JLab proposal PR 12-06-112.* (2006).
27. A. Airapetian *et al.*, *Phys. Lett.* **B562**, 182 (2003).
28. H. Avakian, P. E. Bosted, V. Burkert and L. Elouadrhiri, *AIP Conf. Proc.* **792**, 945 (2005).

EXPLORING CONFINEMENT WITH SPIN

J.P. Ralston

Department of Physics & Astronomy, University of Kansas, Lawrence, KS 66045 USA

A confining gauge theory violates the completeness of asymptotic states held as foundation points of the S-matrix. Spin-dependent experiments can yield results that appear to violate quantum mechanics. The point is illustrated by violation of the Soffer bound in QCD. Experimental confirmation that the bound is violated would be a discovery of immense importance, sweeping away fundamental assumptions of strong interaction physics held for the past 50 years.

1. A Completeness Paradox of Confinement

Confinement was not anticipated in early days formulating the S-matrix. Let $|p,\, s\rangle$ be a set of asymptotic hadron *in*-states. The S-matrix elements $S_{p'p}$ are defined by

$$S_{p's';ps} = lim_{t\to\infty} \langle p',\, s'|\, e^{-iH_{tot}t}\, |p,\, s\rangle \,.$$

When the exact Hamiltonian H_{tot} maps states out of the Hilbert space of asymptotic states - meaning that the hadrons are not complete for their own interactions - then a classic S-matrix assumption is violated.

QCD poses a definite paradox with hadronic completeness. Let Q_a be a *global* color generator in equal-time quantization,

$$[Q_a,\, Q_b\,] = if_{abc}Q_c,$$

where f_{abc} are the structure constants of the color group. Let $1_h = \sum_h |h\rangle\, \langle h|$ be a complete set of all orthogonal hadrons of all momenta and spin. Insert the complete set, and impose they are color singlets, yielding

$$[Q_a,\, Q_b\,] = Q_a \sum_h |h\rangle\, \langle h|\, Q_b - Q_b \sum_h |h\rangle\, \langle h|\, Q_a \to 0.$$

The "paradox of colored completeness" is *how do physical states support the algebra?* Resolving this paradox leads to phenomena that superficially appear to violate quantum mechanics, while providing a fascinating probe of confinement.

2. Two Kinds of Gauge Transformation

We explore the paradox by looking more deeply into QCD. Let $\psi_a(x)$ be a matter field ("type 1") in the fundamental color representation. Let $A_a(x)$ be a gauge potential ("type 2") in the adjoint representation. Fix $A_0 = 0$; there remains space-dependent local gauge transformations of type 1 and 2. Under a gauge transformation at equal times

$$[Q_a^{(1)}(x), \psi_b(x')] = i\tau_{bc}^a \psi_c \delta(x - x');$$
$$[Q_a^{(2)}(x), A_b(x')] = if_{abc}\delta A_c \delta(x - x'),$$

by which

$$A(x) \to U(\theta(x)) A U^\dagger(\theta(x)) - igU(\theta(x)) \partial U^\dagger(\theta(x)).$$

Here

$$[Q_a^{(1)}(x), Q_b^{(1)}(x')] = if_{abc}Q_c^{(1)}(x)\delta(x - x');$$
$$[Q_a^{(2)}(x), Q_b^{(2)}(x')] = if_{abc}Q_c^{(2)}(x)\delta(x - x');$$
$$[Q_a^{(1)}(x), Q_b^{(2)}(x')] = 0.$$

If the color group is $SU(N)$, the gauge transformations of type 1 and 2 generate $SU(N) \times SU(N)$. The breakdown of this group will explain the paradox.

2.1. *Gauss' Law*

Physical states are *gauge invariant* under the diagonal subgroup $SU(N) \subset SU(N) \times SU(N)$ generated by $Q^{(1)}(x) + Q^{(2)}(x)$. The invariant subspace $|all\rangle$ is defined by

$$(Q_a^{(1)}(x) + Q_a^{(2)}(x))\,|all\rangle = 0;$$
$$((D \cdot E)_a(x) - \rho_a(x))\,|all\rangle = 0,$$

where $\rho_a(x)$ is the charge operator on the matter fields, and $(D \cdot E)_a(x)$ is the Gauss-law operator on the gauge fields, with D the gauge-covariant derivative. Under Gauss' Law certain gluonic configurations are attached to the matter fields to make *singlet operators invariant under the joint transformations of type 1 and 2:*

$$\psi_a(x) \otimes A_b \to \psi_a(x)\bar{e}_a^\mu(x; A);$$
$$[Q_b(x), \psi_a(x')\bar{e}_a^\mu(x'; A)] = 0.$$

Index μ (the "solution number") is a composite index including Lorentz properties and internal parameters.

2.1.1. *Dirac's Solution for QED*

For QED pure gauge-fields, Dirac[1] solved $e_a^\mu(x;\, A) \to e_{Dirac}(x;\, A)$

$$e_{Dirac}(x;\, A) = e^{ig\frac{\nabla \cdot A}{\nabla^2}(x)}.$$

Under a type-2 gauge transformation $A(x) \to A(x) + \nabla\theta(x)$ we find:

$$e_{Dirac}(x;\, A + \nabla\theta(x)) = e^{ig\frac{\nabla\cdot(A+\nabla\theta)}{\nabla^2}(x)} = e^{ig\theta(x)}\, e_{Dirac}(x;\, A).$$

Dirac's "dressing operator" transforms like an electron. With $E = i\delta/\delta A$ the electric field operator, we check that Gauss' Law was solved by

$$\langle 0|\, \bar{e}_{Dirac}(x;\, A)\, i\frac{\delta}{\delta A(y)} e_{Dirac}(x;\, A)\, |0\rangle = -g\frac{\nabla}{\nabla_{xy}^2}.$$

But Dirac's pure-gauge construction is only part of the story.

2.2. *Embedded Frames and Connections*

The geometrical interpretation of e_a^μ is a *frame* for color polarizations on a "big" (embedding) space labeled with index μ. By Gauss the frame e_a^μ must transform like a quark - the fundamental rep - on index a:

$$e_a^\mu(x) \to U_{ab}(x) e_a^\mu(x); \quad U_{ab}(x) U_{bc}^\dagger(x) = \delta_{ac}. \tag{1}$$

Linear combinations of solutions are expressed with index μ. The frame is "normalized" by its inverse $\bar{e}_{a\mu}$,

$$\bar{e}_{a\mu} e_b^\mu = \delta_{ab}. \tag{2}$$

The range $\mu = 1...M$ depends on the geometry of the embedding space. Choosing e_a^μ a square, unitary matrix will only generate pure gauge configurations. By a theorem of John Nash one can always choose M large enough so that an arbitrary manifold can be embedded in a geometrically-trivial *Euclidean* space.

The *connection* A_{ab} is

$$A_{ab}(e;\, \bar{e};\, x) = \frac{-i}{g} \sum_\mu e_{a\mu}(x)\, \partial \bar{e}_b^\mu(x).$$

Here ∂ means $\partial/\partial x_\beta$. The symbol A_{ab} transforms locally like type-2:

$$A_{ab}(e;\, \bar{e};\, x) \to A_{ab}(Ue;\, \bar{e}U^\dagger;\, x) = (U(x) A(e;\, \bar{e};\, x) U^\dagger)_{ab} - \frac{i}{g} U_{ac}(x)\, \partial\, U_{cb}^\dagger(x).$$

The whole theory can be expressed using the frames.

2.2.1. *Gauge Links and Dressed Partons*

Parallel transport of frames is described by choosing a path and an associated gauge transformation

$$e_P(A;\, x,\, x') = \mathcal{P}e^{ig\int_x^{x'} dx\cdot A(z)}.$$

Parallel transport is neither an efficient nor necessary way to set up frames. Instead, path dependence under parallel transport is a symptom that the frames describe a curved space. The Gauss' Law frames $e_a^\mu(x)$ and connections $A_{ab}(e;\,\bar{e};\,x)$ are ordinary field that exist "everywhere" with no inherent path dependence.

The gauge invariant coordinate-space parton distribution is

$$\Phi(x^-,\, x_T) = \sum_\mu \langle p,\, s|\, \bar{\psi}_a(x)e_a^\mu(x)\Gamma\bar{e}_b^\mu(0)\psi_b(0)\, |p,\, s\rangle\, ,$$

where Γ represents spin projections. By Gauss' Law the frame-factors dress the bare quark operators with gluons. Early "leading twist" QCD set $x^+ = x_T = 0$, implementing the impulse approximation and integrating over parton transverse momenta. The assumption of short distance trivializes dressing to an unphysical light-like line along $x_T = 0$. Beyond leading twist, and in discussion of initial state interactions such as the Sivers effect, the correlations of dressing with spin and momentum are just what we seek to learn from experiments.

3. Two Kinds of Completeness

We now return to the paradox of colored completeness. The full QCD interaction comes into play between overlapping hadrons at finite range. The QCD Hamiltonian is only invariant under $SU(3) \subset SU^{(1)}(3) \times SU^{(2)}(3)$. The breakdown of invariance of the type-1 and type-2 groups allows exchange of local color and dressing while maintaining the gauge symmetry.

Consider the local generator relations

$$[Q_a^{(1)}(x),\, Q_b^{(1)}(x')\,] = if_{abc}Q_c^{(1)}(x)\delta(x-x');$$
$$[Q_a^{(2)}(x),\, Q_b^{(2)}(x')\,] = if_{abc}Q_c^{(2)}(x)\delta(x-x')$$

Let $|all\rangle$ be a complete set of *all* states of hadrons under interactions. Inset $1_{all} = \sum_{all} |all\rangle\langle all|$, giving

$$[(\,Q_a^{(1)}(x) + Q_a^{(2)}(x)\,)\sum_{all} |all\rangle\langle all|\,(\,Q_b^{(1)}(x') + Q_b^{(2)}(x')\,)\,]$$
$$= if_{abc}Q_c^{(tot)}(x)\delta(x-x');$$
$$Q^{(tot)}(x) = Q_a^{(1)}(x) + Q_a^{(2)}(x).$$

We made it explicit that interacting states transform under the type-1 and type-2 groups. But these transformations are not good symmetries, so we do not get new quantum numbers. In comparison the asymptotic $|p,\, s\rangle$ only transform under the Lorentz group and global symmetries that commute. It follows that the two spaces $|p,\, s\rangle$ and $|all\rangle$ are not equivalent. Completeness of $|all\rangle$ is "bigger" than completeness of trivial asymptotic states (lacking color labels):

$$1_h = \sum_h |h\rangle \langle h| \subset 1_{all} = \sum_{all} |all\rangle \langle all| .$$

This resolves the paradox.

4. Reduction

Without much discussion parton phenomenology implements the more-complete completeness of interacting states by giving color indices to perturbative quark and gluon lines at intermediate steps in the calculation. Summation over the color indices produces *reduction*, a step decreasing the dimension of a direct-product space.

Reduction requires density matrix formalism. It is convenient to suppress all dimensions except color and spin. Let $|\chi_\alpha\rangle$ be a basis element on the spin space, and $|\phi_i\rangle$ be a basis element on the color space, which spans type-1 and type-2. Take a particular state $|\psi\rangle$ on the interacting space, and expand:

$$|\psi\rangle = \sum_{i\alpha} |\chi_\alpha\rangle \, |\phi_i\rangle \; \psi_{i\alpha}, \tag{3}$$

Any pure state $|\psi\rangle$ is equivalent to a density matrix ρ_ψ made from the outer product with itself. The density matrix on asymptotic states comes from tracing out the color:

$$\rho_h = tr_{color}(\rho) = tr_{color}(|\psi\rangle \langle\psi|) = \sum_i |\chi_\alpha\rangle \, \psi_{i\alpha}\psi^*_{i\beta} \, \langle\chi_\beta| .$$

After reduction it is not generally possible to represent ρ_h as the outer product of a pure state with itself. Even more interesting, ρ_h will not describe hadron dynamics when color is exchanged. *Hadrons are incomplete to describe their own interactions*

Transitions of a general operator Ω are given by

$$S = tr(\rho_{all,\, in}\Omega\rho_{all,\, out}\Omega).$$

Since neither ρ_h nor $|p,\, s\rangle$ appear, properties of superposition assuming a wave function can fail. It is not quantum mechanics that is violated, however, it is the naive use of quantum mechanics developed for pure states that is violated.

I illustrate this by showing that the Soffer bound can be violated in QCD.

5. The Soffer Bound

Using the rules of quantum mechanics for pure states Soffer[2] produced a bound on $h_T(x,\, Q^2)$, the distribution of transversely polarized quarks in a transversely polarized target.[3] Soffer's bound is

$$h_T(x,\, Q^2) < |q(x,\, Q^2) + \Delta q(x,\, Q^2)|/2,$$

where $q(x,\, Q^2)$ and $\Delta q(x,\, Q^2$ are the unpolarized and longitudinally polarized quark distributions. Soffer's bound is more restrictive than positivity. A recent review by Artru *et al*[4] presents the bound and many related spin inequalities. Barone *et al*[5] handsomely review the general subject of transverse polarization. Experiments are in flux, but indicate that transversity is large.[6] The relation of experiments to matrix elements is the important goal of the transversity project reviewed in these Proceedings. Let us re-examine the use of quantum mechanics in the reduced system.

Eq. 3 can be simplified using the singular value decomposition:

$$|\psi\rangle = \sum_{\alpha} |\tilde{\chi}_\alpha\rangle\, \Lambda_\alpha \left|\tilde{\phi}_\alpha\right\rangle. \tag{4}$$

The tilde-basis are orthonormal, and can be found from diagonalizing two reduced density matrices on color and spin. Notice that the number of terms cannot exceed the dimension of the smaller space. The interpretation of Eq .4 is a basis in which each tilde-spin and color state are strictly correlated, with weights represented by Λ_α Correlation of color and spin would ordinarily seem to violate gauge invariance, but we have done the work to show it is an outcome of gauge invariance. The tilde-basis is particular to the particular state $|\psi\rangle$, which is time-dependent, and dynamical.

Consider an asymptotic spin 1/2 hadron helicity basis $|h\pm\rangle$, for which transversely polarized states $|hT_\pm\rangle$ are related by ordinary superposition:

$$|hT_\pm\rangle = \frac{1}{\sqrt{2}}(|h+\rangle \pm |h-\rangle).$$

A measurement in a "pure " transversely polarized state is described by

$$\begin{aligned} P(\Omega) &= tr(|T_+\rangle\,\langle T_+|\,\Omega), \\ &= \frac{1}{2}(tr(|h+\rangle\,\langle h+|\,\Omega) + tr(|h+\rangle\,\langle h-|\,\Omega) + ... \end{aligned}$$

Soffer's bound comes by letting Ω be the correlations of transversely polarized quarks. In reality an experiment measures

$$P(\Omega) = \sum_{\alpha,\beta} \Lambda_\alpha \Lambda_\beta (\left\langle \tilde{\phi}_\alpha \right| \langle \tilde{\chi}_\alpha |) \, \Omega \, (|\tilde{\chi}_\beta\rangle \left| \tilde{\phi}_\beta \right\rangle).$$

It is not possible to proceed without detailed information about the time evolution in QCD.

Some relations can by developed by making unrealistic restrictions. Suppose an operator does not correlate color with spin: $\Omega \to \Omega_{color} \otimes \Omega_{spin}$. Then

$$\begin{aligned} \langle \Omega \rangle &\to tr(\rho_{h;\,\Omega} \Omega_{spin}); \\ \rho_{h;\,\Omega} &= \sum_\alpha |\tilde{\chi}_\alpha(t)\rangle \, \rho_{\alpha\beta}(t) \, \langle \tilde{\chi}_\alpha(t)| \, , \\ where \quad \rho_{\alpha\beta}(t) &= \Lambda_\alpha(t) \left\langle \tilde{\phi}_\alpha(t) \right| \Omega_{color} \left| \tilde{\phi}_\beta(t) \right\rangle \Lambda_\beta(t). \end{aligned}$$

In this case an effective hadronic theory exists, but it is driven from outside by the color interactions the same theory cannot express. Requiring that the interacting hadronic system remain pure at all times would be completely arbitrary. QCD has no such rules, so we again conclude that hadrons are incomplete for their own interactions. For instance chiral models, which are expressed entirely by hadronic degrees of freedom, can't possibly represent QCD, which happens to be consistent with the fact nobody uses chiral models at high energies. It is much more surprising to realize *there is not supposed to be a local effective hadronic theory of any kind* representing QCD.

Soffer's bound fails because the final state "complete sets" depend on how the system was prepared and interacts. We can be quite sure that the final state hadrons of a transversely polarized reaction will differ from those of longitudinally polarized one. One would expect detailed calculations to reveal the general effect. The Q^2 dependence of parton distributions calculated perturbatively confirms the general argument. The anomalous dimensions of h_T, Δq, and q have been calculated to leading and next to leading order.[7] Since the functions scale by different rules, Soffer's bound must fail under evolution in one direction (increasing Q^2) or the other (decreasing Q^2). Neither outcome is acceptable under the strict logic used to make the bound. Consistency problems were noticed before.[8] However the bound was compared to the Callan-Gross relation, which is a correct zeroth order kinematic relation disturbed by radiative corrections. Our statement here is much stronger: the Soffer bound is based on premises that became obsolete with confintement. Examining the size and the direction of $pQCD$ effects is a side issue that does not address faults of the non-perturbative starting point.

The breakdown of hadronic completeness is the central issue. It would be good if studies extracting transversity distributions would assume *positivity*, rather than the Soffer bound, in order not to prejudice the analysis, Experimental confirmation that the bound is violated would be a discovery of immense importance, sweeping away fundamental assumptions of strong interaction physics held for the past 50 years.

Acknowledgments

I thank the workshop organizers for their generosity, and Mauro Anselmino, Leonard Gamberg, Gary Goldstein, Kloya Nikolaev and Oleg Teryaev for helpful discussions. Research supported in part under DOE Grant Number DE-FG02-04ER14308.

References

1. P. A. M. Dirac, *The Principles of Quantum Mechanics*, 4th Edition (Oxford University Press, 1958).
2. J. Soffer, Phys. Rev. Lett. **74**, 1292 (1995) [arXiv:hep-ph/9409254].
3. J. P. Ralston and D. E. Soper, Nucl. Phys. B **152**, 109 (1979); J. L. Cortes, B. Pire and J. P. Ralston, Z. Phys. C **55**, 409 (1992). See also AIP Conf. Proc. **223**, 184 (1991).
4. X. Artru, M. Elchikh, J. M. Richard, J. Soffer and O. V. Teryaev, arXiv:0802.0164 [hep-ph].
5. V. Barone, A. Drago and P. G. Ratcliffe, Phys. Rept. **359**, 1 (2002) [arXiv:hep-ph/0104283].
6. For the recent experimental situation see these Proceedings. See also M. Anselmino, M. Boglione, U. D'Alesio, A. Kotzinian, F. Murgia, A. Prokudin and C. Turk, Phys. Rev. D **75**, 054032 (2007) [arXiv:hep-ph/0701006] and references therein; M. Anselmino *et al.*, arXiv:0807.0173 [hep-ph].
7. A. V. Efremov and O. V. Teryaev, Sov. J. Nucl. Phys. **39**, 962 (1984) [Yad. Fiz. **39**, 1517 (1984)]; S. Kumano and M. Miyama, Phys. Rev. D **56**, 2504 (1997) [arXiv:hep-ph/9706420].
8. G. R. Goldstein, R. L. Jaffe and X. D. Ji, Phys. Rev. D **52**, 5006 (1995); B. Kamal, A. P. Contogouris and Z. Merebashvili, Phys. Lett. B **376**, 290 (1996); [arXiv:hep-ph/9501297]; V. Barone, Phys. Lett. B **409**, 499 (1997) [arXiv:hep-ph/9703343].

SIDIS ASYMMETRIES IN QUARK-DIQUARK MODEL

A. Kotzinian

CEA-Saclay, IRFU/Service de Physique Nucléaire, 91191 Gif-sur-Yvette, France
On leave in absence from YerPhI, Yerevan, Armenia and JINR, Dubna, Russia

Some properties of intrinsic transverse momentum dependent nucleon distribution functions are considered in the simple quark-diquark model. The transverse target polarization dependent asymmetries for SIDIS are calculated and compared with recent results of COMPASS. The model describes well the measured asymmetries. Generalization of quark-diquark model for Sivers function is discussed.

Keywords: distribution function; transverse momentum; asymmetry; azimuthal.

1. Introduction

The production of one unpolarized hadron in polarized lepton-nucleon DIS can be described using eighteen structure functions[1,2]. Eight of them are describing azimuthal asymmetries proportional to target transverse polarization. Within factorized QCD parton model, these structure functions can be expressed as convolutions of transverse momentum dependent (TMD) quark distribution and fragmentation functions (DFs and FFs). To evaluate these structure functions only six T-even twist-two DFs and two twist-two FFs together with $\mathcal{O}(1/Q)$ kinematic corrections were taken into account in[1]. The full expressions for the structure functions in terms of convolutions of twist-two and twist-three DFs and FFs are given in[2].

Since the TMD DFs and FFs are nonperturbative objects and cannot be calculated from first principles, the standard way to study them is to use some parameterizations and then, using expressions for the cross-sections and asymmetries, fit the parameters to existing data[3,4]. Other approach — develop nonperturbative models for nucleon structure and quark fragmentation, see for example[5–8].

Here I'll discuss some properties of DFs calculated in a simple quark-diquark model[5] and present the comparison of results for target transverse spin azimuthal asymmetries with COMPASS preliminary data[9,10].

We use the model of Ref.[5] with a modified formfactor for the proton-quark-diquark vertex. Namely, instead of power-like dependence on quark virtuality, we use the Gaussian dependence:

$$\frac{1}{|k^2 - \Lambda^2|^\alpha} \Rightarrow \exp\left(k^2/2\Lambda^2\right). \tag{1}$$

With this choice of formfactor the six leading order (LO) T-even DFs for the scalar ($R = S$) and axial-vector ($R = A$) diquark look like

$$\begin{aligned}
f_1^R(x, \mathbf{k}_T^2) &= f_0^R(x)\left((xM+m)^2 + \mathbf{k}_T^2\right) G(\mathbf{k}_T^2, x), \\
g_1^R(x, \mathbf{k}_T^2) &= a_R\, f_0^R(x)\left((xM+m)^2 - \mathbf{k}_T^2\right) G(\mathbf{k}_T^2, x), \\
h_1^R(x, \mathbf{k}_T^2) &= a_R\, f_0^R(x)(xM+m)^2 G(\mathbf{k}_T^2, x), \\
h_{1T}^{\perp R}(x, \mathbf{k}_T^2) &= -2a_R\, f_0^R(x) M^2 G(\mathbf{k}_T^2, x), \\
g_{1T}^{\perp R}(x, \mathbf{k}_T^2) &= 2a_R\, f_0^R(x) M (xM+m)^2 G(\mathbf{k}_T^2, x), \\
h_{1L}^{\perp R}(x, \mathbf{k}_T^2) &= -g_{1T}^{\perp R}(x, \mathbf{k}_T^2),
\end{aligned} \tag{2}$$

where

$$\begin{aligned}
G(\mathbf{k}_T^2, x) &= \frac{1}{\pi \mu^2(x)} \exp\left(\mathbf{k}_T^2/\mu^2(x)\right), \\
f_0^R(x) &= N_R \exp\left(\frac{x(1-x)M^2 - xM_R^2}{\mu^2(x)}\right), \\
\mu^2(x) &= \Lambda^2(1-x),
\end{aligned} \tag{3}$$

$a_S = 1$ and $a_A = -1/3$. The DFs for u (d) quarks in the proton are given by

$$\begin{aligned}
f_1^u(x, \mathbf{k}_T^2) &= \frac{3}{2} f_1^S(x, \mathbf{k}_T^2) + \frac{1}{2} f_1^A(x, \mathbf{k}_T^2), \\
f_1^d(x, \mathbf{k}_T^2) &= f_1^A(x, \mathbf{k}_T^2)
\end{aligned} \tag{4}$$

and similarly for other DFs. Since the model is describing only the valence quarks, the normalization factors N_S and N_A are obtained by requiring that f_1^S and f_1^A are normalized to unity upon integration over x and $\mathbf{k}_T^2$. In expressions (1–4), M is the proton mass and $m = 0.36$ GeV/c, $M_S = 0.6$ GeV/c, $M_A = 0.8$ GeV/c and $\Lambda = 0.5$ GeV/c are free parameters of the model suggested in[5].

Let us stress here that since the model is based on tree diagrams, the T-odd Sivers, $f_{1T}^{\perp}(x, \mathbf{k}_T^2)$, and Boer-Mulders, $h_1^{\perp}(x, \mathbf{k}_T^2)$, functions are equal to zero.

2. Helicity distribution function

In many phenomenological applications for TMD DFs the x-k_T factorized form: $f_1^q(x,\mathbf{k}_T^2) = f_1^q(x)\exp\left(\mathbf{k}_T^2/\mu_1^2\right)/\pi\mu_1^2$, $g_1^q(x,\mathbf{k}_T^2) = g_1^q(x)\exp\left(\mathbf{k}_T^2/\mu_2^2\right)/\pi\mu_2^2$ with x-independent width of transverse momentum distribution ($\mu_{1,2}^2 = Const$) is used. With this choice the quark longitudinal polarization, $P_q(x,\mathbf{k}_T^2) = g_1^u(x,\mathbf{k}_T^2)/f_1^u(x,\mathbf{k}_T^2)$, also has a x-k_T factorized form. Thus, for different fixed transverse momenta, the quark polarization has similar (re-scaled depending on these transverse momenta factors) x-dependence.

In contrast, with our choice of quark-diquark model, not only does the x-dependence of $P_q(x,\mathbf{k}_T^2)$ changes its form for different fixed k_T but it changes sign as well, see Fig. 1.

This behavior is tightly related to different quark orbital momentum contributions to DFs with scalar and vector diquark. Indeed, from Eqs. (2) the positive and negative helicity DFs $q_\pm^R(x,\mathbf{k}_T^2)$ look as

$$\begin{aligned}
q_+^S(x,\mathbf{k}_T^2) &= f_0^S(x)(xM+m)^2 G(\mathbf{k}_T^2,x),\\
q_-^S(x,\mathbf{k}_T^2) &= f_0^S(x)\mathbf{k}_T^2 G(\mathbf{k}_T^2,x),\\
q_+^A(x,\mathbf{k}_T^2) &= \frac{1}{3}f_0^A(x)\left(2(xM+m)^2+\mathbf{k}_T^2\right)G(\mathbf{k}_T^2,x),\\
q_-^A(x,\mathbf{k}_T^2) &= \frac{1}{3}f_0^A(x)\left((xM+m)^2+2\mathbf{k}_T^2\right)G(\mathbf{k}_T^2,x).
\end{aligned}\tag{5}$$

The DF q_+^S has the relative angular momentum $L=0$ and for q_-^S we have $L=1$, $l_z=1$. For the axial-vector diquark case $L=0$ and $L=1$ orbital momenta contribute both in q_+^A and q_-^A (for more discussion, see[11,12]).

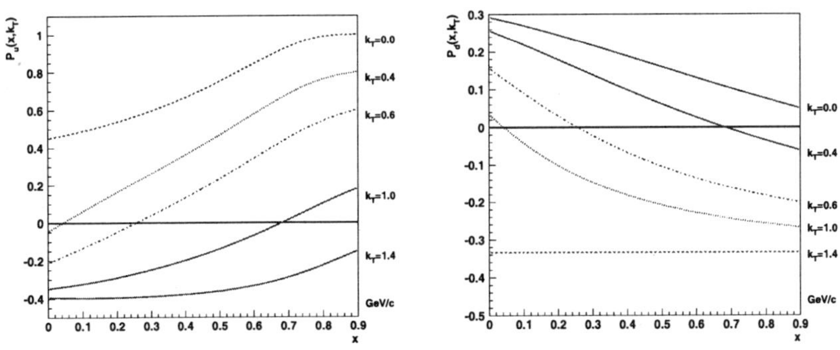

Fig. 1. The u- (left) and d-quark (right) polarization for different values of k_T.

The x-k_T factorization assumption was used in some analyzes of high-p_T hadron production asymmetries where scattering on quark is considered as a background process in polarized gluon distribution extraction. It is clear that the effective range of intrinsic k_T depends on the lower cut in hadron p_T and one can conclude from Fig. 1 that the magnitude of this background contribution can be different depending on the choice of k_T-dependence of TMD DFs. This consideration demonstrates that it is very important to obtain a reliable information on unpolarized and helicity TMD DFs by precise measurement of x-, z-, p_T- and azimuthal dependences of unpolarized SIDIS cross section and A_{LL} asymmetry and perform a flavor analysis (see also[13]).

3. Azimuthal asymmetries on transversely polarized target

Let us start by asymmetries which are described in the parton model by convolutions of twist-two DFs and FFs. The Sivers function is equal to zero in quark-diquark model, thus $A_{UT}^{\sin(\phi_h-\phi_S)} = 0$[a].

The Collins asymmetry can be expressed as

$$A_{UT}^{\sin(\phi_h+\phi_S)} \propto \frac{h_1^q \otimes H_{1q}^{\perp h}}{f_1^q \otimes D_{1q}^h}, \tag{6}$$

where $\otimes$ denotes integration over intrinsic transverse momentum of quark and hadron z- and p_T-variables over the kinematic domain of experiment. For details see Refs.[1,2,16]. The results of calculations for charged hadron production asymmetry on deuteron[14] and proton[15] target are presented in left and right panels of Fig. 2. For unpolarized FFs, we are using the recent DSS LO set[17] and for Collins FFs the parametrization from the recent global analysis.[3,16]

The DF $g_{1T}^{\perp}$ gives origin to double spin asymmetry

$$A_{LT}^{\cos(\phi_h-\phi_S)} \propto \frac{g_{1T}^{\perp q} \otimes D_{1q}^h}{f_1^q \otimes D_{1q}^h} \tag{7}$$

and the DF $h_{1T}^{\perp}$ — to

$$A_{UT}^{\sin(3\phi_h-\phi_S)} \propto \frac{h_{1T}^{\perp q} \otimes H_{1q}^{\perp h}}{f_1^q \otimes D_{1q}^h}. \tag{8}$$

The results are presented in left and right panels of Fig. 3.

---

[a]For notations and definition of azimuthal asymmetries see[9,10]

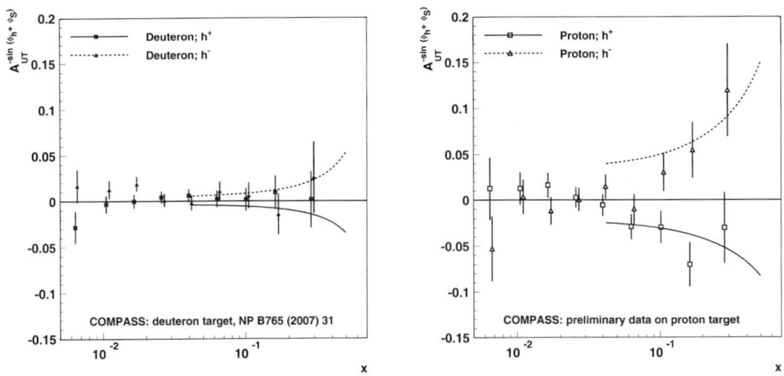

Fig. 2. Collins asymmetry for charged hadron production.

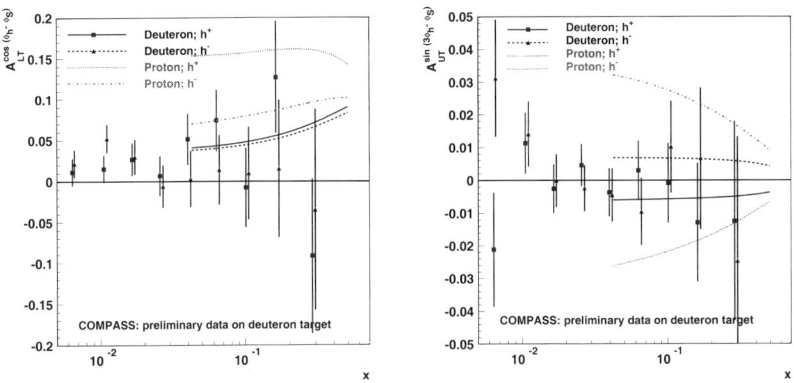

Fig. 3. The $A_{LT}^{\cos(\phi_h-\phi_S)}$ (left) and $A_{UT}^{\sin(3\phi_h-\phi_S)}$ (right) asymmetries.

Other four target transverse spin dependent asymmetries can be treated as kinematic $\mathcal{O}(1/Q)$ corrections to twist-two contributions.[1,2] One has

$$A_{LT}^{\cos(\phi_S)} \propto \frac{M}{Q}\frac{g_{1T}^{\perp q} \otimes D_{1q}^{h}}{f_1^q \otimes D_{1q}^{h}}, \tag{9}$$

$$A_{LT}^{\cos(2\phi_h\phi_S)} \propto \frac{M}{Q}\frac{g_{1T}^{\perp q} \otimes D_{1q}^{h}}{f_1^q \otimes D_{1q}^{h}}, \tag{10}$$

$$A_{UT}^{\sin(\phi_S)} \propto \frac{M}{Q}\frac{\left(h_1^q \otimes H_{1q}^{\perp h} + f_{1T}^{\perp q} \otimes D_{1q}^{h}\right)}{f_1^q \otimes D_{1q}^{h}}, \tag{11}$$

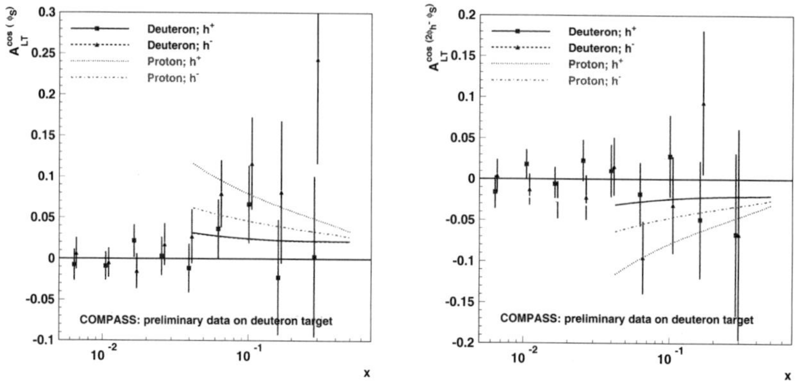

Fig. 4. Twist-three asymmetries $A_{LT}^{\cos(\phi_S)}$ (left) and $A_{LT}^{\cos(2\phi_h-\phi_S)}$ (right).

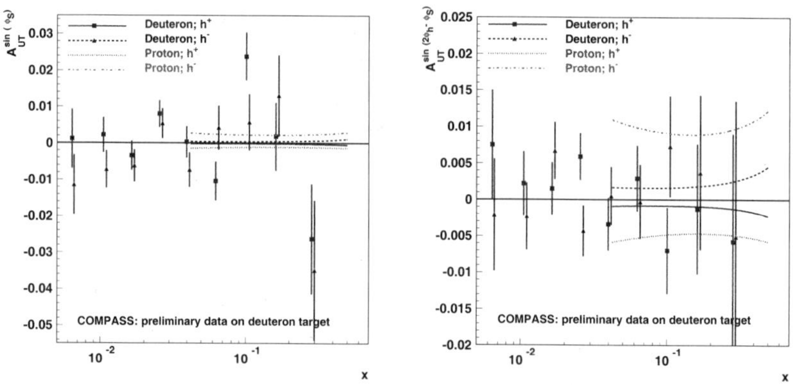

Fig. 5. Twist-tree asymmetries $A_{UT}^{\sin(\phi_S)}$ (left) and $A_{uT}^{\sin(2\phi_h-\phi_S)}$ (right).

$$A_{LT}^{\sin(2\phi_h\phi_S)} \propto \frac{M}{Q} \frac{\left(h_{1T}^{\perp q} \otimes H_{1q}^{\perp h} + f_{1T}^{\perp q} \otimes D_{1q}^{h}\right)}{f_1^q \otimes D_{1q}^h}. \tag{12}$$

The results are presented in Figs. 4 and 5.

Note that, because the simple quark-diquark model is developed for valence quarks, the calculations are presented for not too small values of Bjorken variable x. Calculations of z- and p_T-dependences of asymmetries for COMPASS kinematics include x-integration starting from very small $x \approx 0.003$. This means that the main contribution to unpolarized cross-section z- and p_T-dependence comes from sea quark region. For this reason, the quark-diquark model is not well adapted to describe these dependences of asymmetries at COMPASS kinematics.

4. Remarks on Sivers function

The mechanism of final state interaction (FSI) for generating the Sivers asymmetry in SIDIS was proposed in Ref.[18]. Application of this mechanism to quark-diquark model was considered in[6–8] and[19]. With Sivers DFs calculated within this approach, one can rather well describe the x-dependences of pion asymmetries observed by HERMES and COMPASS. However, we have observed that these functions are violating the naive positivity constraint for Sivers analyzing power:

$$\left|A_{Siv}(x, \mathbf{k}_T^2)\right| \leq 1, \qquad A_{Siv}(x, \mathbf{k}_T^2) = \frac{k_T}{M} \frac{f_{1T}^{\perp}(x, \mathbf{k}_T^2)}{f_1(x, \mathbf{k}_T^2)}. \tag{13}$$

Namely, with dipole-like formfactor in the proton-quark-diquark vertex in models[6,8,19] one obtains for high-k_T $A_{Siv}(x, \mathbf{k}_T^2) \propto k_T \to \infty$, with Gaussian choice Eq. (1) $A_{Siv}(x, \mathbf{k}_T^2) \propto 1/k_T \exp\left(\mathbf{k}_T^2/2\mu(x)^2\right) \to \infty$ and even stronger divergence in approach of[7]. The similar positivity violation for polarizing power, $P_{BM}(x, \mathbf{k}_T^2) = (k_T/M)\left(h_1^{\perp}(x, \mathbf{k}_T^2)/f_1(x, \mathbf{k}_T^2)\right)$, holds for the Boer-Mulders function calculated in[7] — $P_{BM}(x, \mathbf{k}_T^2) \to \infty$ at high k_T.

5. Conclusions

The quark-diquark model provides a good tool to study nonperturbative spin dynamics. The predictions of target transverse spin dependent asymmetries obtained using this model are in good agreement with COMPASS measurement.[9,10]

The interesting property of this model is the violation of the x-k_T factorization and dependence on x of width of transverse momentum distribution. It is be very important to measure the SIDIS cross-sections and asymmetries with high precision and perform a global analysis of data to get reliable two-dimensional parameterizations of different TMD DFs.

Sivers and Boer-Mulders functions obtained by applying the FSI mechanism to quark-diquark model do not satisfy the positivity bounds. Therefore, the questions arise: is the FSI mechanism[18] a universal explanation for nonzero Sivers function? Is it possible to avoid the violation of positivity by adding a formfactor in the gluon-diquark vertex or by a re-summation of higher order diagrams? More studies are needed to answer these questions.

References

1. A. Kotzinian, Nucl. Phys. B **441**, 234 (1995) [arXiv:hep-ph/9412283].
2. A. Bacchetta, M. Diehl, K. Goeke, A. Metz, P. Mulders and M. Schlegel, JHEP **0702**, 093 (2007) [arXiv:hep-ph/0611265].
3. A. Prokudin, in this proceedings.
4. M. Boglione, in this proceedings.
5. R. Jakob, P.J. Mulders, and J. Rodrigues, Nucl. Phys. A **626**, 937 (1997).
6. A. Bacchetta, A. Schaefer and J. J. Yang, Phys. Lett. B **578**, 109 (2004) [arXiv:hep-ph/0309246].
7. L. P. Gamberg, G. R. Goldstein and M. Schlegel, Phys. Rev. D **77**, 094016 (2008) [arXiv:0708.0324 [hep-ph]],
   L. Gamberg, in this proceedings.
8. M. Radici, in this proceedings.
9. A. Kotzinian [for COMPASS collaboration], arXiv:0705.2402 [hep-ex].
10. B. Parsamyan [for COMPASS Collaboration], arXiv:0709.3440 [hep-ex].
11. S. J. Brodsky, D. S. Hwang, B. Q. Ma and I. Schmidt Nucl. Phys. B **593**, 311 (2001).
12. H. Avakian, S. J. Brodsky, A. Deur and F. Yuan, Phys. Rev. Lett. **99**, 082001 (2007) [arXiv:0705.1553 [hep-ph]].
13. M. Anselmino, A. Efremov, A. Kotzinian and B. Parsamyan, Phys. Rev. D **74**, 074015 (2006) [arXiv:hep-ph/0608048].
14. E. S. Ageev *et al.* [COMPASS Collaboration], Nucl. Phys. B **765**, 31 (2007) [arXiv:hep-ex/0610068];
15. S. Levorato [for COMPASS collaboration], in this proceedings.
16. M. Anselmino, M. Boglione, U. D'Alesio, A. Kotzinian, F. Murgia, A. Prokudin and C. Turk, Phys. Rev. D **75**, 054032 (2007) [arXiv:hep-ph/0701006].
17. D. de Florian, R. Sassot and M. Stratmann, Phys. Rev. D **75**, 114010 (2007) [arXiv:hep-ph/0703242].
18. S. J. Brodsky, D. S. Hwang and I. Schmidt, Phys. Lett. B **530**, 99 (2002) [arXiv:hep-ph/0201296].
19. J. Ellis, D. S. Hwang and A. Kotzinian, work in progress.

PARTON DENSITIES IN A SPECTATOR MODEL WITH AXIAL-VECTOR DIQUARKS

M. Radici

INFN - Sezione di Pavia
via Bassi 6, I-27100 Pavia, Italy

We use a spectator model of the nucleon, including scalar and axial-vector diquarks, to investigate all its leading-twist spin structures. We interpret the results in terms of overlaps of light-cone wave functions, and we identify the non-trivial role of the quark-diquark relative orbital angular momentum. The model parameters are fixed by reproducing the phenomenological parametrization of unpolarized and helicity parton distributions at the lowest available scale. Predictions for the other parton densities, and for some of the related spin asymmetries, are given and, whenever possible, compared with available phenomenological parametrizations.

Keywords: Spectator diquark model; unintegrated parton distribution functions; spin densities.

1. Introduction

Partonic transverse-momentum distributions (TMD), alternatively called unintegrated parton distribution functions (PDF), describe the probability to find in a hadron, with momentum P, mass M and spin S, a parton with longitudinal momentum fraction x and transverse momentum p_T with respect to the direction of P. They can be extracted from the quark-quark correlator $\Phi(x, \mathbf{p}_T, S)$.

In recent years, a lot of theoretical activity was performed on TMD (for brevity, see the introduction of Ref.[1] and references quoted therein). On the experimental and phenomenological side, however, not much information is available as yet. New parametrizations (with large errors) have been recently released for the chiral-odd transversity h_1,[2] and the T-odd Sivers function $f_{1T}^{\perp}$.[3] But all their $\mathbf{p}_T$ dependences are very poorly constrained. These analyses are biased by neglecting evolution effects in extracting TMD from azimuthal asymmetries at very different scales. Moreover, the assumed x and $\mathbf{p}_T$ dependences are not very general: the former does not allow for sign changes; the latter is usually represented by a gaussian, while there is no compelling reason for this choice. At the same time, knowledge of TMD is becoming a central issue for the physics program of new accelerators like LHC,[4] where multiple-jet events are sensitive to QCD initial-state radiation that depends on the p_T tail of TMD.

Therefore, it is very useful to build a relatively simple and calculable model of TMD. In the literature, there are several approaches, including the one based on the diquark spectator representation of the nucleon (see, again, the introduction of Ref.[1] and references quoted therein). However, nowhere a systematic description of all leading-twist T-even and T-odd parton densities was given including also the axial-vector diquark, which is needed to describe the d quark in the nucleon. This task was accomplished in Ref.[1] and those results are briefly summarized here.

2. Formalism

The diquark spectator approximation amounts to insert a completeness relation in $\Phi(x, \mathbf{p}_T, S)$ and to truncate the sum over final states to a single on-shell spectator state with mass M_D and momentum $P-p$. By expanding the gauge link operator in the increasing number of exchanges of gluons and retaining just the first two terms (i.e., up to the single gluon-exchange), we get the leading-twist $O(\alpha_s)$ analytic approximation of the quark-quark correlator as[1]

$$\Phi(x, \boldsymbol{p}_T, S) \approx \frac{1}{(2\pi)^3} \frac{1}{2(1-x)P^+} \Big\{ \overline{\mathcal{M}}^{(0)}(S)\, \mathcal{M}^{(0)}(S) \tag{1}$$

$$+ \overline{\mathcal{M}}^{(0)}(S)\, \mathcal{M}^{(1)}(S) + \overline{\mathcal{M}}^{(1)}(S)\, \mathcal{M}^{(0)}(S) \Big\}\Big|_{p^2=\tau(x,\mathbf{p}_T)} \tag{2}$$

where

$$p^2 \equiv \tau(x, \mathbf{p}_T) = -\frac{\mathbf{p}_T^2 + L_D^2(m^2)}{1-x} + m^2 \tag{3}$$

$$L_D^2(m^2) = xM_D^2 + (1-x)m^2 - x(1-x)M^2 \tag{4}$$

is the off-shell condition for the parton with momentum p and mass m, induced by the spectator on-shell condition $(P-p)^2 = M_D^2$.

The contribution (1) to Φ corresponds to the left panel of Fig. 1. The calculation of the tree-level amplitude involves a nucleon-quark-diquark vertex $\mathcal{Y}_D$, which assumes different forms depending if the diquark has spin 0 (scalar $D = s$) or spin 1 (axial-vector $D = a$). A suitable form factor $g_D(p^2)$ is attached. In Ref.,[1] three choices were explored for point-like, dipolar, and exponential form factors. Here, we present results for the dipolar choice,

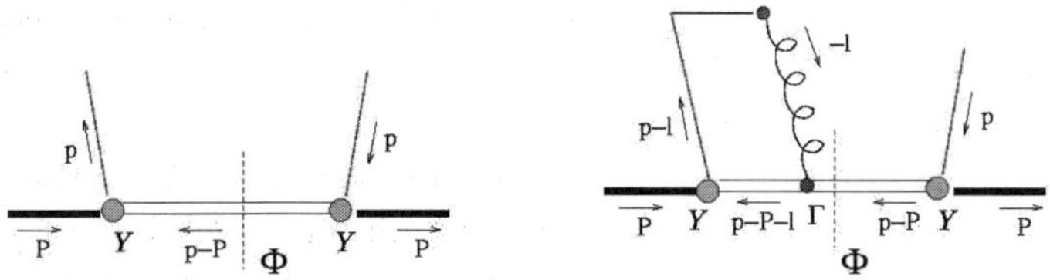

Fig. 1. Left panel: tree-level cut diagram for the calculation of T-even leading-twist parton densities. Right panel: interference between the one-gluon exchange diagram in eikonal approximation and the tree level diagram for the calculation of T-odd parton densities (the Hermitean conjugate diagram is not shown). The double line indicates both scalar and axial-vector diquarks.

because it grants the correct asymptotic behaviour. For the spin-1 diquark several choices are possible for the corresponding propagator across the cut in Fig. 1 (see Ref.[1]). Here, we will consider just the so-called "light-cone" one, where only the light-cone transverse polarization states are propagated. The reason for this choice mainly relies on its "minimal impact" on the parton interpretation of DIS, where F_T still can be interpreted as the charge-weighted sum of parton densities.[1] The contribution (2) to Φ corresponds to the right panel of Fig. 1. The amplitude for the one gluon-exchange can be calculated in the eikonal approximation using Cutkosky rules.[1] The final result is still analytic.

Both T-even and T-odd leading-twist TMD are extracted from suitable projections of the contributions (1) and (2) to $\Phi(x, \mathbf{p}_T, S)$. For example, the unpolarized parton density $f_1(x, \mathbf{p}_T)$ is defined by

$$f_1(x, \mathbf{p}_T) = \frac{1}{4} \frac{1}{(2\pi)^3} \frac{1}{2(1-x)P^+} \\ \mathrm{Tr}\left[\left(\overline{\mathcal{M}}^{(0)}(S)\,\mathcal{M}^{(0)}(S) + \overline{\mathcal{M}}^{(0)}(-S)\,\mathcal{M}^{(0)}(-S)\right)\gamma^+\right] + \text{h.c.}\,. \tag{5}$$

Expressions for all the other leading-twist TMD can be found in Ref.[1] Here, we just note that with scalar diquarks, $D = s$, the transversity saturates the Soffer bound, the Boer-Mulders function equals the Sivers one, and the so-called Pretzelosity function $h_{1T}^{\perp}$ satisfies the relation suggested in Ref.[5] However, the latter does not hold for axial-vector diquarks and, consequently, it must not be considered as a general one.[1]

The TMD can conveniently be represented as overlaps of light-cone wave functions (lcwf):

$$\psi^{\lambda_N}_{\lambda_q \lambda_D}(x, \mathbf{p}_T) = \sqrt{\frac{p^+}{(P-p)^+}}\, \frac{\bar{u}(p, \lambda_q)}{p^2 - m^2}\, \mathcal{Y}_D\, U(P, \lambda_N)\,, \tag{6}$$

where the dependence upon the diquark helicity λ_D is implicitly contained in the vertex $\mathcal{Y}_D$. For scalar diquarks we assume $\lambda_D \equiv \lambda_s = 0$, for axial-vector diquarks we have $\lambda_D \equiv \lambda_a = \pm 1$.

Angular momentum conservation constrains the helicities of the nucleon, the diquark, and the quark, implying the "spin sum rule" $\lambda_N = \lambda_q + \lambda_D + L_z$, where L_z is the projection of the relative orbital angular momentum between the quark and the diquark. It turns out that the lowest-energy configurations do not necessarily correspond to $L_z = 0$. Hence, in this version of the spectator diquark model the nucleon wave function does not show a SU(4)=SU(2)⊗SU(2) spin-isospin symmetry, contrary to what is usually assumed.[6] Moreover, the components with $|L_z| = 2$ vanish.[1]

3. Numerical results

The nine free parameters of the model — nucleon-quark-diquark couplings, diquark masses, and vertex cutoffs, for scalar, vector-isoscalar, vector-isovector diquarks — have been fixed by fitting the ZEUS parametrization at $Q_0^2 = 0.3$ GeV^2 for $f_1^{u,d}$,[7] and the leading-order GRSV2000 at $Q^2 = 0.26$ GeV^2 for $g_1^{u,d}$.[8] The results are reported in Fig. 2. The fitted parame-

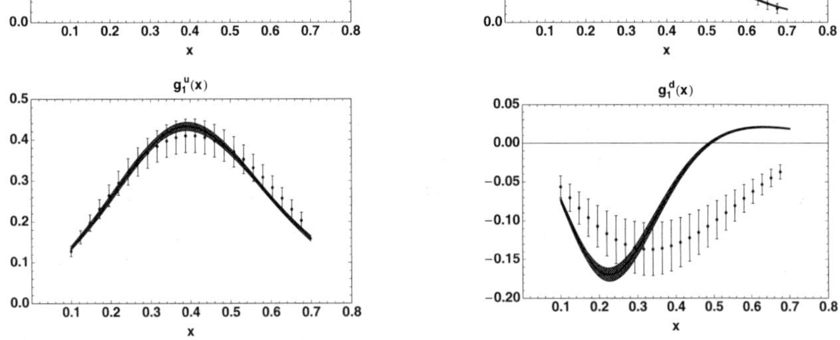

Fig. 2. The distribution functions $f_1(x)$ (above) and $g_1(x)$ (below) for the up quark (left panel) and the down quark (right panel). Data from Ref.[7] and Ref.[8] at LO, respectively. The curves represent the best fit (χ^2/d.o.f. = 3.88) obtained with our spectator model. The statistical uncertainty bands correspond to $\Delta\chi^2 = 1$.

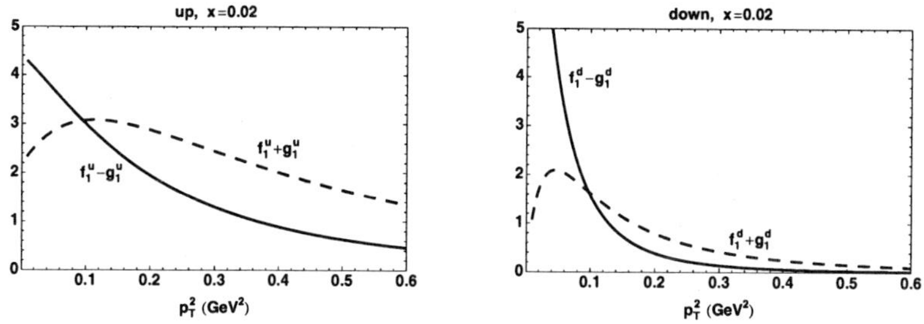

Fig. 3. The $f_1(x, \mathbf{p}_T^2) - g_1(x, \mathbf{p}_T^2)$ (solid line) and $f_1(x, \mathbf{p}_T^2) + g_1(x, \mathbf{p}_T^2)$ (dashed line) for up (left panel) and down quark (right panel), at $x = 0.02$.

ters induce the proton momentum fraction carried by valence quarks to be 0.584 ± 0.010, to be compared with the ZEUS result of 0.55.[7]

Evidence of the effect from orbital angular momentum can come from the combined analysis of unpolarized and helicity distributions. An example is given in Fig. 3, where the combinations $f_1(x = 0.02, \mathbf{p}_T^2) - g_1(x = 0.02, \mathbf{p}_T^2)$ and $f_1(x = 0.02, \mathbf{p}_T^2) + g_1(x = 0.02, \mathbf{p}_T^2)$ are displayed for the up (left panel) and down quarks (right panel). For brevity, let's consider just the down quark which is determined by the axial-vector isovector diquark alone. The $f_1 + g_1$ combination is related to $\psi^+_{+\pm}$, which carries $L_z = \pm 1$; hence, it drops down to 0 for $\mathbf{p}_T^2 \to 0$. At variance, the $f_1 - g_1$ one is given by ψ^+_{-+} with $L_z = 0$, which has a completely different behaviour in the same limit.

In Fig. 4, the model predictions for $xh_1(x)$ (upper panels) and $xh_1(x, \mathbf{p}_T^2)$ at $x = 0.1$ (lower panels) are presented for the up (left panels) and down quark (right panels). The results at the assumed original scale $Q_0^2 = 0.3$ GeV^2 are represented by the dashed line. The solid line indicates the result after applying the LO evolution up to the scale $Q^2 = 2.5$ GeV^2, where the only available parametrization of Ref.[2] is displayed with its uncertainty band due to parameter errors.

In Fig. 5, the $xf_{1T}^{\perp(1)}(x)$ moment of the Sivers function is predicted for the up (left panel) and down quark (right panel). It is compared with the parametrizations of Ref.[3] (darker shaded area) and of Ref.[9] (lighter shaded area). It must be stressed that no evolution was applied in the displayed model results.

In Fig. 6, we plot the spin density $f_{q/p^\uparrow}(x = 0.1, \mathbf{p}_T)$ of unpolarized quarks with flavor q in protons transversely polarized along $\hat{\mathbf{x}}$. The size of the asymmetry is driven by the Sivers function. The deformation is in

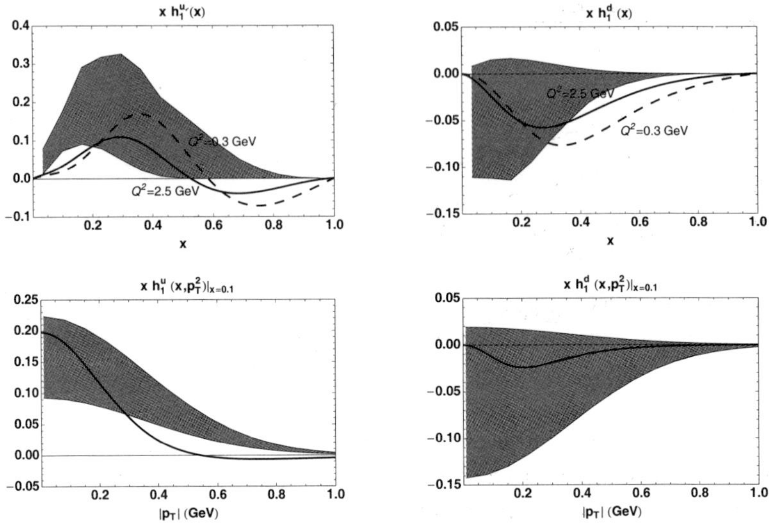

Fig. 4. The transversity distribution $xh_1(x)$ (upper panels) and $xh_1(x, \mathbf{p}_T^2)$ at $x = 0.1$ (lower panels) for up (left panel) and down quark (right panel). In the upper panels, dashed (solid) line for the model result before (after) evolution at LO up to the scale of the parametrization from Ref.[2], whose uncertainty band due to errors in the fit parameter is represented by the shaded area.

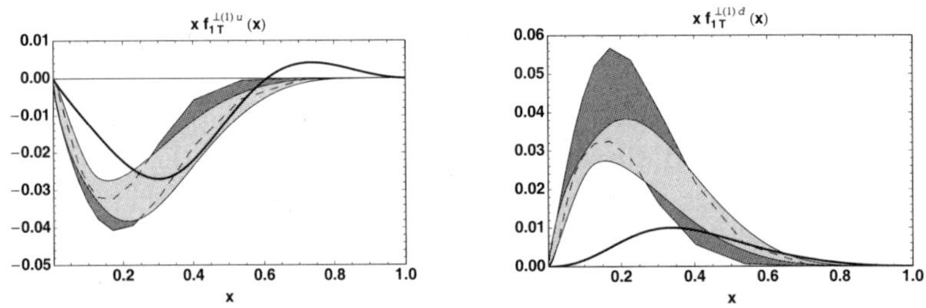

Fig. 5. The $xf_{1T}^{\perp(1)}(x)$ for up (left panel) and down quark (right panel). Solid line for the spectator diquark model. Darker shaded area for the parametrizations from Ref.[3], lighter one from Ref.[9].

agreement with the lattice results of Ref.[10] Analogous results have been worked out for the Boer-Mulders function and the related spin density of transversely polarized quarks in unpolarized protons.[1]

Preliminary results for azimuthal spin asymmetries have already been discussed in Ref.[11] for the $\bar{p}p^{(\uparrow)} \to \mu^+\mu^- X$ Drell-Yan reaction at

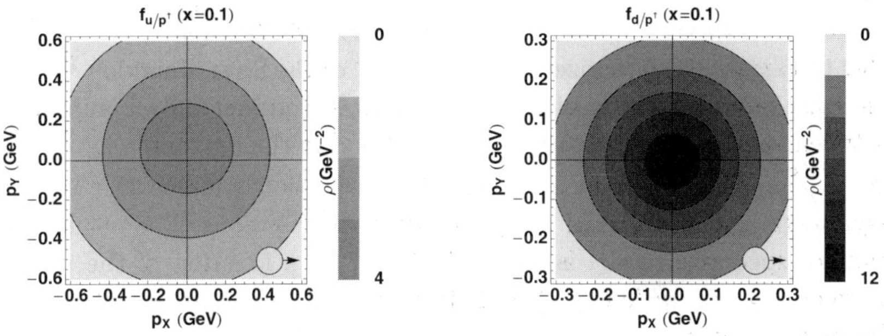

Fig. 6. The spin density of unpolarized quarks in transversely polarized protons (see text) in $\mathbf{p}_T$ space at $x = 0.1$. Left panel for up quark, right panel for down quark. The circle with the arrow indicates the direction of the proton polarization.

$s = 200$ GeV^2 with muon pair invariant masses in the range $4 < q < 9$ GeV and with zenithal angle range $60 < \theta < 120$ deg.

Here, we consider the $\pi^- p^\uparrow \to \mu^+\mu^- X$ reaction in the kinematic conditions explored in Ref.,[12] that could be measured at COMPASS with highly energetic pion beams hitting transversely polarized proton targets at $s = 400$ GeV^2, and producing muon pairs with squared invariant masses

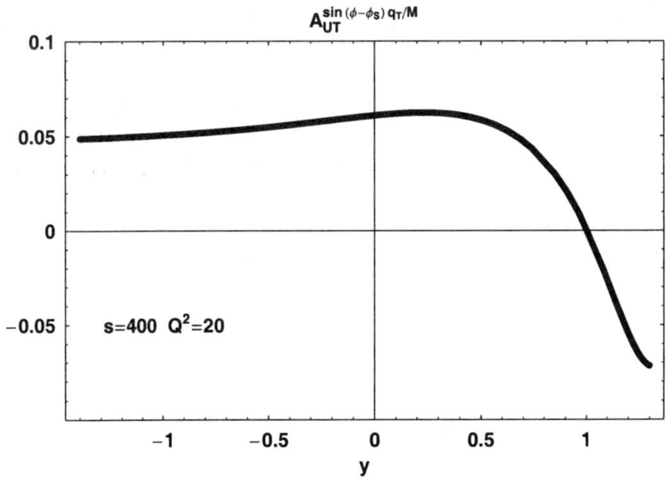

Fig. 7. The spin asymmetry for the $\pi^- p^\uparrow \to \mu^+\mu^- X$ reaction at COMPASS kinematics, as a function of rapidity y.

around $q^2 = 20$ GeV^2. The weighted spin asymmetry $A_{UT}^{\sin(\phi-\phi_{S_p})\, q_T/M_p}$ allows to extract the factorized moment $f_{1T}^{\perp(1)}$ of the Sivers function.[12] Using the same approximation of Ref.,[12] namely to consider only annihilations between valence components in the target and in the beam, the flavor sums in $A_{UT}^{\sin(\phi-\phi_{S_p})\, q_T/M_p}$ reduce to the $u\bar{u}$ contribution, making the weighted asymmetry independent of the pion parton distributions. The result of the diquark spectator model is shown in Fig. 7 as a function of the rapidity y, where $x_p = \sqrt{q^2/s}\, \exp[y]$. It is in very reasonable agreement with the findings of Ref.[12]

Acknowledgments

I warmly thank A. Bacchetta, F. Conti, and A. Bianconi, without whom this presentation would not have been possible.

References

1. A. Bacchetta, F. Conti and M. Radici (2008).
2. M. Anselmino *et al.*, *Phys. Rev.* **D**75, p. 054032 (2007).
3. M. Anselmino *et al.* (2008).
4. F. Hautmann and H. Jung (2008).
5. H. Avakian, A. V. Efremov, P. Schweitzer and F. Yuan (2008).
6. R. Jakob, P. J. Mulders and J. Rodrigues, *Nucl. Phys.* **A**626, 937 (1997).
7. S. Chekanov *et al.*, *Phys. Rev.* **D**67, p. 012007 (2003).
8. M. Glück, E. Reya, M. Stratmann and W. Vogelsang, *Phys. Rev.* **D**63, p. 094005 (2001).
9. J. C. Collins *et al.* (2005).
10. M. Gockeler *et al.*, *Phys. Rev. Lett.* **98**, p. 222001 (2007).
11. M. Radici, F. Conti, A. Bacchetta and A. Bianconi (2007).
12. A. V. Efremov, K. Goeke, S. Menzel, A. Metz and P. Schweitzer, *Phys. Lett.* **B**612, 233 (2005).

GLUONIC POLE MATRIX ELEMENTS IN SPECTATOR MODELS

A. Mukherjee[a], L.P. Gamberg[b], P.J. Mulders[c]

[a]*Physics Department, Indian Institute of Technology, Powai, Mumbai 400076, India.*
[b]*Department of Physics, Penn State University-Berks, Reading, Pennsylvania 19610, USA*
[c]*Department of Physics and Astronomy, VU University NL-1081 HV Amsterdam, The Netherlands*

We investigate the gluonic pole matrix element contributing to the first p_T moment of the distribution and fragmentation functions in a spectator model. By performing a spectral analysis, we find that for a large class of spectator models, the contribution of gluonic pole matrix elements is non-zero for the distribution correlators, whereas in fragmentation correlators they vanish. This outcome is important in the study of universality for fragmentation functions.

Keywords: Distribution function, fragmentation function, gluonic pole

1. Introduction

In order to access intrinsic transverse momenta, one needs to do a careful study of the azimuthal dependencies in high energy processes. Azimuthal imbalance can generate single spin asymmetries. These effects are not suppressed by powers of hard scale in comparison with the leading order terms. But it requires measuring hadronic scale quantities (transverse momenta) in a high momentum environment. Various symmetries, in particular, time reversal invariance play a key role here.

The proper gauge invariant definitions of transverse momentum dependent correlators involves path ordered exponentials (gauge links).[1] Recently, it has been found that the part of the path due to the transverse gluon field at lightcone $\pm\infty$ are not necessarily suppressed in light-cone gauge.[2,3] It is to be noted that the gauge invariance of collinear correlators also require a gauge link, however, in this case the bilocality in the operator is only in the lightcone direction and the gauge link is universal, independent of the process. Moreover, the link can be set to unity by choosing an axial gauge. This is not the case for in the transverse momentum dependent (TMD) correlators, as the bilocality is not restricted to the lightcone direction but exists also in the transverse direction. The resulting full color gauge invariant matrix element becomes process dependent due to the process dependent gauge link.

As QCD is time reversal invariant, it is possible to distinguish quantities and observables according to their T-behavior. For distribution correlators $\Phi(x)$, involving plane wave hadronic states in their definition, T-reversal and Hermiticity implies that the collinear correlators are T-even. However the TMD distribution correlators $\Phi(x, p_T)$ are process dependent due to the gauge link as stated above and time reversal relates $\Phi^{[+]}(x, p_T)$ and $\Phi^{[-]}(x, p_T)$, where the link is in two opposite light cone direction. Thus one can construct T-even and T-odd combinations and in general, TMD correlators can be parametrized in terms of both T-even and T-odd functions. The T-odd distribution functions, such as Sivers function, are non-zero due to the presence of the gauge link in the correlator.

The situation regarding T-invariance is different for the fragmentation TMD correlator. These involve hadronic out state in the definition and thus one cannot use the T-invariance as a constraint while parametrizing them. One always has to allow T-even and T-odd parts of the correlator. For spin 0 and spin 1/2 hadrons, in the collinear case, no T-odd functions appear at leading twist. Including TMD, for fragmentation functions there are now two mechanisms for T-odd terms:[2]

- That due to the operator including the gauge link;
- That due to the final state.

A nice feature, however, is that the two mechanisms leading to T-odd functions can be distinguished. The T-odd operator structure can be traced back to the color gauge link that necessarily appears in correlators to render them color gauge-invariant and it is process dependent. That due to the final state is independent of the process.

Relating T-odd effects in different processes, requires azimuthal weighting, which projects out the transverse momentum weighted parts of the correlators Φ and Δ, referred to as (first) transverse moments Φ_∂ and Δ_∂, respectively. After azimuthal weighting of the cross sections, one simply finds that the T-odd features originating from the gauge link lead to specific factors with which the T-odd functions appear in observables. Comparing T-odd effects in distribution functions in semi-inclusive DIS (SIDIS) and Drell-Yan (DY) processes, one finds a relative minus sign, which means that the Sivers function in these two processes differ by a sign. The T-odd operator parts of Φ_∂ and Δ_∂ are precisely the soft limits ($k_1 \to 0$) of the gluonic pole matrix elements,[2] denoted by $\phi_G(k_1 = 0)$ and $\Delta_G(k_1 = 0)$. In this work, we investigate the spectral properties of these matrix elements, which play an important role in the universality issue of the TMD distribution and fragmentation functions.

Fig. 1. The graphical representation of the correlator in the case of distributions of partons with momentum k in a hadron with momentum P and the spectator model description.

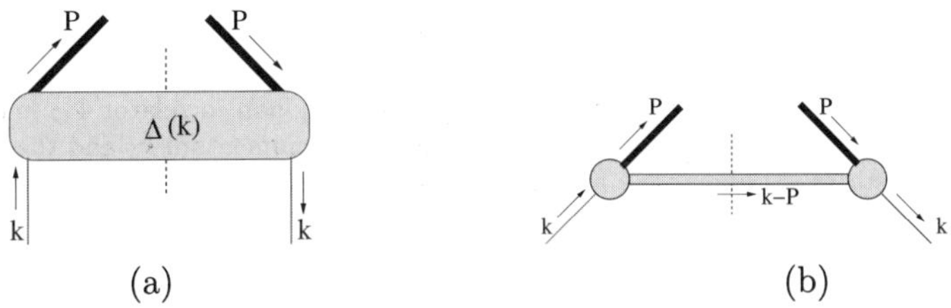

Fig. 2. The graphical representation of the correlator in the case of fragmentation of partons with momentum k into a hadron with momentum P (a) and the spectator model description (b).

For the correlators, depicted in Figs 1 and 2, the expressions in terms of bilocal matrix elements are frequently used as a starting point in modeling distribution and fragmentation functions. In particular the spectator model has become fairly popular, because it is easy, flexible and intuitively attractive. On the other hand, one should be very careful, because the predictive power depends on limiting oneself in the choice of spectator (e.g. a diquark with fixed mass for the nucleon) and using simple vertices. In fact making a spectral analysis of the spectator and allowing for the most general vertices one would lose all predictive power. Here we will investigate differences between distribution and fragmentation functions using a spectral analysis and using physical intuition in restricting the momentum dependence and asymptotic behavior of the vertices.

Rather than looking at the full process in the model and carefully studying the cuts,[4,5] we look at the soft part only,[6] but do this for the quark-quark-gluon correlators Φ_G and Δ_G, as shown in Figs 3 and 4, respectively. Our approach starting directly with the multi-parton correlator has the advantage that we can work at tree-level and just do a spectral analysis, unlike a one loop calculation.[7]

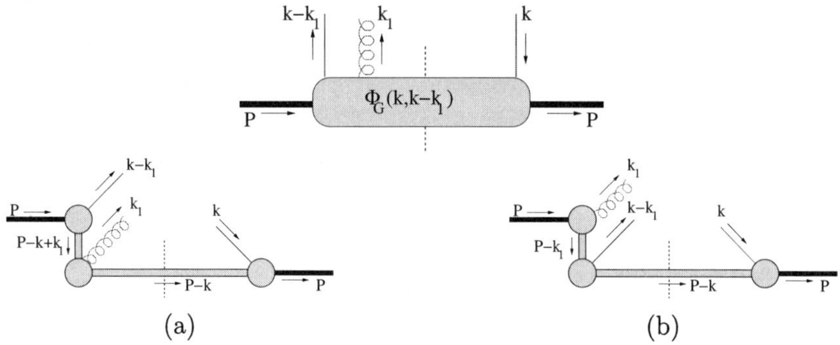

Fig. 3. The graphical representation of the quark-quark-gluon correlator Φ_G in the case of distributions including besides the quark a gluon with momentum k_1 and the possible intermediate states (a) and (b) in a spectator model description.

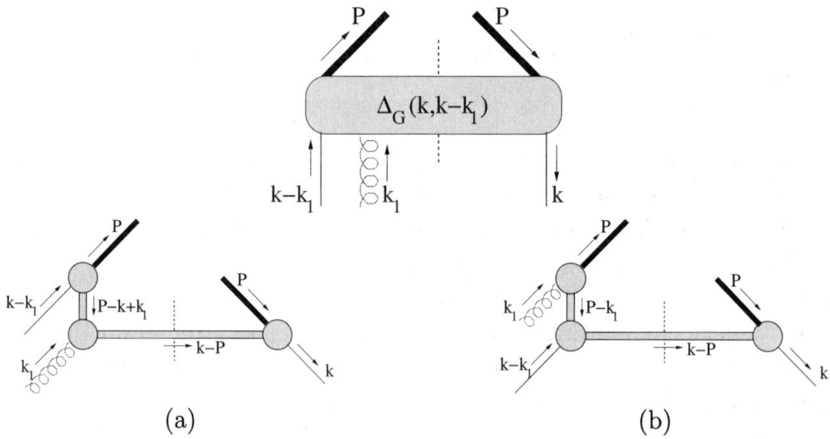

Fig. 4. The graphical representation of the quark-quark-gluon correlator Δ_G in the case of fragmentation including besides the quark a gluon with momentum k_1 and the possible intermediate states (a) and (b) in a spectator model description.

2. Gluonic pole matrix elements

We make a Sudakov decomposition of the momenta of active partons, $k = x\,P + \sigma\,n + k_T$. The Sudakov vector n is an arbitrary light-like four-vector $n^2 = 0$ that has non-zero overlap $P \cdot n$ with the hadron's momentum P. We will simply choose $P \cdot n = 1$.

The TMD distribution correlators are given by :

$$\Phi_{ij}^{[\mathcal{U}]}(x,k_T) = \int \frac{d(\xi{\cdot}P)\,d^2\xi_T}{(2\pi)^3}\; e^{ik\cdot\xi}\,\langle P|\,\overline{\psi}_j(0)\,\mathcal{U}_{[0;\xi]}\,\psi_i(\xi)\,|P\rangle\big]_{\rm LF}\,. \tag{1}$$

The *Wilson line* or *gauge link* $\mathcal{U}_{[\eta;\xi]} = \mathcal{P}\exp\big[-ig\int_C ds{\cdot}A^a(s)\,t^a\,\big]$ is a path-ordered exponential along the integration path C with endpoints at η and ξ.

In azimuthal asymmetries one needs the transverse moments contained in the correlator

$$\Phi_{\partial}^{\alpha\,[\mathcal{U}]}(x) = \int d^2k_T\; k_T^\alpha\,\Phi^{[\mathcal{U}]}(x,k_T)\,. \tag{2}$$

This can be written as

$$\Phi_{\partial}^{\alpha\,[\mathcal{U}]}(x) = \tilde{\Phi}_{\partial}^{\alpha}(x) + C_G^{[\mathcal{U}]}\,\pi\Phi_G^\alpha(x,x), \tag{3}$$

with calculable process-dependent gluonic pole factors $C_G^{[\mathcal{U}]}$ and process (link) independent correlators $\tilde{\Phi}_\partial$ (T even) and Φ_G (T-odd). The latter is precisely the soft limit $x_1 \to 0$ of a quark-gluon correlator $\Phi_G(x,x_1)$ of the type

$$\Phi_G^\alpha(x,x{-}x_1) = n_\mu \int \frac{d(\xi{\cdot}P)}{2\pi}\frac{d(\eta{\cdot}P)}{2\pi}\; e^{ix_1(\eta\cdot P)}e^{i(x-x_1)(\xi\cdot P)}$$
$$\langle P|\,\overline{\psi}(0)\,U^n_{[0;\eta]}\,gG^{\mu\alpha}(\eta)\,U^n_{[\eta;\xi]}\,\psi(\xi)\,|P\rangle\,\big]_{\rm LC}\,. \tag{4}$$

The TMD fragmentation correlator is given by[2]

$$\Delta_{ij}^{[\mathcal{U}]}(z,k_T) = \sum_X \int \frac{d(\xi\cdot P_h)\,d^2\xi_T}{(2\pi)^3} e^{i\,k\cdot\xi}\langle 0|\mathcal{U}_{[0,\xi]}\psi_i(\xi)|P,X\rangle$$
$$\langle P,X|\bar{\psi}_j(0)|0\rangle|_{LF}\,. \tag{5}$$

In the transverse moments obtained after k_T-weighting,

$$\Delta_{\partial}^{\alpha\,[\mathcal{U}]}(z) = \int d^2k_T\; k_T^\alpha\Delta^{[\mathcal{U}]}(z,k_T) = \tilde{\Delta}_\partial^\alpha\left(\tfrac{1}{z}\right) + C_G^{[\mathcal{U}]}\,\pi\Delta_G^\alpha\left(\tfrac{1}{z},\tfrac{1}{z}\right); \tag{6}$$

the two link independent correlators $\tilde{\Delta}_\partial$ and Δ_G contain again a T-even and T-odd operator combination, respectively, however, parametrizations of both of them contain T-odd functions as the final state is involved in both. The gluonic pole correlator is the soft limit, $z_1^{-1} = x_1 \to 0$, of the quark-gluon correlator

$$\Delta^{\alpha}_{G\,ij}(x, x - x_1) = \sum_X \int \frac{d(\xi{\cdot}P)}{2\pi}\frac{d(\eta{\cdot}P)}{2\pi}\; e^{i\,x_1(\eta\cdot P)} e^{i\,(x-x_1)(\xi\cdot P)} \times \langle 0|\mathcal{U}^n_{[0,\eta]}\, gG^{n\alpha}(\eta)\,\mathcal{U}^n_{[\eta,\xi]}\psi_i(\xi)|P,X\rangle\langle P,X|\overline{\psi}_j(0)|0\rangle\Big|_{LC}. \tag{7}$$

With $\Delta_G(x,x)$ being zero, T-odd FFs in Δ_∂ only come from the T-even operator combination in $\tilde{\Delta}_\partial$, which is due to the hadron-jet state and they are process independent, for instance the T-odd Collins function. In contrast T-odd DFs in Φ_∂ only can come from $\Phi_G(x,x)$. They can still be universal but acquire process dependent gluonic pole factors.[8,9] We calculate Φ_G and Δ_G in a spectator model approach.

3. The spectator model approach

In a typical spectator model approach to distribution or fragmentation correlators one considers a spectator with mass M_s. The result for the cut, but untruncated, diagrams, such as in Figs. 1 and 2 are of the form

$$\Phi(x, k_T) \sim \int d(k\cdot P)\; \frac{F(k^2, k\cdot P)}{(k^2 - m^2 + i\epsilon)^2}\,\delta\left((k-P)^2 - M_s^2\right). \tag{8}$$

The details of the numerator function depend on the details of the model, including the vertices, polarization sums, etc. These must be chosen in such a way as to not produce unphysical effects, such as a decaying proton if $M \geq m + M_s$. Thus m in Eq. 8 must represent some constituent mass in the quark propagator, rather than the bare mass. The useful feature of the model is its ability to produce reasonable valence and even sea quark distributions using the freedom in the model connected to an intuitive picture. The results for the fragmentation function in the spectator model is identical upon the substitution of $x = 1/z$.

The quark-gluon correlators as shown in Figs. 3 and 4 can be calculated in the spectator model and the gluonic pole matrix elements can be extracted by taking the limit $x_1 \to 0$. Assuming that the numerator does not grow with k_1^- one can easily perform the k_1^- integrations.[6] We obtain (for $x \geq 0$)

$$\Phi_G(x,x) = -\int d^2k_T\, d^2k_{1T}\; \frac{(1-x)\,F_1(x,0,k_T,k_{1T})\theta(1-x)}{(\mu^2 - k_T^2)\left(x\,B_1 + (1-x)\,A_2\right)A_1}. \tag{9}$$

and for fragmentation functions ($x = 1/z \geq 1$)

$$\Delta_G(x, x) = 0\,. \tag{10}$$

4. Conclusion

We have investigated the gluonic pole contributions to the distribution and fragmentation functions. Instead of doing a quantitative analysis involving details of a phenomenological model, we limit ourselves to a spectral analysis within the spectator framework, in order to understand the basic features of these quantities. We simply assumed that masses and vertices do not spoil our analysis. We find that under realistic assumptions, the gluonic pole contributions for fragmentation correlators vanish whereas these contributions do not vanish for distribution correlators. The result for fragmentation correlators at nonzero gluon momentum is nonzero. A full proof that the gluonic pole contributions to the fragmentation correlators vanish is important as it eliminates a whole class of matrix elements parameterized in terms of T-odd fragmentation functions besides the T-odd fragmentation functions in the parameterization of the two-parton correlators.

Acknowledgments

AM would like to thank the organizers of Transversity 2008 for the kind invitation and support.

References

1. J. C. Collins, D. E. Soper, Nucl. Phy. **B** 194, 445 (1982).
2. D. Boer, P. J. Mulders, F. Pijlman, Nucl. Phys. **B** 667, 201 (2003).
3. A. V. Belitsky, X. Ji, F. Yuang, Nucl. Phys. **B** 656, 165 (2003).
4. A. Metz, Phys. Lett. **B** 549, 139 (2002).
5. J. C. Collins, A. Metz, Phys. Rev. Lett. **93**, 252001 (2004).
6. L.P. Gamberg, A. Mukherjee, P. J. Mulders, Phys. Rev. **D** 77, 114026 (2008).
7. A. Bacchetta, L.P. Gamberg, G. Goldstein, A. Mukherjee, Phys. Lett. **B** 659, 234 (2008).
8. C. J. Bomhof, P. J. Mulders, F. Pijlman, Phys. Lett. **B** 596, 277 (2004).
9. C. J. Bomhof, P. J. Mulders, JHEP, **02**, 029 (2007).

TMD MEASUREMENTS AT CLAS6 AND CLAS12

H. Avakian
(On behalf of the CLAS6 AND CLAS12 Collaborations)
Jefferson Lab,
Newport News, VA, 23601, USA

We present recent results on studies of spin orbit correlations using the Jefferson Lab CLAS detector and discuss both near term and long-term future plans with CLAS, including measurements with longitudinally and transversely polarized targets. Single Spin Asymmetries (SSAs) were measured with CLAS for unpolarized and longitudinally polarized targets using a longitudinally polarized beam. Measurements of SSA for charged and neutral pions indicate that there are significant spin and azimuthal asymmetries, which can be interpreted in terms of transverse momentum dependent (TMD) distributions.

Keywords: SSA,SIDIS,polarized electroproduction

1. Introduction

Semi-inclusive deep inelastic scattering (SIDIS), for which a hadron is detected in coincidence with the scattered lepton, gives access to the transverse momentum distributions of quarks, which is not accessible in inclusive scattering. Azimuthal distributions of final state particles in semi-inclusive deep inelastic scattering are sensitive to the orbital motion of quarks and play an important role in the study of transverse momentum distributions of quarks in the nucleon.

Large Single Spin Asymmetries (SSAs), appearing as azimuthal moments of the cross section, observed for decades in hadronic reactions, have been among the most difficult phenomena to understand from first principles in QCD. Two fundamental mechanisms have been identified leading to SSAs in hard processes; the Sivers mechanism,[1–5] which generates an asymmetry in the distribution of quarks due to orbital motion of partons, and the Collins mechanism,[4,6] which generates an asymmetry during the hadronization of quarks.

Significant progress has been made recently in understanding the role of partonic initial and final state interactions.[3–5] Accounting of interaction between the active parton in the hadron and the spectators leads to the introduction of gauge-invariant transverse momentum dependent parton distributions .[3–5,7–9] Furthermore, QCD factorization for semi-inclusive deep inelastic scattering at low transverse momentum in the current-fragmentation region has been established in Refs.[10,11] This new framework provides a rigorous basis to study the TMD parton distributions from SIDIS data using different spin-dependent and independent observables.

SSAs, giving direct access to TMD distributions, have been reported for pion production in semi-inclusive deep-inelastic scattering by the HERMES collaboration[12–15] for unpolarized, as well as longitudinally and transversely polarized targets, by the COMPASS collaboration at CERN for a transversely polarized target[16] and by the CLAS Collaboration for longitudinally polarized beam and target.[17,18]

2. SIDIS with unpolarized target

The beam-spin asymmetries in single-pion production off the unpolarized target are higher-twist by their nature.[19,20] The higher twist observables are important for understanding the long-range quark-gluon dynamics and they may also be accessible as leading contributions through the measurements of certain asymmetries[21–24] and in particular beam SSAs. Recently, higher twist effects in SIDIS were interpreted in terms of average transverse force acting on the active quarks in the instant after being struck by the virtual photon.[25]

Different contributions to the beam SSAs, discussed so far, provide information on different leading and sub-leading parton distribution and fragmentation functions, related both to Collins and Sivers production mechanisms. Large beam SSAs were predicted[20] based on the Sivers mechanism as the dynamical origin of the observable. Within this framework, the asymmetry generated at the distribution level, is given by either the convolution of the T-odd parton distribution $h_1^\perp$ with the twist-3 fragmentation function $E(x)$,[21] or the convolution of the twist-3 T-odd distribution function $g^\perp$ with unpolarized fragmentation function D_1.[26] Parton model analyzes[19,20,26,27] of sub-leading single-spin asymmetries observed at HERMES[28] and CLAS[17] lead to introduction of a new twist-3 T-odd distribution function.[9,11] Despite the fact that partonic formalism is much better suited for higher energy reactions, there is a reasonable agreement in size and behavior of different predicted SSAs with the data.

Latest measurements at HERMES,[12,14] COMPASS[16] and Belle,[29] indicate that the Collins function describing the fragmentation of polarized quarks[30] is significant. The favored and unfavored Collins functions appeared to be roughly equal and have opposite signs. That puts the π^0 in a unique position in SSA studies. Since the π^0 fragmentation function (FF) is the sum of π^+ and π^- FFs, its favored and unfavored FFs will be equal and the Collins FF will be strongly suppressed. The contributions related to Sivers mechanism can thus be studied without a significant background from the Collins mechanism.

The multiplicity of π^0, defined as a ratio of counts of π^0s to the DIS events, in the valence region, where the sea contribution is suppressed, will coincide with its fragmentation function. The π^0 multiplicity measured using the latest CLAS data obtained with a CLAS configuration including the inner calorimeter (e1dvcs)[31] indeed indicates a good agreement (see Fig. 1) with the data at much higher energies from SLAC,[32] EMC,[33] HERMES,[34] as well as the fragmentation function of the π^0 extracted recently from fits to the world data.[35]

The measured beam SSA for π^0 appeared to be comparable with π^+ beam SSA from older CLAS data sets[36] both in the sign and value (Fig. 1), indicating that contribution from Collins mechanism could not be the dominating one.

A similar measurement has been performed by the HERMES collaboration at higher beam energy (27.6 GeV). After taking into account the kinematical factor in the expression of the beam spin dependent contri-

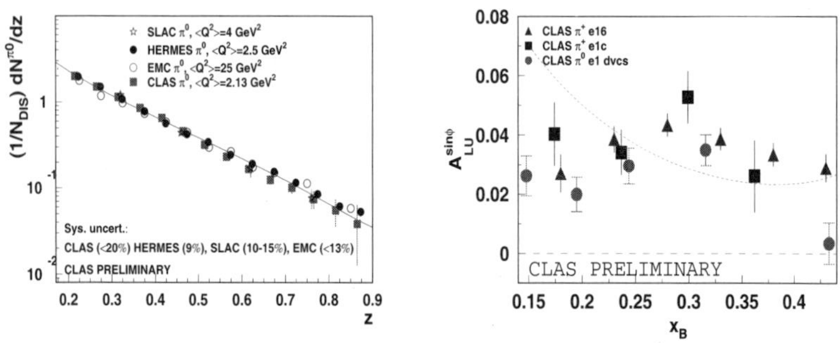

(a) (b)

Fig. 1. The π^0 multiplicities (a) from e1dvcs (CLAS@5.8 GeV). The curve is for DSS fragmentation function,[35] which was also used to scale the EMC data to 2.5 GeV^2. The beam SSA A_{LU} vs x_B (b) for π^+ from e1c (CLAS@4.3 GeV), e16 (CLAS@5.7 GeV) and π^0 from e1dvcs (CLAS@5.8 GeV) datasets (b). The curve is the theory prediction for the Sivers contribution.[19]

bution, CLAS and HERMES measurements were found to be consistent, indicating that at energies as low as 4-6 GeV the behavior of spin asymmetries is consistent with higher energy measurements.

3. SIDIS with longitudinally polarized target

For a longitudinally polarized target the only azimuthal asymmetry arising at leading order is the $\sin 2\phi$ moment,[6,23,37] involving the Collins fragmentation function $H_1^\perp$ and the Ralston-Soper-Mulders-Tangerman (RSMT) distribution function $h_{1L}^\perp$,[6,38] describing the transverse polarization of quarks in a longitudinally polarized proton.[6,23,37,39] The same distribution function is accessible in double polarized Drell-Yan, where it gives rise to a $\cos 2\phi$ azimuthal moment in the cross section.[22]

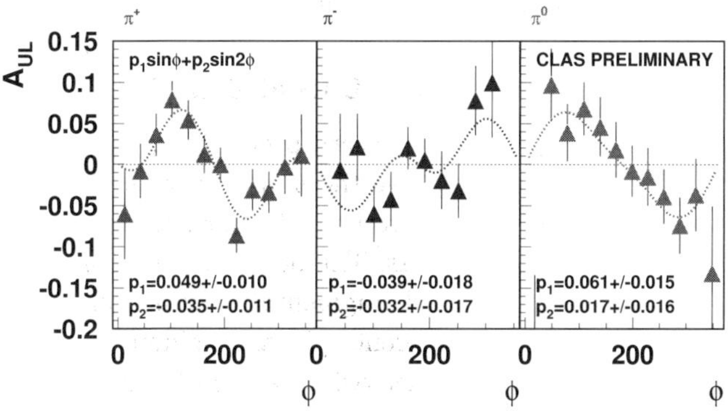

Fig. 2. Azimuthal modulation of the target spin dependent asymmetry $A_{UL} = 1/P_T(N^+ - N^-)/(N^+ + N^-)$, where $N^\pm$ are luminosity weighted counts for 2 target spin orientations. The dotted line shows the fit by $(p_1 \sin\phi + p_2 \sin 2\phi)/(1 + p_3 \cos\phi)$. Two dashed lines for π^+ show separately contributions from $\sin\phi$ and $\sin 2\phi$ moments.

Since the $\sin 2\phi$ asymmetry, first discussed by Kotzinian and Mulders[37] doesn't have a Sivers type contribution in the leading order, it can provide a clean measurement of the Collins fragmentation. The observed azimuthal dependence of the asymmetry for π^0 (see Fig. 2) is indeed consistent with a small $\sin 2\phi$ in agreement with observation of roughly equal favored and unfavored Collins FFs by HERMES.[14] Amplitudes of $\sin\phi$ modulations for π^+ and π^0 are comparable as in the case of beam SSA, also supporting relatively small contribution from Collins effect in general. The $\sin 2\phi$ moment of the π^+ SSA at large x_B is dominated by the u-quarks; therefore with some assumption about the ratio of unfavored to favored Collins fragmentation functions, it can provide a first glimpse of the twist-2 TMD function $h_{1L}^{\perp}$. The extraction, of the RSMT TMD from π^+ SSA alone, however, will require the knowledge of the unknown ratio of the Collins favored and unfavored fragmentation functions as well as the $h_{1L}^{\perp d}/h_{1L}^{\perp u}$ ratio.

The $\sin\phi$ moment of the cross section, measured with CLAS at 5.7 GeV, is in agreement with the HERMES measurement at 27.5 GeV for a longitudinally polarized target,[12] indicating a bigger asymmetry for the higher twist contribution compared to the leading twist $\sin 2\phi$ moment. The $P_\perp$ dependence of the $\sin\phi$ moment for π^+ is consistent with an increase with increasing $P_\perp$,[18] as expected for the TMD function.[40] A non-zero $\sin\phi$ moment for the longitudinally polarized target SSA was also observed for π^-.[18] The $\sin\phi$ and $\sin 2\phi$ moments of the SIDIS cross section for longitudinally polarized targets can be an important source of independent information on the Collins fragmentation mechanism.

4. Future measurements with CLAS at 6 GeV

New measurements of SSAs in electroproduction, approved by JLab PAC,[41–43] will significantly improve the statistics with the longitudinally polarized target and will provide important information in the valence region for the transverse target SSAs. The new data will constrain the ratio of favored to unfavored polarized fragmentation functions and also allow studies of Q^2 dependence of different spin and azimuthal asymmetries to define the leading-twist and higher-twist nature of the corresponding observables.[11,19–21,24,26,44]

The impact parameter space displacement of unpolarized quarks in the transversely polarized proton is described by the Generalized Parton Distribution E (GPD-E),[45] and has recently been calculated in lattice QCD.[46,47] The relations between nucleon Generalized Parton Distributions (GPDs) and TMDs were recently discussed in the framework of so called " Generalized TMDs", including also the possible nontrivial relations between them.[48,49] New CLAS measurements (see Fig. 3) of exclusive photon production (DVCS), giving access to the GPD E,[41] and single spin asymmetries in pion production giving access to the Sivers function,[42] will provide crucial input for testing possible relations between GPDs and TMDs.

The measurements with transversely polarized target offer a variety of observables sensitive to different combinations of GPDs. The wide phase space coverage of CLAS will allow us to bin observables in different bins for relevant kinematical variable. Shown below are some projections for representative measurements. The corresponding projections for CLAS data

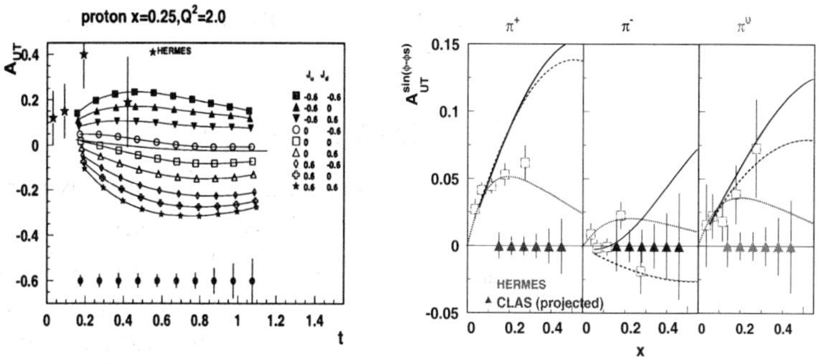

(a) (b)

Fig. 3. Expected DVCS TTSA amplitudes $A_{UT}^{\sin(\phi-\phi_S)\cos\phi}$ as a function of t (a), as predicted by the dual parameterization for GPDs H and E[50] ($\widetilde{\mathcal{H}} = \widetilde{\mathcal{E}} = 0$). The solid line (extending to t=1.2) corresponds to the case E=0. Expected Sivers TTSA $A_{UT}^{\sin(\phi-\phi_S)}$ as a function of x (b). The projected statistical error for hydrogen target. Curves were calculated using two parameterizations from Ref.[51] (solid and dashed) and Ref.[52,53] (dotted).The systematic error is expected to not exceed the statistical one in both cases.

set, based on 25 days of data taking using the HD-Ice target,[41] are shown in Fig. 3 for few different values of the total angular momentum carried by quarks (J_u, J_d).

Measured single and double spin asymmetries for all pions in a large range of kinematic variables (x_B, Q^2, z, $P_\perp$ and ϕ) combined with measurements with unpolarized target, will provide detailed information on flavor and polarization dependence of transverse momentum distributions of quarks in valence region and in particular on the x_B and k_T dependence of leading TMD parton distribution functions of u and d quarks.

The measurement of the transverse asymmetry from the Sivers effect with π^0 in valence region, in particular, will provide a model-independent extraction of the Sivers function. Furthermore, measurements with proton and neutron targets will provide model-independent information on flavor partners of the Sivers function.

The transversely polarized target measurements also provide access to the leading-twist TMD $g_{1T}^q(x)$, appearing in convolution with the unpolarized fragmentation function $D_1^q(z)$ in a $\cos\phi$ moment of the cross section. Significant asymmetries were predicted recently for CLAS[54] providing access also to $g_{1T}^q(x)$, describing longitudinally polarized quarks in the transversely polarized nucleon. New asymmetry with transversely polarized target was recently discussed, related to the leading-twist "pretzelosity" TMD,[55–58] $h_{1T}^{\perp}q(x)$ distribution, which appears in convolution with the Collins fragmentation function.[6]

4.1. *TMD Measurements with JLab at 12 GeV*

Measurements of semi-inclusive processes combined with inclusive and exclusive measurements with upgraded JLab will allow to study the quark structure of the nucleon with unprecedented detail. Significantly higher statistics from CLAS12 data, especially in the large x_B region, will also enable the extraction of the x_B and Q^2 dependencies for different azimuthal moments in a wide kinematic range allowing extraction of the underlying distribution functions.

Measurement of the P_T dependence of the Kotzinian-Mulders-asymmetry will also allow for checking of the predictions of a unified description of SSA by Ji and collaborators[10,59] and for study of the transition from a non-perturbative to a perturbative description at large P_T. The $\sin 2\phi$ asymmetry for semi-inclusive deep inelastic scattering in the kinematic regions of CLAS12 is predicted to be significant (a few percent on average) and tends to be larger in the large-x and large-z region.

The comparison of the future CLAS12 data on different azimuthal moments with the previous HERMES measurements in the high-Q^2 domain, where higher-twist effects are less significant, will clarify the role of higher twists in observed SSAs. The RSMT distribution function and transversity can be extracted from global analysis with Belle data following the procedure for extraction of the transversity from $A_{UT}^{\sin\phi}$ developed by Anselmino et al.[60]

The projections for the measurements with CLAS12 longitudinally polarized target are shown in the Fig. 4 in comparison with published HERMES[12] and preliminary CLAS data at 5.7 GeV. The projected transverse target single-spin asymmetries for CLAS12 kinematics at 11 GeV are plotted in Fig. 5 compared with theory predictions. The projected error bars have been calculated assuming a luminosity of 2×10^{34} cm^{-2}s^{-1}, with an NH_3 target polarization of 85% and a dilution factor of 0.14, with 100 days of data taking (or HD-Ice with luminosity of 5×10^{33} cm^{-2}s^{-1}) . The asymmetry is integrated over all hadron transverse momenta.

The extraction of the Sivers function from the transverse asymmetry is more straight forward and doesn't require any additional knowledge of polarized fragmentation functions. Combination of measurements of Sivers asymmetry for pions and Kaons at CLAS (see Fig. 5) will help to clarify the role of the sea in the Sivers effect in the significant difference between Sivers

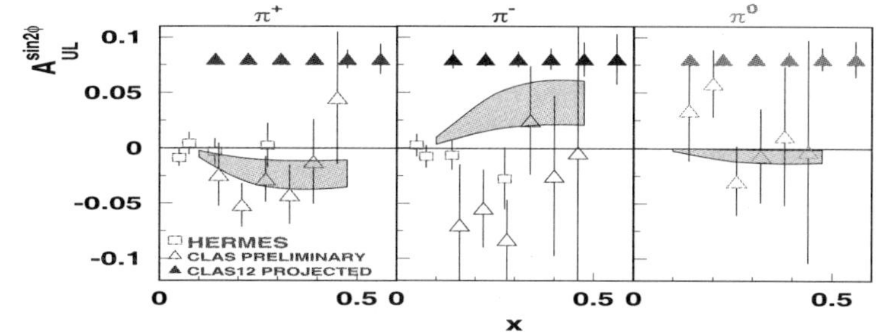

Fig. 4. The projected x-dependence of the longitudinal target SSA at 11 GeV. The triangles illustrate the expected statistical accuracy. The open squares and triangles show the existing measurement of the RSMT TMD from HERMES and the preliminary results from CLAS 5.7 GeV EG1 data sets, respectively. The curves are calculated using Ref.[61] The band shows the existing uncertainty in theory predictions due to the (1/2)-moment of the Collins function.[62]

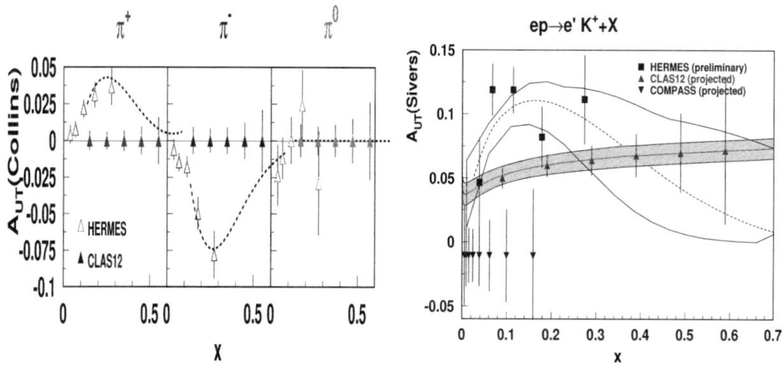

(a) (b)

Fig. 5. Measurements of the Collins asymmetry for pions (a). The curves indicate existing predictions.[63] The expected Kaon Sivers TTSA $A_{UT}^{\sin(\phi-\phi_S)}$ as a function of x for CLAS12 (b). The projected statistical error for hydrogen target. The projected statistical error for 100 days with hydrogen polarization 75% (2 bins in x, Q^2) is shown. The dashed[64] and dotted[65] lines show predictions with corresponding bands estimated from fit errors.

asymmetries of pions and Kaons first observed by the HERMES Collaboration.[66] Semi-inclusive Kaon production have a very different background from exclusive processes and its comparison with pions will help to clarify also the role of exclusive processes,which is one of the missing components in the interpretation of large SSAs observed in different processes.

References

1. D. W. Sivers, *Phys. Rev.* **D**43, 261 (1991).
2. M. Anselmino and F. Murgia, *Phys. Lett.* **B**442, 470 (1998).
3. S. J. Brodsky, D. S. Hwang and I. Schmidt, *Phys. Lett.* **B**530, 99 (2002).
4. J. C. Collins, *Phys. Lett.* **B**536, 43 (2002).
5. X. Ji and F. Yuan, *Phys. Lett.* **B**543, 66 (2002).
6. P. J. Mulders and R. D. Tangerman, *Nucl. Phys.* **B**461, 197 (1996).
7. A. V. Belitsky, X. Ji and F. Yuan, *Nucl. Phys.* **B**656, 165 (2003).
8. D. Boer, S. J. Brodsky and D. S. Hwang, *Phys. Rev.* **D**67, p. 054003 (2003).
9. D. Boer, P. J. Mulders and F. Pijlman, *Nucl. Phys.* **B**667, 201 (2003).
10. X. Ji, J. Ma and F. Yuan, *Phys. Rev.* **D**71, p. 034005 (2005).
11. J. C. Collins and A. Metz, *Phys. Rev. Lett.* **93**, p. 252001 (2004).
12. A. Airapetian *et al.*, *Phys. Rev. Lett.* **8**4, 4047 (2000).
13. A. Airapetian *et al.*, *Phys. Rev.* **D**64, p. 097101 (2001).
14. A. Airapetian *et al.*, *Phys. Rev. Lett.* **9**4, p. 012002 (2005).
15. A. Airapetian *et al.*, *Phys. Lett.* **B**648, 164 (2007).
16. V. Y. Alexakhin *et al.*, *Phys. Rev. Lett.* **9**4, p. 202002 (2005).
17. H. Avakian *et al.*, *Phys. Rev.* **D**69, p. 112004 (2004).
18. H. Avakian *et al.*, *AIP Conf. Proc.* **7**92, 945 (2005).
19. A. Afanasev and C. E. Carlson (2003).
20. F. Yuan, *Phys. Lett.* **B**589, 28 (2004).
21. R. L. Jaffe and X.-D. Ji, *Nucl. Phys.* **B**375, 527 (1992).
22. R. D. Tangerman and P. J. Mulders, *Phys. Rev.* **D**51, 3357 (1995).
23. A. Kotzinian, *Nucl. Phys.* **B**441, 234 (1995).
24. J. Levelt and P. J. Mulders, *Phys. Lett.* **B**338, 357 (1994).
25. M. Burkardt, *hep-ph* **0**807.2599 (2008).
26. A. Metz and M. Schlegel, *Eur. Phys. J.* **A**22, 489 (2004).
27. A. V. Efremov, K. Goeke and P. Schweitzer, *Nucl. Phys.* **A**711, 84 (2002).
28. A. Airapetian *et al.*, *Phys. Rev.* **D**71, p. 012003 (2005).
29. K. Abe *et al.*, *Phys. Rev. Lett.* **96**, p. 232002 (2006).
30. J. C. Collins, *Nucl. Phys.* **B**396, 161 (1993).
31. F. X. Girod *et al.*, *Phys. Rev. Lett.* **100**, p. 162002 (2008).
32. T. P. McPharlin *et al.*, *Phys. Lett.* **B**90, p. 479 (1980).
33. J. J. Aubert *et al.*, *Z. Phys.* **C**18, p. 189 (1983).
34. A. Airapetian *et al.*, *Eur. Phys. J.* **C**21, 599 (2001).
35. D. de Florian, R. Sassot and M. Stratmann, *J. Phys. Conf. Ser.* **110**, p. 022045 (2008).
36. H. Avakian and L. Elouadrhiri, *High Energy Spin Physics Proc., Editors A.V. Efremov and O.V. Teryaev* , p. 239 (2003).
37. A. M. Kotzinian and P. J. Mulders, *Phys. Rev.* **D**54, 1229 (1996).
38. J. P. Ralston and D. E. Soper, *Nucl. Phys.* **B**152, p. 109 (1979).
39. E. Di Salvo, *Int. J. Mod. Phys.* **A**22, p. 2145 (2007).
40. X.-d. Ji, J.-P. Ma and F. Yuan, *Nucl. Phys.* **B**652, 383 (2003).
41. H. Avakian *et al.*, *JLab Experiment E-08-021* (2008).
42. H. Avakian *et al.*, *JLab Experiment E-08-015* (2008).
43. H. Avakian, *et al.*, *JLab Experiment E-05-113* (2005).

44. A. M. Kotzinian, K. A. Oganesian and H. R. Avakian (1999).
45. M. Burkardt, *Phys. Rev.* **D**72, p. 094020 (2005).
46. M. Gockeler *et al.*, *Nucl. Phys. Proc. Suppl.* **1**53, 146 (2006).
47. M. Gockeler *et al.*, *Phys. Rev. Lett.* **9**8, p. 222001 (2007).
48. M. Burkardt and D. S. Hwang, *Phys. Rev.* **D**69, p. 074032 (2004).
49. S. Meissner, A. Metz and M. Schlegel (2008).
50. V. Guzey and T. Teckentrup, *Phys. Rev.* **D**74, p. 054027 (2006).
51. W. Vogelsang and F. Yuan, *Phys. Rev.* **D**72, p. 054028 (2005).
52. A. V. Efremov *et al.*, *Phys. Lett.* **B**612, 233 (2005).
53. J. C. Collins *et al.*, *Phys. Rev.* **D**73, p. 014021 (2006).
54. A. Kotzinian, B. Parsamyan and A. Prokudin, *Phys. Rev.* **D**73, p. 114017 (2006).
55. G. A. Miller, *Phys. Rev.* **C**76, p. 065209 (2007).
56. M. Burkardt, *hep-ph* **0**709.2966 (2007).
57. H. Avakian, A. V. Efremov, P. Schweitzer and F. Yuan, *hep-ph* **0**805.3355 (2008).
58. B. Pasquini, S. Cazzaniga and S. Boffi, *hep-ph* **0**806.2298 (2008).
59. X. Ji, J.-w. Qiu, W. Vogelsang and F. Yuan, *Phys. Rev.* **D**73, p. 094017 (2006).
60. M. Anselmino *et al.*, *hep-ph* **0**807.0173 (2008).
61. A. V. Efremov, K. Goeke and P. Schweitzer, *Czech. J. Phys.* **5**5, A189 (2005).
62. H. Avakian *et al.*, *Phys. Rev.* **D**77, p. 014023 (2008).
63. M. Anselmino *et al.*, *Phys. Rev.* **D**75, p. 054032 (2007).
64. S. Arnold *et al.*, *hep-ph* **0**805.2137 (2008).
65. M. Anselmino *et al.*, *hep-ph* **0**807.0166 (2008).
66. M. Diefenthaler, *Proceedings of the XV International Workshop on Deep Inelastic Scattering* **M**unich,Germany, April 16th - April 20th (2008).

NUCLEON TO PION TRANSITION DISTRIBUTION AMPLITUDES IN A LIGHT-CONE QUARK MODEL

M. Pincetti, B. Pasquini, S. Boffi

Dip. di Fisica Nucleare e Teorica, Università degli Studi di Pavia and INFN, Sezione di Pavia, Pavia, 27100, Italy

We present a general representation for the nucleon distribution amplitudes and for the nucleon to pion transition distribution amplitudes in terms of light-cone wave functions. We apply our formalism to a light-cone constituent quark model giving some numerical results for both the classes of observables.

Keywords: Distribution Amplitudes; Transition Distribution Amplitudes; Meson Cloud.

1. Introduction

Until few years ago all matrix elements taken into account in the description of exclusive reactions at high momentum transfer were those between initial and final hadron states, giving rise to the so-called Generalized Parton Distributions (GPDs),[1] or those involving an hadronic and a vacuum state, describing the well-known Distribution Amplitudes (DAs).[2] The generalization of such distributions to the case where the initial and final states correspond to different particles has recently been proposed. This new mathematical objects, initially called skewed distribution amplitudes,[3] are now established as Transition Distribution Amplitudes[4] (TDAs), since they describe a transition between two different hadronic states.

In this article, we perform a Fock-state decomposition of the nucleon state with the aim to understand the structure of the nucleon in term of its Light-Cone Wave Function (LCWF) representation. First, we derive the general LCWF representation for the nucleon DAs taking into account the lower Fock-state component made up of three valence quarks only. Then, for the first time in the literature we apply the LCWF formalism to the nucleon to pion TDAs for which we are forced to consider at least the state composed of five partons, *i.e.* three valence quarks and a quark-antiquark pair describing the mesonic cloud surrounding the nucleon. The nucleon to pion TDAs are believed to be the most direct observables to test the pion-cloud contribution to the nucleon structure. Their experimental study is ongoing (JLab) or is planned (GSI, CERN).

The outline of the paper is as follows. In section 2 we give a detailed description of the LCWF representation for the DAs in the valence sector and we give some numerical results for the DAs and their moments within a light-cone quark model, in comparison with other model calculations and lattice predictions. Then in section 3 in the framework of the convolution model of Ref. 5 we present the representation of the nucleon to pion TDAs in terms of LCWFs giving also some numerical predictions for three of them.

2. DAs in the LCWFs formalism

2.1. *Definitions and properties*

In general the notion of DAs refers to hadron to vacuum matrix elements of nonlocal operators built up of quark and gluon fields at light-like separation. In this paper we deal with the three-quark matrix element

$$\langle 0|\epsilon^{ijk} u_\alpha^{i'}(z_1 n)[z_1; z_0]_{i'i} u_\beta^{j'}(z_2 n)[z_2; z_0]_{j'j} d_\gamma^{k'}(z_3 n)[z_3; z_0]_{k'k}|P(p_1, \lambda)\rangle, \quad (1)$$

where $|P(p_1, \lambda)\rangle$ denotes the proton state with momentum p_1 ($p_1^2 = M^2$) and helicity λ. The Latin letters i, j and k refer to colour, while the Greek letters α, β and γ stand for Dirac indices. We define the light-cone vectors p and n ($p^2 = n^2 = 0$) such that $2p \cdot n = 1$. The most general decomposition of the matrix element in Eq. (1) involves 24 invariant functions,[6] but to leading twist accuracy only three of them survive, *i.e.*

$$4\mathcal{F}\Big(\langle 0|\epsilon^{ijk} u_\alpha^i(z_1 n) u_\beta^j(z_2 n) d_\gamma^k(z_3 n)|P(p_1, \lambda)\rangle\Big) = f_N\Big[V^p(\not{p}_1 C)_{\alpha\beta}(\gamma^5 N^+)_\gamma$$
$$+A^p(\not{p}_1\gamma^5 C)_{\alpha\beta}(N^+)_\gamma + T^p(\sigma_{p_1\mu}C)_{\alpha\beta}(\gamma^\mu\gamma^5 N^+)_\gamma\Big], \quad (2)$$

where $\sigma^{\mu\nu} = 1/2[\gamma^\mu, \gamma^\nu]$, $\sigma^{p_1\mu}$ is a shorthand notation for $p_{1\nu}\sigma^{\nu\mu}$, C is the charge conjugation matrix, N^+ is the "good" or "large" component of the nucleon spinor N, $\mathcal{F}$ represents a Fourier transform[a], and f_N is the value

[a]The symbol $\mathcal{F}$ is a shorthand notation for the Fourier transform

$$(p \cdot n)^3 \int \frac{\prod_j \mathrm{d}z_j}{(2\pi)^3} \exp[i x_k z_k p \cdot n].$$

of the nucleon wave function at the origin. The three functions V^p, A^p and T^p depend on the variables x_i ($0 < x_i < 1$, $\sum_i x_i = 1$) which correspond to the longitudinal momentum fractions carried by the quarks inside the nucleon. Because of the symmetry properties of the operator in Eq. (2) it is easy to see that the three invariant functions can be expressed in term of a single function Φ. Indeed, labelling with 1, 2 or 3 the arguments of the DAs the following relations hold:

$$V(1,2,3) = V(2,1,3), \;\; A(1,2,3) = -A(2,1,3), \;\; T(1,2,3) = T(2,1,3), \quad (3)$$

and

$$2T(1,2,3) = \Phi(1,3,2) + \Phi(2,3,1), \;\; \Phi(1,2,3) = V(1,2,3) - A(1,2,3). \quad (4)$$

Defining with the symbol $M^{\uparrow/\downarrow}_{\alpha\beta,\gamma}$ the matrix element in the left-hand side of Eq. (2), we can express the three DAs as[b]

$$V^p = \frac{1}{f_N}\frac{1}{\sqrt[4]{2}}(p_1^+)^{-\frac{3}{2}}\left(M^{\uparrow}_{12,1} + M^{\uparrow}_{21,1}\right), \quad (5)$$

$$A^p = \frac{1}{f_N}\frac{1}{\sqrt[4]{2}}(p_1^+)^{-\frac{3}{2}}\left(M^{\uparrow}_{21,1} - M^{\uparrow}_{12,1}\right), \quad (6)$$

$$T^p = -\frac{1}{f_N}\frac{1}{\sqrt[4]{2}}(p_1^+)^{-\frac{3}{2}} M^{\uparrow}_{11,2}. \quad (7)$$

Thus to calculate the leading twist DAs we must give an expression for the $M^{\uparrow/\downarrow}_{\alpha\beta,\gamma}$ matrix elements.

2.2. *LCWFs representation*

The representation in terms of LCWFs has been applyed to a large range of observables like form factors,[7] the transversity distribution,[8] GPDs[9] and transverse momentum dependent distributions[10] as a uselful formalism to disentangle the contribution of the different partonic configurations in the nucleon. Here we sketch the derivation of the LCWF representation for the DAs. Starting from the left-hand side of Eq. (2) and substituting the

---

[b]In principle one has to face with 128 matrix elements ($4 \times 4 \times 4 \times 2$) because there are three Dirac indices and two helicity states, but for symmetry reasons only three of them are independent.

general Fourier expansion in momentum space of the free quark field one finds,

$$
\langle 0|4(p_1^+)^3\epsilon^{ijk}\int\frac{\mathrm{d}k_1^+\mathrm{d}^2\mathbf{k}_{1\perp}}{16\pi^3k_1^+}\int\frac{\mathrm{d}k_2^+\mathrm{d}^2\mathbf{k}_{2\perp}}{16\pi^3k_2^+}\int\frac{\mathrm{d}k_3^+\mathrm{d}^2\mathbf{k}_{3\perp}}{16\pi^3k_3^+}\Theta(k_1^+)\Theta(k_2^+)\Theta(k_3^+)
$$

$$
\times\sum_{\lambda_1,\lambda_2,\lambda_3} b_1^i(\tilde{k}_1,\lambda_1)b_2^j(\tilde{k}_2,\lambda_2)b_3^k(\tilde{k}_3,\lambda_3)u_{+\alpha}(k_1^+,\lambda_1)u_{+\beta}(k_2^+,\lambda_2)u_{+\gamma}(k_3^+,\lambda_3)
$$

$$
\times\delta\Big(x_1p_1^+-k_1^+\Big)\delta\Big(x_2p_1^+-k_2^+\Big)\delta\Big(x_3p_1^+-k_3^+\Big)|P(p_1,s_1)\rangle, \tag{8}
$$

where $\tilde{k}_i\equiv(k_i^+,\mathbf{k}_{i\perp})$ is a shorthand notation for the plus and transverse parton momentum components, λ_i represents the parton helicity and b is the annihilator of the "good" component of the quark fields. Truncating the Fock expansion to the minimal configuration corresponding to three valence quarks, the proton state in Eq. (8) can be written as

$$
|P(p_1,s_1)\rangle=\sum_{\lambda_i,\tau_i,c_i}\int\prod_{i=1}^{3}\frac{\mathrm{d}y_i}{\sqrt{y_i}}\int\frac{\prod_{i=1}^3\mathrm{d}^2\boldsymbol{k}_{i\perp}}{[2(2\pi)^3]^2}\delta\Big(1-\sum_{i=1}^3y_i\Big)\delta^{(2)}\Big(\sum_{i=1}^3\boldsymbol{k}_{i\perp}\Big)
$$

$$
\times\tilde{\Psi}_\lambda^{[f]}(\{y_i,\boldsymbol{k}_{i\perp};\lambda_i,\tau_i,c_i\})\prod_{i=1}^{3}|y_ip_1^+,\boldsymbol{k}_{i\perp}+y_i\mathbf{p}_{1\perp},\lambda_i,\tau_i,c_i;q\rangle, \tag{9}
$$

where λ_i,τ_i and c_i are spin, isospin and colour variables of the quarks, respectively. Then, after some algebra one obtains[c]

$$
M_{\alpha\beta,\gamma}^{\uparrow/\downarrow}=-\frac{24}{\sqrt{x_1x_2x_3}}\sum_{\lambda_1,\lambda_2,\lambda_3}u_{+\alpha}(x_1p_1^+,\lambda_1)u_{+\beta}(x_2p_1^+,\lambda_2)u_{+\gamma}(x_3p_1^+,\lambda_3)
$$

$$
\times\int\frac{\prod_{i=1}^3\mathrm{d}^2\boldsymbol{k}_{i\perp}}{[2(2\pi)^3]^2}\delta^{(2)}\Big(\sum_{i=1}^3\boldsymbol{k}_{i\perp}\Big)
$$

$$
\times\tilde{\Psi}_\lambda^{[f]}(\{x_1,\boldsymbol{k}_{1\perp};\lambda_1,1/2\}\{x_2,\boldsymbol{k}_{2\perp};\lambda_2,1/2\}\{x_3,\boldsymbol{k}_{3\perp};\lambda_3,-1/2\}). \tag{10}
$$

The previous equation is a formal general representation for the matrix element involved in the DAs definition. To get the final representation for the matrix element what is missing is an expression for the LCWFs. In this work we follow the same procedure adopted in Refs. 11, 12. We refer to

[c]The colour dependence in the LCWF $\tilde{\Psi}_\lambda^{[f]}$ has already been taken into account in the numerical prefactor.

these works for the details and here we give the final results for the DA Φ

$$\Phi = \frac{8\sqrt{2}}{f_N}\left[\frac{1}{M_0}\frac{\omega_1\omega_2\omega_3}{x_1x_2x_3}\right]^{\frac{1}{2}} \int \frac{\prod_{i=1}^{3} \mathrm{d}^2\boldsymbol{k}_{i\perp}}{16\pi^3}\delta^{(2)}\left(\sum_{i=1}^{3}\boldsymbol{k}_{i\perp}\right)\psi(\boldsymbol{k}_1,\boldsymbol{k}_2,\boldsymbol{k}_3)$$

$$\times \frac{\left\{\left[a_1\kappa_2^R\kappa_3^L\right] + \frac{1}{2}\left[a_1a_2a_3\right] - \frac{1}{2}\left[\kappa_1^L\kappa_2^R a_3\right]\right\}}{\sqrt{\left[a_1^2+\boldsymbol{k}_{1\perp}^2\right]\left[a_2^2+\boldsymbol{k}_{2\perp}^2\right]\left[a_3^2+\boldsymbol{k}_{3\perp}^2\right]}}, \tag{11}$$

where $a_i = m + x_iM_0$ and $\kappa_i^{L,R} = \kappa_{ix} \mp \kappa_{iy}$, with m being the quark mass and M_0 the eigenvalue solution of the free Hamiltonian. For the description of the momentum dependence part $\psi(\boldsymbol{k}_1,\boldsymbol{k}_2,\boldsymbol{k}_3)$ of the LCWF we adopt the parameterization of Ref. 21.

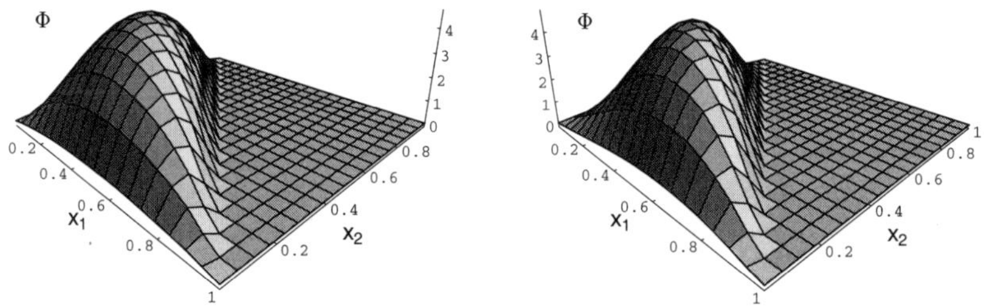

Fig. 1. Comparison of the DA Φ between our model prediction (left panel) and the data fit performed in Ref.[16] (right panel).

2.3. *Numerical results*

In the literature there are many different papers trying to give a good description of the nucleon DAs.[13–20] In particular, in Fig. 1 the DA Φ from the data fit of Ref. 20 is compared with our model calculation. Our result is in good agreement with the data fit of Bolz and Kroll. In Table 1 we report our predictions for some moments of the DA Φ in comparison with the results from Refs. 13, 14, 18, 19, 20, 22. The moments are defined as

$$\Phi^{(l,m,n)} = \int_0^1 [dx]x_1^l x_2^m x_3^n \Phi(x_1,x_2,x_3) \Big/ \int_0^1 [dx]\Phi(x_1,x_2,x_3), \tag{12}$$

where $[dx] = \delta(1-\sum_{i=1}^{3} x_i)\prod_{i=1}^{3} dx_i$. Our results are pretty close to the BK data parameterization and to the very recent lattice predictions.

Table 1. Results for the moments of the DA Φ in our light-cone quark model in comparison with other model calculations: COZ[13], KS[14], DF[18], SB[19], data parameterization BK[20] and lattice predictions LAT[22].

(l, m, n)	COZ	KS	DF	SB	BK	Our	LAT
0 0 0	1	1	1	1	1	1	1
1 0 0	0.54	0.46	0.582	0.572	0.381	0.346	0.394
0 1 0	0.18	0.18	0.213	0.184	0.309	0.331	0.302
0 0 1	0.20	0.22	0.207	0.244	0.309	0.323	0.304
2 0 0	0.32	0.27	0.367	0.338	0.179	0.152	0.18
0 2 0	0.065	0.08	0.085	0.066	0.125	0.142	0.132
0 0 2	0.09	0.10	0.83	0.170	0.125	0.137	0.138

3. Nucleon to pion TDAs

In this section we apply the same formalism used for the DAs to the study of the TDAs in the baryonic sector and in particular for the description of the nucleon to pion transition. Such quantities can be analyzed in reactions such as backward pion electroproduction, e.g., $ep \to e'p'\pi_0$, and nucleon-antinucleon annihilation, e.g., $N\bar{N} \to \gamma^*\pi$. As in Ref. 4 we rewrite the general matrix element describing a transition from a nucleon to a pion in terms of eight TDAs: two vectorial $V_{1,2}^{p\pi^0}(x_i, \xi, \Delta^2)$, two axial $A_{1,2}^{p\pi^0}(x_i, \xi, \Delta^2)$ and four tensorial $T_{1,\ldots,4}^{p\pi^0}(x_i, \xi, \Delta^2)$, *i.e.*

$$
\begin{aligned}
&4\mathcal{F}\Big(\langle\pi^0(p_\pi)|\epsilon^{ijk}u_\alpha^i(z_1n)u_\beta^j(z_2n)d_\gamma^k(z_3n)|P(p_1,s_1)\rangle\Big) = \\
&i\frac{f_N}{f_\pi}\Big[V_1^{p\pi^0}(\not{p}C)_{\alpha\beta}(N^+)_\gamma + A_1^{p\pi^0}(\not{p}\gamma^5C)_{\alpha\beta}(\gamma^5N^+)_\gamma \\
&+T_1^{p\pi^0}(\sigma_{p\mu}C)_{\alpha\beta}(\gamma^\mu N^+)_\gamma + M^{-1}V_2^{p\pi^0}(\not{p}C)_{\alpha\beta}(\not{\Delta}_T N^+)_\gamma \\
&+M^{-1}A_2^{p\pi^0}(\not{p}\gamma^5C)_{\alpha\beta}(\gamma^5\not{\Delta}_T N^+)_\gamma + M^{-1}T_2^{p\pi^0}(\sigma_{p\Delta_T}C)_{\alpha\beta}(N^+)_\gamma \\
&+M^{-1}T_3^{p\pi^0}(\sigma_{p\mu}C)_{\alpha\beta}(\sigma^{\mu\Delta_T}N^+)_\gamma + M^{-2}T_4^{p\pi^0}(\sigma_{p\Delta_T}C)_{\alpha\beta}(\not{\Delta}_T N^+)_\gamma\Big]. \quad (13)
\end{aligned}
$$

Here, $p_1 = (1+\xi)p + \frac{M^2}{1+\xi}n$, $p_\pi = (1-\xi)p + \frac{m_\pi^2+\Delta_T^2}{1-\xi}n + \Delta_T$, and $\Delta = p_\pi - p_1$, where we have defined $\xi = -\Delta^+/2P^+$ with $P = \frac{1}{2}(p_1 + p_\pi)$. The TDAs depend on the light-cone momentum fractions x_i ($\sum_i x_i = 2\xi$), on the skewedness parameter ξ that describes the change of longitudinal momentum from the initial and the final state, and on Δ^2. As for the DAs, defining the left-hand side of Eq. (13) with the symbol $D_{\alpha\beta,\gamma}^{\uparrow/\downarrow}$ it is possible to express the TDAs in terms of eight of these matrix elements. The calculations of the LCWF representation follows exactly the same line we used for the DAs. However to match the final one-pion state the nucleon state must now be composed at least by five partons with their dynamics encoded into two LCWFs, one for the $3q$ structure and one for $q\bar{q}$ pair, convoluted with a splitting function, $\phi_{\lambda' 0}^{\lambda(N,N\pi)}$, that weighs the different possible configurations. For the details about the description of the nucleon as a bound state of a baryon surronded by its meson cloud we refer to Ref. 5, while we postpone all the details regarding the calculation to a forthcoming paper.[23] We report here the general LCWF representation for the matrix element $D^{\lambda}_{\alpha\beta,\gamma}$,

$$-\frac{24}{\sqrt{x_1 x_2 x_3}}\left(\frac{1}{2\xi}\right)^{\frac{3}{2}} \sum_{\lambda_1,\lambda_2,\lambda_3} u_{+\alpha}(x_1 P^+,\lambda_1) u_{+\beta}(x_2 P^+,\lambda_2) u_{+\gamma}(x_3 P^+,\lambda_3)$$

$$\times \sum_{\lambda'} \int \mathrm{d}y \mathrm{d}^2\mathbf{k}_\perp \phi_{\lambda' 0}^{\lambda(N,N\pi)}(y,\mathbf{k}_\perp)\sqrt{y(1-y)}\,\delta\left(1-y-\frac{p_\pi^+}{p_1^+}\right)$$

$$\times \int \frac{\prod_i \mathrm{d}^2\boldsymbol{k}_{i\perp}}{[2(2\pi)^3]^2}\delta^{(2)}(\mathbf{p}_{\pi\perp}+\mathbf{k}_\perp)\delta^{(2)}\left(\sum_{i=1}^{3}\boldsymbol{k}_{i\perp}\right)$$

$$\times \tilde{\Psi}^{[f]}_{\lambda'}\left(\left\{\frac{x_1}{2\xi},\boldsymbol{k}_{1\perp};\lambda_1,1/2\right\}\left\{\frac{x_2}{2\xi},\boldsymbol{k}_{2\perp};\lambda_2,1/2\right\}\left\{\frac{x_3}{2\xi},\boldsymbol{k}_{3\perp};\lambda_3,-1/2\right\}\right), \quad (14)$$

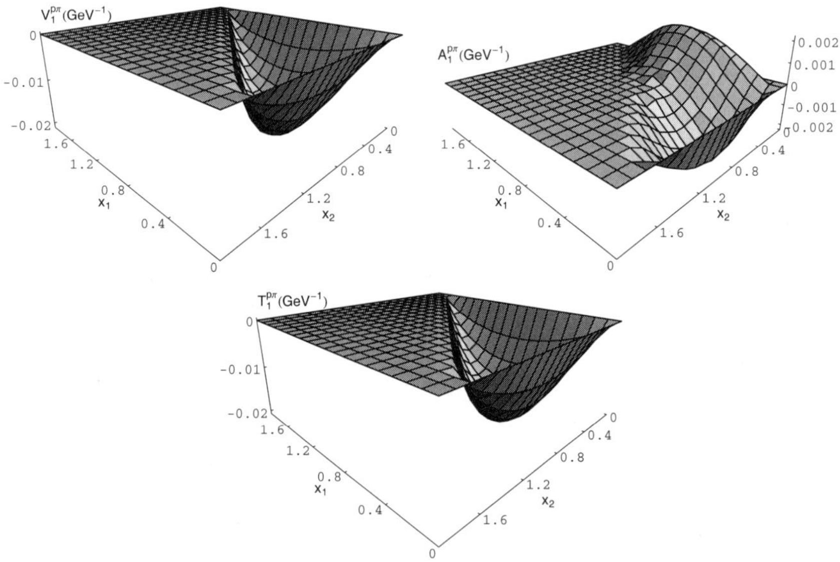

Fig. 2. Results for the TDAs $V_1^{p\pi^0}$, $A_1^{p\pi^0}$ and $T_1^{p\pi^0}$ at Δ^2=-1GeV^2 and ξ=0.9.

where, apart from some kinematical factors, it is easy to recognize the same structure found for the DAs. In Eq. (14) we note that the dependence from the light-cone momentum fraction in the LCWF is rescaled by a factor 2ξ with respect to the DAs case, as found in Ref. 4 in the soft pion limit[24,25] ($\Delta_T = 0, \xi \to 1$) and for the pion-nucleon GDAs in Ref. 26. In Fig. 2 the results for $V_1^{p\pi^0}$, $A_1^{p\pi^0}$ and $T_1^{p\pi^0}$ are shown.

For a detailed numerical analysis of all the TDAs we refer to Ref. 23.

Acknowledgments

M.P. acknowledges useful discussions with F. Conti, A. Courtoy and O. Teryaev.

References

1. For recent reviews on the subject see: M. Diehl, Phys. Rep. **388**, 41 (2003); Xiangdong Ji, Ann. Rev. Nucl. Part. Sci. **54**, 413 (2004); A.V. Belitsky and A.V. Radyushkin, Phys. Rept. **418**, 1 (2005); S. Boffi and B. Pasquini, Riv. Nuovo Cim. **30**, 387 (2007), arXiv:0711.2625 [hep-ph].
2. V.L. Chernyak and A.R. Zhitnitsky, Phys. Rep. **112**, 173 (1984); G.P. Lepage and S.J. Brodsky, Phys. Rev. D **22**, 2157 (1980).
3. L.L. Frankfurt *et al.*, Phys. Rev. D **60**, 014010 (1999); hep-ph/9808449.
4. B. Pire and L. Szymanowski, Phys. Rev. D **71**, 111501 (2005); Phys. Lett B **622**, 83 (2005); J.P. Lansberg, B. Pire and L. Szymanowski, Phys. Rev. D **75**, 074004 (2007); arXiv:0709.2567 [hep-ph]; Phys. Rev. D **76**, 111502 (2007).
5. B. Pasquini and S. Boffi, Phys. Rev. D **73**, 094001 (2006).
6. V. Braun *et al.*, Nucl. Phys. B **589**, 381 (2000); *ibidem* **607**, 433 (2001).
7. S.D. Drell and T. Yan, Phys. Rev. Lett. **24**, 181 (1970); G.B. West, Phys. Rev. Lett. **24**, 1206 (1970).
8. S. Boffi, B. Pasquini and M. Pincetti, Phys. Rev. D **76**, 034020 (2007).
9. M. Diehl *et al.*, Nucl. Phys. B **596**, 33 (2001); *ibidem*, 99 (2001).
10. X. Ji, J.-P. Ma, F. Yuan, Nucl. Phys. B **652**, 383 (2003); B. Pasquini, S. Cazzaniga and S. Boffi, arXiv:0806.2298 [hep-ph].
11. S. Boffi, B. Pasquini and M. Traini, Nucl. Phys. B **649**, 243 (2003).
12. S. Boffi, B. Pasquini and M. Traini, Nucl. Phys. B **680**, 147 (2004); *ibidem* A **755**, 545 (2005); Phys. Rev. D **71**, 034022 (2005); S. Boffi, B. Pasquini and M. Pincetti, Phys. Rev. D **72**, 094029 (2005).
13. V.L. Chernyak, A.A. Ogloblin, and I.R. Zhitnitsky, Yad. Fiz. **48**, 841 (1988).
14. I.D. King and C.T. Sachrajda, Nucl. Phys B **279**, 785 (1987).
15. M.F. Gari and N.G. Stefanis, Phys. Rev. D **35**, 1074 (1987).
16. Z. Dziembowski, Phys. Rev. D **37**, 2030 (1988).
17. A. Schäfer, Phys. Lett B **217**, 545 (1989).
18. Z. Dziembowski and J. Franklin, Phys. Rev. D **42**, 905 (1990).
19. N.G. Stefanis and M. Bergmann, Phys. Rev. D **47**, R3685 (1993).
20. J. Bolz and P. Kroll, Z. Phys A **356**, 327 (1996).

21. F. Schlumpf, J. Phys. G **20**, 237 (1994).
22. QCDSF/UKQCD collaborations, Göckeler *et al.*, arXiv:0804.1877 [hep-lat].
23. B. Pasquini, M. Pincetti and S. Boffi, *forthcoming.*
24. S. Adler and R. Dashen, *Currents Algebra* (W.A. Benjamin, New York, 1968).
25. P.V. Pobylitsa, M.V. Polyakov and M. Strikman, Phys. Rev. Lett. **87**, 022001 (2001); M.V. Polyakov and S. Stratman, hep-ph/0609045.
26. V.M. Braun *et al.*, Phys. Rev. D **75**, 014021 (2007).

ANTIPROTON POLARIZATION STUDIES FOR FAIR (HOW TO POLARIZE ANTIPROTONS AND WHAT TO USE THEM FOR?)

H. Ströher and F. Rathmann

Institut für Kernphysik, Forschungszentrum Jülich,
Jülich, 52425, Germany

P. Lenisa

Università di Ferrara and INFN,
44100 Ferrara, Italy

A brief report of the ongoing experimental efforts to produce a beam of polarized antiprotons, which may be used in a second stage of the FAIR/HESR-facility currently under construction in Darmstadt (Germany), is given.

Keywords: antiproton polarisation, spin filtering.

1. Introduction

The physics case for antiproton beams – if polarized – is enormous, including a first time measurement of double-spin observables in proton-antiproton Drell-Yan reactions.[1] Since these are dominated by annihilations of valence quark – valence antiquark in the proton and in the antiproton, respectively, the double spin asymmetry A_{TT} will give direct access to transversity, the important but largely unknown piece of the nucleon spin structure,[2] which is the subject of this meeting. Unfortunately, the preparation of polarized antiprotons is very demanding, and in fact has not been achieved with sufficient intensity and/or degree of polarization up to now. Some time ago, it has proposed to use *spin filtering* for polarization build-up in a stored beam of antiprotons to achieve the required parameters.[3] Recently, a different method has been suggested, *viz.* exploiting the interaction of antiprotons with polarized positrons at small relative velocity, where the expected cross sections are gigantic.[4] The PAX-collaboration has taken on these issues, and has started a systematic experimental search for the best method and operation parameters for producing a polarized beam of antiprotons, which could be used at FAIR ("Facility for Antiproton and Ion Research", to be build at GSI, Darmstadt, Germany)[5] in an upgrade of the "High Energy Storage Ring" (HESR) into an asymmetric collider.

2. Strategies Towards Polarized Antiprotons

For more than two decades, physicists have tried to produce beams of polarized antiprotons. Recent workshops at Daresbury (UK)[6] and Bad Honnef (Germany)[7] have reviewed the different possibilities. In principle the following two approaches can be applied: (i) flipping of the spins in an ensemble of particles to one preferred state, and (ii) selective removal of "wrong" particles from the beam. Conventional methods that work very reliably for protons, like atomic beam sources (ABS), cannot be used for antiprotons due to their annihilation with matter. Polarized antiprotons are being produced in the in-flight decay of antihyperons. At Fermilab, polarizations of ~35% and intensities up to $1.5{\cdot}10^5\,\mathrm{s}^{-1}$ have been achieved.[8] Scattering of energetic antiprotons off liquid hydrogen yielded $P \sim 0.2$ and $I \sim 2 \cdot 10^3\,\mathrm{s}^{-1}$.[9] It must be stated, however, that both approaches cannot be used for an efficient accumulation in a storage ring, which would greatly increase the luminosity. The spacial separation by the Stern-Gerlach effect ("Spin splitting") in an inhomogeneous magnetic field could also be used, as proposed in 1985,[10] but due to the enormous technological expense this method has never been tested experimentally. Two more methods have been proposed: (i) "spin filtering" and (ii) "spin transfer" – in the following, both are described in more detail.

(1) Spin-Filtering Method:
The spin-filtering technique is based on a repeated traversal of a beam of particles through a polarized hydrogen gas target: since the scattering cross sections for "spin-up" and "spin-down" are somewhat different, the beam will slowly acquire an increasing degree of polarization, however, at the expense of a reduction in beam intensity – this immediately implies that it makes no sense to filter for too long time, because no beam would be left. The optimum filtering time in a storage ring is about twice the beam lifetime. Therefore it is of utmost importance to design or use a polarizer ring such that it keeps scattered particles in the ring acceptance. The proof-of-principle of this method has been conducted with protons in the "FILTEX" (filter experiment) at the TSR (Test Storage Ring) of the Max-Planck Institute for Nuclear Physics in

Heidelberg (Germany). Since TSR was not optimized for acceptance, the achieved polarization build-up was about 1% per hour, resulting in a polarization of the proton beam of a few percent.[11]

(2) Spin-Transfer Method:
A different way of polarizing an (anti-)/proton beam was recently put forward by a group from Mainz (Germany):[4] they propose to use the spin-transfer from a polarized (positron)/electron beam to the hadrons, since, according to their calculations, the corresponding cross sections become very large for small relative velocities between leptons and hadrons. It should be mentioned that a scientific dispute is ongoing about the size of this cross section – a different group claims that the cross section is much smaller,[12] which would make this method impracticable. The excitement about this potential technique comes because there would be no dramatic beam loss in a storage ring. However, up to now, no experimental verification of whether this method actually works, has been achieved.

The PAX-collaboration has taken on this proposal and performed an experiment, which hopefully should provide such a test – in the following section, this experiment will be described.

3. Spin-Transfer Measurement by Depolarization

The polarization build-up experiment of a hadron beam in a storage ring by interaction with a polarized electron beam is not possible for the time being, since no intense polarized electron beam is available at a hadron storage ring. One might want to use a polarized hydrogen target, where the electron makes the polarization, but due to the much larger relative velocities, a signal would be very hard to find. Instead, we have looked for the time-reversed process, i.e. the depolarization of an originally polarized proton beam in COSY-Jülich (Germany),[13] which is interacting with the unpolarized electron beam of our electron cooler. The corresponding set-up is schematically given in Fig. 1.

The experiment consists of two parts:

(1) Interaction of the polarized proton beam with the electrons of the cooler; this should depolarize the protons, i.e. $\bar{p} + e \rightarrow p + e$, and
(2) Analysis of the remaining proton beam polarization; to do this, the ANKE deuteron cluster jet target, located immediately behind the interaction region, has been used. The reaction exploited was pd elastic

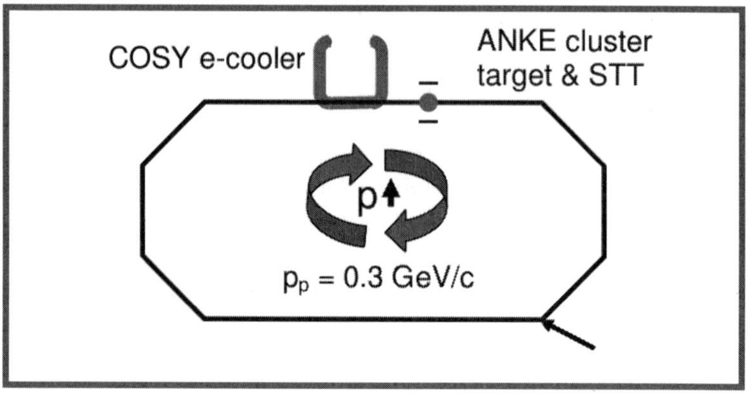

Fig. 1. Schematic layout of the experiment at COSY-Jülich: The circulating polarized proton beam of COSY interacts with electrons of the electron cooler at small relative velocity. Subsequently the polarization is determined with the help of the ANKE cluster jet target and silicon tracking telescopes (STT) after some effect "build-up" time.

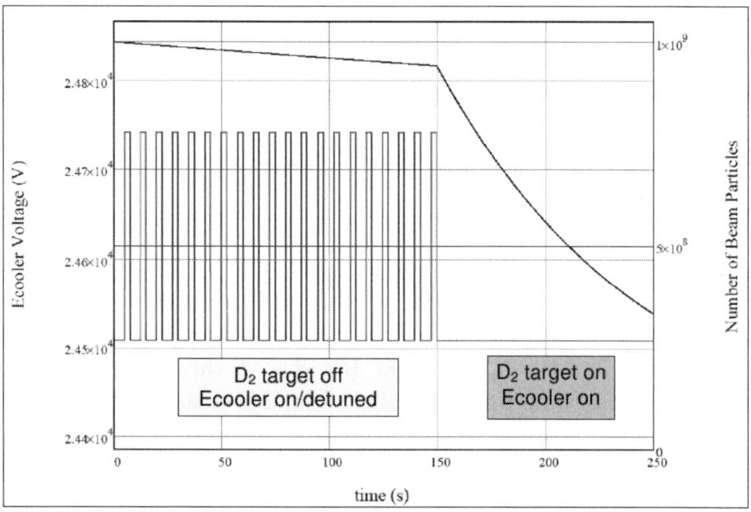

Fig. 2. Scheme of the depolarization measurement conducted at COSY. The electron cooler was repeatedly (in cycles of 5 s) switched from the nominal energy, for which electron and proton velocity match, to the detuned voltage, corresponding to the wanted velocity mismatch, for about 150 seconds (see red histogram). During this time, the D_2 cluster jet target was switched off. After that – with the electron voltage now at its nominal value – the cluster target was switched on to determine the remaining degree of polarization of the proton beam. While doing this, the beam intensity (blue curve) continuously drops, during the depolarization part slowly, and faster during the polarization analysis.

scattering, using two silicon tracking telescopes left and right of the beam.

These measurements were conducted for different relative velocities between electrons and protons. Actually, the experiment is more complicated, because electron cooling tends to remove the necessary velocity mismatch. Therefore, we used a measurement scheme as sketched in Fig. 2.

The experiment was carried out in February 2008, and the detailed analysis is still going on. Preliminary results of the measurement have been shown in conference talks, which indicate that no proton beam depolarization has been observed. Such a result would indicate that the spin-transfer cross section must be much smaller than expected by the Mainz-group, and thus make this method inapplicable for achieving our goal, the effective polarization of a beam of antiprotons.

Still, we have the spin-filtering method, and corresponding investigations are foreseen for the near future at COSY (with proton beams) and at CERN/AD (with antiprotons).

References

1. See e.g.: Yuji Koike *et al.*, arXiv: 0805.2289v2 (August 2008).
2. See e.g.: Daniel Boer, arXiv: 0808.2886v1 (August 2008).
3. F. Rathmann *et al.*, Phys. Rev. Lett. **94**, 014801 (2005).
4. Th. Walcher *et al.*, Eur. Phys. J. **A34**, 447 (2007).
5. H.H. Gutbrod (Ed. in Chief), FAIR Baseline Technical Report, ISBN: 3-9811298-0-6, 2006, see also: `http://www.gsi.de/fair/reports/btr.html`.
6. Polarized Antiproton Beams – How? An International Workshop Cockcroft Institute, Daresbury, U.K., 29-31 August 2007, AIP Conf. Proc., Vol. 1008, Chattopadhyay, S.; Barber, D.P.; Buttimore, N.; Court, G.; Steffens, E. (Eds.) 2008, ISBN: 978-0-7354-0527-1.
7. 409.WE-Heraeus Seminar, June 23-25, 2008, Bad Honnef (Germany), see: `http://www.fe.infn.it/heraeus`.
8. see eg.: A. Bravar *et al.*, Phys. Rev. Lett. **77**, 2626 (1996).
9. E.W. Vaandering *et al.*, Nucl. Instr. Meth. in Phys. Res. **A351**, 266 (1994).
10. H. Keiser *at al.*, Polarized Antiprotons with the Spin Splitter, EPAC1988, see: `http://accelconf.web.cern.ch/accelconf/e88/PDF/EPAC1988_0848.PDF`.
11. F. Rathmann *et al.*, Phys. Rev. Lett. **71**, 1379 (1993).
12. A.I. Milstein *et al.*, Nucl. Instr. Meth. in Phys. Res. **B266**, 3453 (2008).
13. R. Maier, Nucl. Instr. Meth. In Phys. Res. **A390**, 1 (1997).

Closing Lecture

NO ONE CAN ENCOMPASS THE UNENCOMPASSABLE: THE HIGHLIGHTS OF TRANSVERSITY-2008

N.N. Nikolaev

Institut f. Kernphysik, Forschungszentrum Jülich, D-52425 Jülich, Germany
and
L.D.Landau Institute for Theoretical Physics, 142432 Chernogolovka, Russia

Keywords: Transverse spin physics.

1. Introduction

Transversity is the last missing leading twist parton distribution in the proton. Without transversity the spin tomography of the proton will be ever incomplete. This Workshop held in Ferrara, Transversity-2008, is a sequel of the previous one, Transversity-2005 held in Como. They were a part of the sequential Progetti di Ricerca Interesse Nazionale sponsored by Italian Ministry of Education, University and Research.[1] Many a participants of the present Workshop took part in Transversity-2005 and would readily testify to a tremendous progress in this frontier physics, reported by 44 speakers in Ferrara. This is a humble effort to review the highlights of Transversity-2008, the title, a citation from the Russian philosopher and poet Kozma Prutkov, serves as an warning to futility of such an attempt which is doomed to be incomplete in many aspects. With due apologies, I will focus on the transversity aspect of the meeting.

2. The first glance at the transversity in the proton

The past 40 years witnessed a fantastic progress with the unpolarized and helicity structure functions of the proton. Elliot Leader explained at this meeting that the transversity is as fundamental, and complementary to, the helicity distribution in the proton, and vindicated the transverse spin sum rule.[2] The early development in the field is well documented in our Bible - the review by Barone, Drago and Ratcliff ([3] and references therein). As a prelude to the Workshop, Daniel Boer ([4] and references therein) reminded us of the seminal 1978 paper by Ralston and Soper[5] where the double transverse spin asymmetry in the Drell-Yan production was shown to persist in the scaling limit. It took however quite a time to classify the full menagerie of possible transverse spin distributions and observables, reviewed to great detail by Boer.

Large single spin asymmetries observed in hadronic interactions were the first to sparkle an interest in our subject and were reviewed in detail by C.Aidala, who presented an wealth of the new data from the PHENIX and RHIC Collaborations at RHIC.[6]

Double-polarized Drell-Yan in the proton-proton collisions has not yet been studied experimentally. At RHIC at BNL the observed signal would couple the quarks in the proton with antiquarks in the second proton and is expected to be very small and difficult to interpret, as was discussed by Les Bland, see also.[4] We all look forward to a golden-plated proton-antiproton Drell-Yan experiment PAX at FAIR at GSI, which would provide a direct measure of the transversity in the most interesting valence region.[7] As of today we are bound to indirect extractions of the transversity: SIDIS (HERMES & COMPASS) provides us with the convolution of the transversity and the Collins function, while at BELLE one measures a convolution of the two Collins functions. With an educated guess for the transverse and longitudinal momentum dependence of all that, and firm belief in the factorization approximation, one can extract the transversity. Such an extraction by the Torino group reported by Alexei Prokudin is the single most important development of the past year[8] and is shown in Fig. 1. The corresponding results for the Collins function are shown in Fig. 2.

Recall the term Gamow-Teller transitions in the theory of nuclear β-decays. This term from the pre-$(V-A)$ era is still with us in the application to the axial-vector matrix elements. And for a good reason: nonrelativistically the tensor, i.e., the transversity, and the axial-vector, i.e., the helicity distributions of nucleons in a nucleus are identical. This feature is shared by nonrelativistic valence quark models: taking the known helicity distributions for the starting point of the QCD evolutions of the transversity sounds as not a bad idea. It would also give the results not far from the Soffer bound. Anselmino et al. parameterizations amount to tensor chagres $\Delta_T u = 0.59^{+0.14}_{-0.13}$, $\Delta_T d = -0.20^{+0.05}_{-0.07}$, in the ball-park of the above expectations. There was quite a discussion on the error bars assigned to the

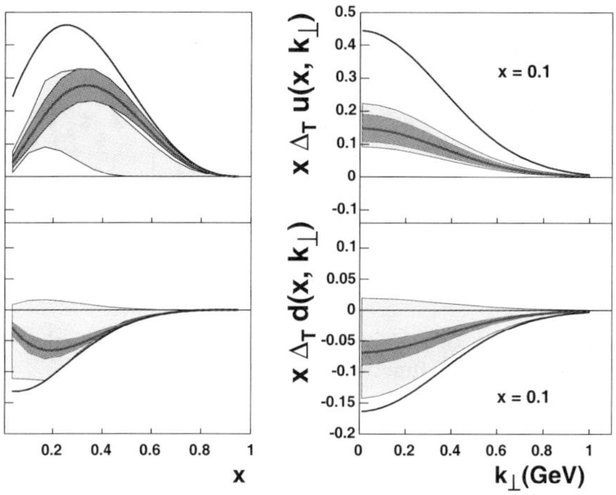

Fig. 1. Left: the 2008 transversity distribution functions for u and d flavors from the global fit by the Torino group. The Soffer bounds are shown by the highest and lowest lines, the wider bands are from the 2007 extraction by the same group. Right:the k_T-dependence using a Gaussian Ansatz.

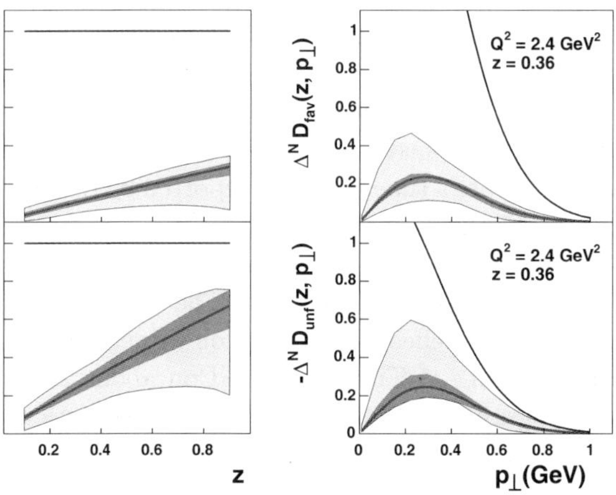

Fig. 2. Favored and unfavored Collins fragmentation functions from the 2008 global fit by the Torino group, showing also the positivity bounds.

so-deduced transversity, to my view the crucial point is that the transversity is large and the expected double spin asymmetry in the PAX experiment will be comfortably large, see the talk by M.Nekipelov.[9]

The first principal ingredient in this extraction is the BELLE analysis of the correlation of fragmentation of the back-to-back jets. While the average

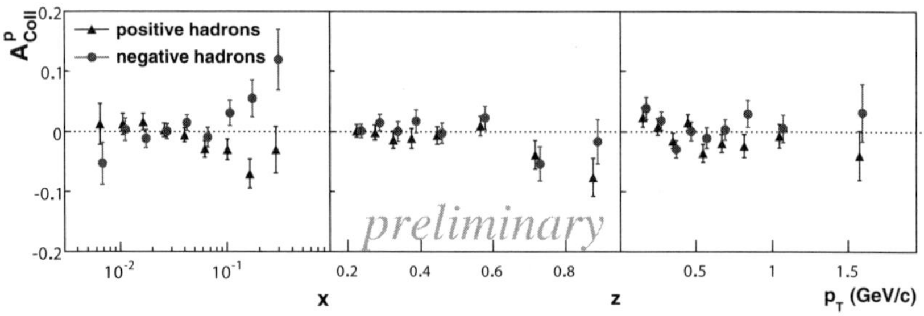

Fig. 3. Collins asymmetries for positive unidentified hadrons (triangles) and for negative unidentified hadrons (circles) as function of x, p_T^h, and z.

polarization of each of the (anti)quark jets in $e^+e^- \to q\bar{q}$ is zero, the quark and antiquark spins are correlated, resulting in the observable coupling of Collins asymmetries in the fragmentation of two jets. The so-induced azimuthal asymmetry is quadratic in the Collins fragmentation function. Due to the fantastic statistics, the BELLE results for the Collins function are of highest accuracy in the filed. This important input from BELLE, reported at this meeting by Grosse-Perdekamp, is with us for some time, their final results have already been published[10] and I jump to the second new ingredient: the Collins asymmetries in SIDIS, which measures a coupling of the Collins function and the transversity. The results from the HERMES have been with us also for quite a time and were reported by L.Pappalordo.[13] The already known COMPASS results for the deuteron target were reported by A.Richter,[12] for the time being they remain the most important input for the isospin properties of the Collins effect. The long awaited results from COMPASS on the Collins effect for the proton target were reported at this meeting for the first time.[14] They are based on about 20 per cent of the collected data. In figure 3 the asymmetry is small, statistically compatible with zero up to $x = 0.05$ while at larger x it increases in module up to 10% and is of opposite sign for positive and negative hadrons. In the p_T^h and z plots the asymmetry amplitude is small due to the dominance of the low x contribution. In the valence region shown in fig. 4 the Collins asymmetry does not exhibit any appreciable z or p_T^h dependence.

Figure 5 shows a very satisfactory comparisons with the Torino theoretical expectations[8] of the Collins asymmetries in the COMPASS kinematical range both for positive and negative hadrons. These Torino predictions are obtained from a global analysis of the HERMES[13] data, COMPASS[12] deuterium data and the BELLE[10] data. A good agreement indicates thus

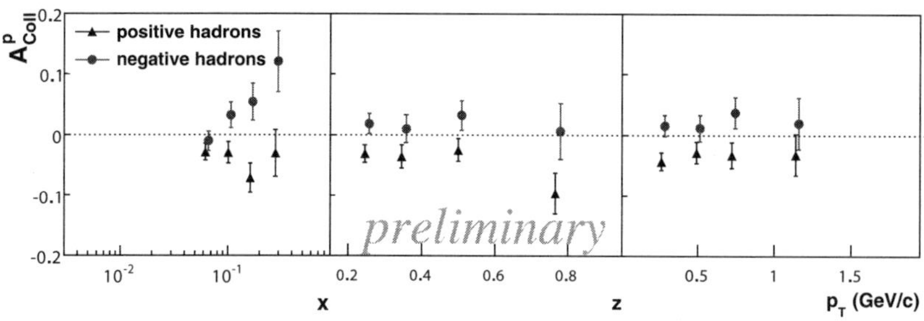

Fig. 4. Collins asymmetries for positive unidentified hadrons (triangles) and for negative unidentified hadrons (circles) with the cut $x > 0.05$ as function of x, p_T^h, and z.

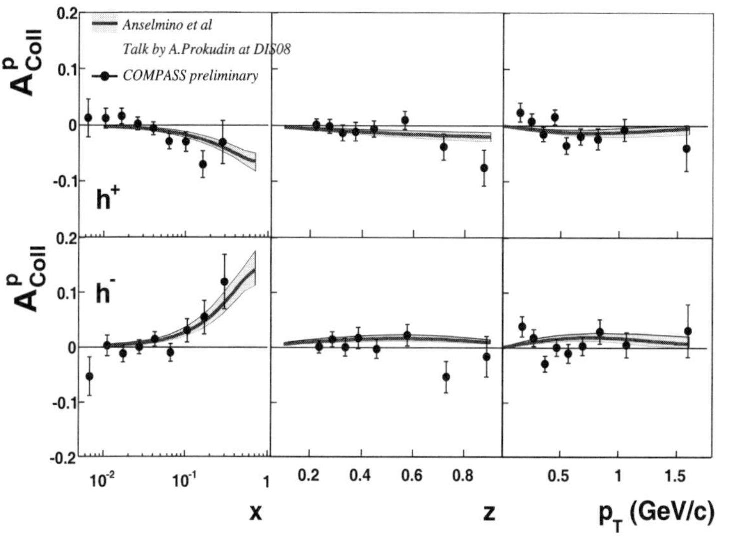

Fig. 5. Compass preliminary Collins asymmetries on proton with the latest prediction of Anselmino et *al.*

a compatibility of the HERMES and COMPASS results for their Collins effect.

The preliminary COMPASS results for the Sivers asymmetries are shown in figure 6 and are compatible with zero. In the case of positive hadrons this is at variance with findings by HERMES.[13] Figure 7 compares the COMPASS Sivers asymmetry on proton data for positive and negative hadrons with the latest prediction of Anselmino et *al.*[8,15] The trends are

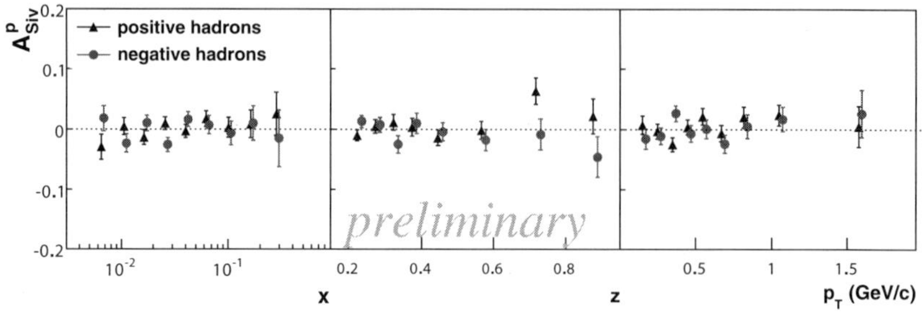

Fig. 6. Sivers asymmetries for positive unidentified hadrons (black) and for negative unidentified hadrons (circles) as function of x, p_T^h, and z.

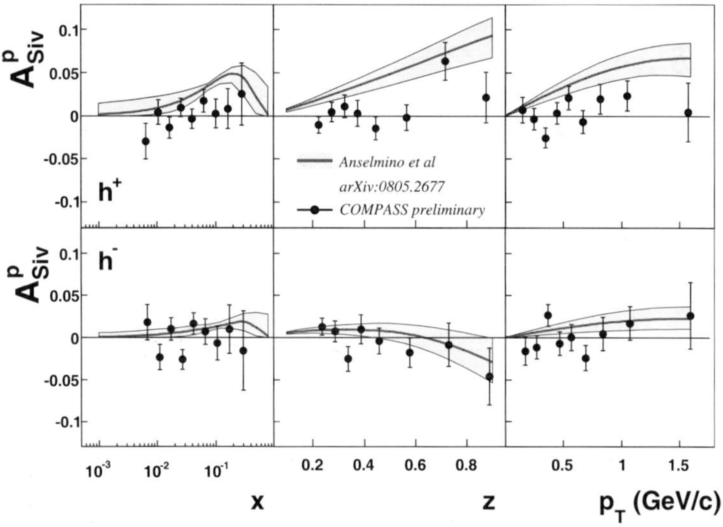

Fig. 7. COMPASS Sivers asymmetry on proton for positive and negative hadrons, with the latest prediction of Anselmino et *al.*

OK. The global fit by Anselmino et al. is driven by the HERMES data, we better await for the full statistics from COMPASS for a more decisive comparison between the COMPASS and HERMES results.

One of the conclusions by F.Murgia from this global analysis is that SSA's will be sufficiently large to be observed in the PAX experiment, where the Sivers, Boer-Mulders and transversity effects would be observable and could be disentangled.[16]

On the theoretical side, the above global analyses hinges on an assumption of the universality of Collins functions, i.e., (soft) finals state interactions in SIDIS and e^+e^- annihilations. Here I would like to make one comment on that, due do D.Musaeva, W.Schäfer and myself.[17] The angular distribution of leptons in the muon decay is asymmetric with respect to the muon polarization. Production of the open charm in e^+e^- annihilations is a hard process and the longitudinal polarizations of the quark and antiquark are correlated.[18] One would naively expect a retention of the polarization of the heavy quark in the course of fragmentation into charmed hadrons. The β-decay of the charmed hadron is a short-distance phenomenon driven by the β-decay of the charmed quark. This decay can well be considered as a final stage of the fragmentation process. The (anti)quark helicity driven forward-backward asymmetries of the positive and negative leptons would couple in complete analogy to the coupling of two Collins asymmetries. This seemingly impeccable chain of arguments fails though. The vector D^* mesons would carry the polarization of the charmed quark but they would decay be strong interaction into spinless D-mesons. The spin of the heavy quark would propagate into the orbital angular momentum of light mesons in $D^* \rightarrow DX$ and for the purposes of the β-decay will be entirely wiped out.[17] Should one worry about anything like that with transverse spins?

For other aspects of the impact on transverse spin phenomena of soft ISI, FSI and absorption corrections in the hadronic basis and the Regge theory approach see the talks by A.Bianconi[19] and G.Goldstein.[20]

3. Polarized antiprotons for PAX

A future of the PAX experiment hinges on whether polarized antiprotons of a sufficiently high polarization will be available at FAIR. The spin filtering of antiprotons in the polarized internal target (PIT) is the only feasible mechanism, and there was quite a confusion on the role of polarized electrons in polarized hydrogen atoms in PIT. Instead of the polarization buildup in the polarized pure electron target one could study the related depolarization of polarized stored protons in the electron cooler as a pure electron target. As H.Ströher reported, for comoving beams at a small relative velocity there are conflicting theoretical calculations with the polarization buildup cross sections differing by 13-14 orders in magnitude. The ongoing experiment at COSY set an upper limit on this cross section which is way below from that of the large-cross-section scenarios.[21]

Misha Nekipelov reported the expectations for double spin asymmetries forseen for the PAX experiment at GSI-FAIR.[9] In the collider mode the energy is sufficiently large to cover the broad valence region in the proton. Daniel Boer emphasized that the double spin asymmetry is remarkably robust against NLO QCD corrections.[4] The possibilities of the fixed target DY experiments at Fermilab Booster and J-PARC we discussed by J.C.Peng,[22] who paid particular attention to the quest for the Boer-Mulders effects. Peng also reminded of the significance of the mesonic cloud effects as evidence by the Gottfried sum rule violations and commented on the pionic contributions to the Boer-Mulders function of the proton. In conjunctions with the COMPASS plans for DY studies reported by M.Colantoni[23] and W.Käfer[24] we have a bright future for TMD's. Add to that future JLAB experiments on TMD's reviewed by H.Avakian.[25]

4. Known knowns, known unknowns, unknown knowns and unknown unknowns

I can not do a full justice to all interesting theoretical reports at the workshop. The time tested constituent quark models are always the first to try and were applied to evaluate the SIDIS asymmetries by A.Kotzinian,[26] S.Scopetta[27] and M.Radici,[28] the emerging results are in the right ball-park set by the Torino analysis.[8] There is also quite a strong model dependence, see also a discussion of Gamberg's talk below.

M.Burkardt gave a nice review on his impact parameter view at the transverse spin phenomena.[29] He outlined a broad programm of quantifying his chromodynamic lensing approach based on the lattice QCD results. In his usual provocative style, John Ralston raised the issue of a possible clash between the confinement property of QCD and the completeness of asymptotic states.[30] He made a point that such a clash would manifest itself as a violation of quantum-mechanical bounds in spin phenomena, citing as a candidate the Soffer bound for the transversity.

On a more conservative QCD side, A.Bacchetta gave quite an extensive review on matching the collinear approximation appropriate at high momentum transfer and the low transverse momentum phenomena in SIDIS described in terms if TMD's. The issue is how the two approaches would meet each other at intermediate transverse momenta. Based on an extensive treatment of specific observables, Bacchetta questioned the completeness of the transverse-momentum-dependent factorization beyond the leading twist approximation.

Still on the same conservative QCD side, several speakers — O.Teryaev,[31] L.Gamberg[32] and A.Mukherjee[33] — went after the standard QCD formalism for the transverse spin structure of the nucleon. Here one of the hot issues is how to match the apparently leading twist property of the Sivers function with the twist-3 property of the quark-gluon correlations. Based on the analysis of the gluon-pole contributions, Teryaev argues that the k_T-dependent Sivers function must be understood as a result of the resummation of the infinite tower of higher twists. He introduces a decomposition of the Sivers function in terms of the Abelian and non-Abelian subprocesses and arrives at a process-dependent color factors for the Sivers effects. The sacred sign-reversal of the Sivers asymmetry from SIDIS to DY survives Teryaev's attack.[31] A closely related and somewhat more specific treatment of the gluon pole effects was presented by Asmitha Mukherjee.[33] Her approach is generic enough, a specific spectator model is invoked only at the last stages. Mukherjee and Teryaev agree on the gluon-pole-induced nonuniversality of the distribution functions, the new point by Mukherjee is that gluon poles do not contribute to the fragmentation functions. The same spectator model has been explored by L.Gamberg who reported numerical studies of the flavor-dependence of the Sivers and Boer-Mulders functions with particular attention to the axial-vector and scalar diquark contributions. An intriguing finding is a near equality of the two functions for the u-quarks, very interesting predictions for the SIDIS on a longitudinally polarized target are reported. I would really love to seen one day a related discussion of the fragmentation of jets in e^+e^- annihilation and a solid proof of the universality of Collins effects from annihilation to SIDIS to hadronic interactions. At the moment I feel somewhat uneasy.

5. Apologies and Acknowledgments

Transversity is a fundamental distribution yet it is very difficult to measure. Still, the past several years the subject of transversity matured much. At Ferrara we heard much more very interesting talks that I was able to cover, I sincerely apologize for being selective. I was indeed excited by the first realistic evaluation of the transversity in the proton and its similarity to the helicity structure of the proton. This was definitely a great step forward and I focused on that particular aspect of the Workshop. Looking forward to our favored Drell-Yan at PAX at GSI-FAIR, should the God Almighty prove not malicious regarding the spin filtering of antiprotons, we shall have large single- and double-spin asymmetries to measure. GPD's and TMD's will be accessible at upgraded JLAB and COMPASS, DY at COMPASS, DY at Fermilab and J-PARC, and we look forward to the final results from HERMES. And we hope our Italian colleagues will succeed with an extension of the Transversity DRIN.

Acknowledgment

I am grateful to the organizers of this extremely productive Workshop for the invitation to review the highlights of the Transversity-2008.

References

1. F. Bradamante, these Proceedings.
2. E. Leader, these Proceedings.
3. V. Barone, A. Drago and P. G. Ratcliffe, Phys. Rept. **3**59 (2002) 1.
4. D. Boer, these Proceedings.
5. J. P. Ralston, D. E. Soper, *Nucl. Phys. B* **152**, 109 (1979).
6. C. Aidala, these Proceedings.
7. V. Barone *et al.* [PAX Collaboration], spokespersons P. Lenisa and F. Rathmann, arXiv:hep-ex/0505054.
8. A. Prokudin, these Proceedings.
9. M. Nekipelov, these Proceedings.
10. R. Seidl *et al.* [Belle Collaboration], Phys. Rev. D **78** (2008) 032011.
11. L.L. Pappalardo, these Proceedings.
12. A. Richter, these Proceedings.
13. L.L. Pappalardo, these Proceedings.
14. S. Levorato, these Proceedings.
15. M. Boglione, these Proceedings.
16. F. Murgia, these Proceedings.
17. D. Musaeva, N. Nikolaev, W. Schaäfer, paper in preparation.
18. R. Dalitz, G. Goldstein and R. Marshall, Phys. Lett. **B** 215 (1988) 783.
19. A. Bianconi, these Proceedings.
20. G. Goldstein, these Proceedings.
21. H. Ströher, these Proceedings.
22. J.C. Peng, these Proceedings.
23. M. Colantoni, these Proceedings.
24. W. Käfer, these Proceedings.
25. H. Avakian, these Proceedings.
26. A. Kotzinian, these Proceedings.
27. S. Scopetta, these Proceedings.
28. M. Radici, these Proceedings.
29. M. Burkardt, these Proceedings.
30. J. Ralston, these Proceedings.
31. O. Teryaev, these Proceedings.
32. L. Gamberg, these Proceedings.
33. A. Mukherjee, these Proceedings.

AUTHOR INDEX